AF580224

Probability and Statistical Inference

Probability and Statistical Inference

RICHARD G. KRUTCHKOFF
Virginia Polytechnic Institute

GORDON AND BREACH SCIENCE PUBLISHERS
New York London Paris

150 Fifth Avenue, New York, N.Y. 10011

Library of Congress catalog card number: 77–99079

Editorial office for the United Kingdom:

Gordon and Breach, Science Publishers Ltd.
12 Bloomsbury Way
London W.C.1

Editorial office for France:

Gordon & Breach
7–9 rue Emile Dubois
Paris 14^e^

Printed in Scotland by Bell and Bain Limited, Glasgow.

Preface

A good textbook on Probability and Statistical Inference, at the level proposed here, should satisfy each of several requirements. It is difficult for a person to learn probability from one text and statistical inference from a second text and, therefore, both topics should be treated under one cover. The text should give a convincing (but not tedious) derivation or justification for everything it asserts. It should be readable at the undergraduate level requiring no statistical prerequisites and only a calculus prerequisite in mathematics. It should treat those topics in probability and statistical inference which are required as preparation for more advanced graduate work. It should contain enough new material to be of interest to researchers and applied statisticians who have already completed their schooling. And finally, it should be detailed enough to be read by those who want to teach themselves without the aid of an instructor. I have tried in this text to satisfy all these requirements.

The first seven chapters treat topics in probability, and the last six chapters treat topics in statistical inference. The text can, therefore, be used either for two separate courses, one in Probability and the other in Statistical Inference, or for a two semester course in Probability and Statistical Inference. Since there are no prerequisites except calculus, the text can be used at the junior or senior undergraduate level or at the first year graduate level. Before being made into a textbook, the manuscript was used as class notes for a two quarter, five hour, first year graduate level course in Probability and Statistical Inference at the Virginia Polytechnic Institute.

There are several more advanced topics that can be more simply treated by using Matrix Theory, Complex Variables, or Measure Theory. These topics are, however, adequately treated without the use of these more advanced mathematical tools. Several exercises which are marked with an asterisk require these more advanced mathematical tools and are included for the more advanced reader only. The exercises which are not marked with an asterisk and which appear in the body of the text should be considered to be a part of the text. Sometimes their solutions are trivial, and sometimes they are very difficult. These solutions, however, are assumed known by the reader in the subsequent text.

There are two appendices at the back of the text which contain tables and other useful information. The reader should become familiar with these appendices.

Answers to some of the problems also appear in the back of the text. It is suggested, however, that the answers be checked only after the problems are solved, never work from the answer to a solution.

This text, as most others, is not the offspring of the author alone. Thanks are due to all my students who made comments on the drafts of the text, which were used as class notes.

Special thanks are due to Mrs. Robert B. Nikodem, who typed the several versions of the manuscript. Thanks are also due to Drs. Hilal Al-Bayyati, Herman Chernoff, Nasser Hadidi, and Norman L. Johnson for their constructive criticism of the manuscript. I would especially like to thank my family and friends who tolerated my neglect while I was engrossed in writing the manuscript.

Virginia Polytechnic Institute RICHARD G. KRUTCHKOFF
January 1970

Contents

CHAPTER 1

Elements of Probability

1-1 Elementary Set Theory

The notation and concepts of Set Theory are very easily adapted to Probability Theory. For this reason we will describe this notation and discuss these concepts in this section. Of course, one section of a text in Probability and Statistical Inference cannot completely describe or discuss the notation and concepts of Set Theory; but we will attempt to include all that is necessary for the subsequent text.

Consider the letters a, b, c, d, e, f, g. They may represent themselves (these seven letters of the alphabet), or they may represent any other seven entities. We call these entities *elements*. The collection of all the elements under consideration is called the *space*. Any collection of these elements is called a *set*. For example, the collection $\{a, b, c\}$ is a set. The space itself $\{a, b, c, d, e, f, g\}$ is also a set. The symbol "Ω" is usually used to represent the space. Brackets "$\{ \}$" are used to contain the elements of a set. The order in which the elements appear within the brackets does not have any significance; a set is completely determined by which elements it contains. For example, the set $\{a, b, c\}$ is the same set as $\{b, c, a\}$, since they have the same elements.

When a set contains several elements, we may abbreviate the notation by using "$\{x$: x has some property$\}$" where the property is explicitly stated. For example, $\{x$: x is an even positive integer less than twenty$\}$ is used to represent $\{2, 4, 6, 8, 10, 12, 14, 16, 18\}$. Another set which will be of particular interest is the set "$\{ \}$", that is, the set with no elements in it. We define this to be a set, even though it contains no elements, and call it the *null* set. The symbol "ϕ" is usually used to represent the null set.

It will be important to distinguish sets from elements. We will use lower case letters to represent elements and upper case letters, as well as Ω and ϕ to represent sets. The symbol "$\in$" is read "is an element of" and is used with an element on the left and a set on the right. For example, if $A = \{a, b, c\}$ then $a \in A$. (Also $b \in A$ and $c \in A$). The symbol "$\subset$" is read "is a subset of" and is used between two sets. For one set A, to be a subset of another set B, all the elements in A must also be in B. For example, if $A = \{a, b\ c\}$ and $B = \{a, b, c\ d\}$ then $A \subset B$. It is not true that $B \subset A$ since B contains an element, namely d, which is not also in A. It is true, however, that $A \subset A$. We will also say that ϕ is a subset of any other set and, therefore, $\phi \subset A$.

Remark *This notation differs from text to text. In texts where the symbol* $\subseteq$ *is used, this corresponds to the symbol* $\subset$ *used here.*

The symbol "$\cup$" is read "union" and is used between two sets. The union of two sets is the set whose elements are in either of the two sets. For example, if $A = \{a, b, c\}$ and $B = \{c, d, e\}$ then $C = A \cup B = \{a, b, c, d, e\}$. Note that the element c is represented only once in the set C. The symbol "$\cap$" is read "intersection" and is used between two sets. The intersection of two sets is a set whose elements are those elements in both of the first two sets. For example, using $A = \{a, b, c\}$ and $B = \{c, d, e\}$ we obtain $D = A \cap B = \{c\}$.

The symbol "$\Rightarrow$" is read "implies" or "*if . . . then . . .*". For example if $A \subset B$ and $a \in A$, then $a \in B$; that is, $A \subset B$, $a \in A \Rightarrow a \in B$. The symbol "$\equiv$" will be read "if and only if," and will be used to replace two $\Rightarrow$ symbols. For example, if A contains the same elements as B, we say $A = B$ and write $a \in A \equiv a \in B$ which is the same as writing both $a \in A \Rightarrow a \in B$ and $a \in B \Rightarrow a \in A$.

Those elements in set B but not in set A comprise the set $B-A$ called the *difference B minus A*. For example, if $A = \{a, b, c\}$ and $B = \{c, d, e\}$ then $E = B-A = \{d, e\}$. A very important difference occurs when the first set is the space. Then $\Omega - A$ is called the *complement* of A, and written $\bar{A}$. The complement of any set is the set consisting of all elements (in Ω) which are not in the set. Two sets are said to be *disjoint* if their intersection is the null set. For example, with $\Omega = \{a, b, c, d, e, f, g\}$, $A = \{a, b, c\}$ and $E = \{d, e\}$ we have $\bar{A} = \{d, e, f, g\}$ and $\bar{E} = \{a, b, c, f, g\}$. Since $A \cap E = \phi$, A and E are disjoint. However, $\bar{A} \cap \bar{E} = \{f, g\}$ and, therefore, $\bar{A}$ and $\bar{E}$ are not disjoint. For any set H it is always true that $H \cap \bar{H} = \phi$ and, therefore, any set is disjoint from its complement.

EXERCISE 1-1

Using $\Omega = \{a, b, c, d, e, f, g\}$, $A = \{a, d, g, e\}$, $B = \{b, e, f, g\}$, and $C = \{c, d, f, g\}$, verify that

(*a*) $A \cup (B \cup C) = (A \cup B) \cup C$

(*b*) $A \cap (B \cap C) = (A \cap B) \cap C$

(*c*) $A \cap (B \cup C) = (A \cap B) \cup (A \cap C)$

(*d*) $A \cup (B \cap C) = (A \cup B) \cap (A \cup C)$

(*e*) $(\overline{A \cup B}) \quad = \bar{A} \cap \bar{B}$

$(f)\ (\overline{A \cap B}) \quad = \bar{A} \cup \bar{B}$

$(g)\ B \cup (A \cap \bar{B}) = A \cup B$

$(h)\ B \cap (A \cap \bar{B}) = \phi$

$(i)\ A \cap \bar{B} \quad = A - B$

$(j)\ A \cup \bar{A} \quad = \Omega$

$(k)\ A \cap \bar{A} \quad = \phi$

$(l)\ \Omega \cup \phi \quad = \Omega$

$(m)\ \Omega \cap \phi \quad = \phi.$

In the text we will need to distinguish two types of infinity, the countably infinite, and the uncountably infinite. The set of integers is said to be a *countably infinite* set. Any set of elements which can be brought into one-to-one correspondence with the integers will be said to be countably infinite. It is interesting to note that the set of rational numbers, (numbers of the form x/y where x and y are integers) is countably infinite. On the other hand, it can be shown that the set of real numbers cannot be brought into one-to-one correspondence with the integers. The set of real numbers is said to be *uncountably infinite*. When a set has either a finite number of elements, or a countably infinite number, we say the number of elements is *countable*.

1-2 Correspondence Between Probability and Set Theory

Consider the pointer in Figure 1-1. Let us say that when we spin this pointer its arrowhead will come to rest in one of the sections labeled a, b, c, d, e, f, g. These sections represented by the labels a through g are possible outcomes of the spinning of the pointer. The only outcomes possible are contained in $\{a, b, c, d, e, f, g\}$ which may be called the sample space. An event, such as "the arrowhead pointing to a vowel" can be represented by $\{a, e\}$. Clearly, an *outcome* of a spin of the pointer (or an outcome of any game or experiment) corresponds to an element of a set. The *sample space* of possible outcomes (also called the *certain event*) corresponds to the space of Set Theory. *Events*, which occur if one or more outcomes occur, correspond to sets.

We read $a \in A$ as "a is an outcome in the event A;" (that is, if a occurs, then the event A has occurred). We read $A \cup B$ as "A or B" meaning that the event $A \cup B$ occurs if either A or B occurs. (That is, if an outcome in either A or B occurs.) For example, if $A = \{a, b, c\}$ and $B = \{c, d, e\}$ then $a \in A$, $b \in A$ and $c \in A$, which means that the event A occurs if the arrowhead falls in one of the sections labeled a, b, or c. Similarly, $A \cup B$ occurs if the arrowhead falls in one of the sections labeled a, b, c, d, or e.

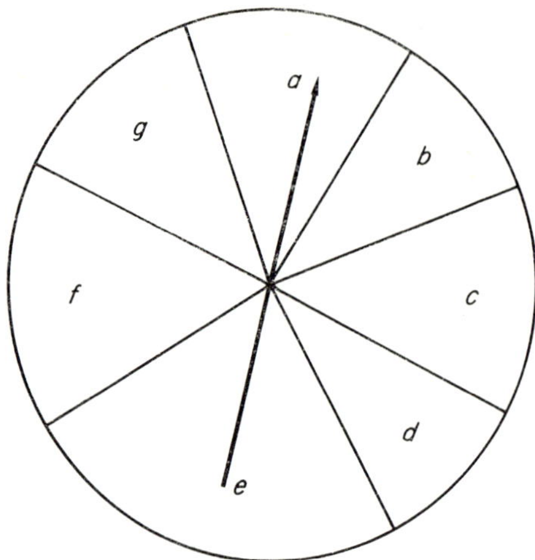

Figure 1-1

We read $A \cap B$ as "A and B" meaning that both the events A and B occur. Using $A = \{a, b, c\}$ and $B = \{c, d, e\}$ then $A \cap B$ can occur only if the arrowhead falls in the section labeled c. That is, $A \cap B = \{c\}$.

Except for the change in the words 'space', 'set', and 'element' to the words 'sample space' (certain event), 'event', and 'outcome', the symbols, notation, and the concept of disjoint remain unchanged. For example, the event $A = \{a, b, c\}$ is a subevent of the event $D = \{a, b, c, d\}$ written $A \subset D$. $D - A = \{d\}$ and $\bar{A} = \{d, e, f, g\}$.

An event which contains no outcomes cannot occur, and is called the *impossible event*, ϕ. As in the case of sets, we will say that ϕ is a subevent of every other event. As an example of a null event consider the situation in which you would win a dollar if you spun the pointer in Figure 1-1, and the arrowhead fell in a section labeled with a vowel other than a or e.

EXERCISE 1-2

Using the events labeled by Ω, A, B, and C given in Exercise 1-1, describe the relations given there in terms of the pointer situation of *Figure* 1-1.

1-3 Definitions of Probability

Probability may be defined or approached from any one of several points of view. The results, however, are generally compatible with one another. Which point of view is most appropriate for a given situation will depend on the details of that situation. We will now present four of these points of view: the Classical definition, the frequency interpretation, the axiomatic development, and the personal approach.

THE CLASSICAL DEFINITION OF PROBABILITY

Probability theory is often thought of as the science behind games of chance. Although games of chance are not the primary concern of the theory of probability, it is often useful to describe probabilities in terms of these games.

Imagine playing a game in which there are *n equally likely* outcomes. This means that there is no reason for one outcome to occur more often than any other. For example, in the tossing of a die, $n = 6$, and in the drawing of a card from a poker deck, $n = 52$. Of course, in practice there may be some slight unbalance which will prevent the outcomes from being precisely equally likely. For this point of view, however, we neglect these slight unbalances and assume that the outcomes are equally likely.

If we define $n(A)$ to be the number of these n equally likely outcomes, which are in the event A, then we can define the *probability* of A by

$$P(A) = n(A)/n.$$

For example, in the tossing of a die there are six equally likely outcomes. The event "an even number of dots facing up" contains three of these outcomes, therefore, $P = 3/6 = 0.5$.

FREQUENCY INTERPRETATION OF PROBABILITY

Consider the situation in which you repeatedly flip a coin, recording after each flip the relative frequency of heads (i.e., the ratio of the current total number of heads divided by the current total number of flips). It would not be surprising if this relative frequency converged to some number near 0.5. The frequency interpretation is based on the apparent existence of this limit.

Let us say that we are about to perform an experiment or play a game, the outcome of which cannot be determined by us in advance. If several of the possible outcomes of the experiment or game comprise the event A, then we can speak of the probability that A will occur, written $P(A)$. Although we may perform the experiment or play the game once, we can imagine recording at the nth repetition the *relative frequency*

$$f_n(A) = \frac{1}{n}\,[\text{total number of times } A \text{ occurred}]$$

then we can define the *probability* of A by

$$P(A) = \lim_{n \to \infty} f_n(A),$$

assuming, of course, that this limit exists. Several properties of this function $P(A)$ should be obvious at this point. However, it will be advantageous to postpone discussion of these properties until we discuss the axiomatic development of probability.

THE AXIOMATIC DEVELOPMENT OF PROBABILITY

This point of view is abstract and, therefore, difficult to visualize. For this reason we will attempt to relate the axiomatic concepts developed to the frequency interpretation.

In this development *probability* is defined to be a function defined on events (certain subevents of Ω) which satisfies the following three axioms.

Axiom I $P(A) \geqq 0, \qquad A$ an event.

Axiom II $P(\Omega) = 1.$

Axiom III If $A_1, A_2, A_3, \ldots$ are a countable number of subevents of Ω such that $A_1 \cup A_2 \cup A_3 \cup \ldots = \Omega$ and $A_i \cap A_j = \phi \quad i \neq j$ (we call this a *subdivision* of Ω) then

$$P(A_i \cup A_j \cup A_k \cup \ldots) = P(A_i) + P(A_j) + P(A_k) + \ldots.$$

In words, this says:

I the probability (relative frequency) of any event is nonnegative;

II the probability (relative frequency) of the certain event (an event which must occur since it contains all the possible outcomes) is unity;

III the probability (relative frequency) of a union of disjoint events is the sum of the probabilities of the events themselves.

This last axiom can be visualized by the following example. Imagine that A_i occurs n_i out of n times. Since the events are all disjoint, only one event can occur at a time. Thus, the total number of times $A_i \cup A_j \cup A_k \cup \ldots$ has occurred is $n_i + n_j + n_k + \ldots$. Therefore,

$$P(A_i \cup A_j \cup A_k \cup \ldots) = \frac{1}{n}[n_i + n_j + n_k + \ldots]$$

However, it is also true that

$$P(A_i) + P(A_j) + P(A_k) + \ldots = \tfrac{1}{2}[n_i + n_j + n_k + \ldots]$$

We now derive several properties of probability which are immediate consequences of these three axioms.

Property a $P(\phi) = 0$.

Proof $\Omega \cup \phi = \Omega$, $\Omega \cap \phi = \phi$, and therefore,

$$P(\Omega \cup \phi) = P(\Omega) = P(\Omega) + P(\phi)$$

and Property *a* follows.

Property b $P(\bar{A}) = 1 - P(A)$

Proof $A \cup \bar{A} = \Omega$, $A \cap \bar{A} = \phi$, and therefore

$$P(A \cup \bar{A}) = P(\Omega) = 1 = P(A) + P(\bar{A})$$

and Property *b* follows.

Property c $A \subset B \Rightarrow P(A) \leqq P(B)$

Proof $A \subset B \Rightarrow A \cup (B \cap \bar{A}) = B \cup A = B$, and since $A \cap (B \cap \bar{A}) = \phi$, we obtain

$$P[A \cup (B \cap \bar{A})] = P(A) + P(B \cap \bar{A}) = P(B).$$

Since $P(B \cap \bar{A}) \geqq 0$, Property *c* follows.

Property d $0 \leqq P(A) \leqq 1 \quad \text{any} \quad A \subset \Omega$

Proof $A \subset \Omega$ and Property *d* follows from Property *c* and Axiom I.

EXERCISE 1-3

The property $A \cap B \cap C = \phi$ is not sufficient to ensure that $P(A \cup B \cup C) = P(A) + P(B) + P(C)$. Demonstrate this by finding an event function which satisfies Axioms I–III and sets A, B, and C, which satisfy the above property, but for which the equation does not hold.

EXERCISE 1-4

Prove that for any two events A and B

$$P(A \cup B) = P(A) + P(B) - P(A \cap B).$$

EXERCISE 1-5

Discuss the meaning of **Properties a** through **d** from the frequency point of view and demonstrate the compatibility of these two definitions of probability.

EXERCISE 1-6

Demonstrate by an example that the classical definition of probability is compatible with both the frequency interpretation and the axiomatic treatment by showing that Axioms I–III are satisfied.

PERSONAL PROBABILITIES

Situations arise in which we can neither assume equally likely outcomes, nor imagine repeating the situation many times. This is often the case in economic problems. For example, assume that a new building contractor has just gone into business, and we are debating whether or not to contract him to build our new home. The stakes are high, and we will only be playing the game once with this new contractor. If, however, we knew the odds on his doing a good job, we could decide more easily whether or not to enter into a contract with him. In order to ascertain these odds, we can speak to him, people who know him, look at the type of work he has done in the past, etc. The impressions obtained by this procedure are, however, subjective. Different people, obtaining the exact same information, might obtain different impressions of this contractor's competence. We will, therefore, call the probabilities, which we will obtain, personal probabilities.

In order to obtain these personal probabilities, imagine the following betting situation. We present two numbers x and y. Our opponent can then bet *for* or *against* the contractor's doing a satisfactory job. If he bets *for*, then he receives y dollars from us if the job is satisfactory, and pays us x dollars if the job is not satisfactory. If he bets *against*, then he pays us y dollars if the job is satisfactory, and receives x dollars from us if it is not satisfactory. The odds (according to us) for the contractor's doing a satisfactory job are then $x:y$. In order for the bet to be fair we should expect, that if the bet could be repeated many times, the relative frequency of the contractor's doing a satisfactory job should be approximately $x/(x+y)$. We can, therefore, define personal probabilities in the following way. If our odds for an event A are $x:y$, then our *personal probability* for the event A is given by

$$P(A) = \frac{x}{x+y} .$$

EXERCISE 1-7

Demonstrate by an example that the personal probability approach is compatible with the other points of view by showing that Axioms I–III can be satisfied.

1-4 Some Methods for Finding Probabilities Using the Classical Approach

Often the sample space involved is a complex combination of simpler sample spaces. For example, if a single die is tossed, the sample space consists of six numbers. If two dice are tossed, the sample space consists of 36 pairs of numbers. If n dice are tossed, the sample space consists of 6^n n-tuples. As another example, consider the drawing of cards from a deck of ten cards. For one draw the sample space contains only ten outcomes. For the second draw the sample space is a 9 outcome subset of the first sample space. For both draws considered at the same time the sample space consists of 90 pairs of cards. Clearly it would be advantageous to obtain probabilities without having to construct the sample space.

Let us assume that the sample space is made up of equally likely outcomes, each of which is in the form of an n-tuple. To find the probability of an event A all we require is the ratio of the number of n-tuples, which comprise property A, to the total number of n-tuples in the sample space. This counting can often be done without explicitly displaying the sample space. In particular, the counting can be done when the n-tuples are constructed by one of the two sampling procedures; sampling with replacement, and sampling without replacement.

WITH REPLACEMENT

Assume that we have three dice which we are to toss. We can obtain the total number of 3-tuples by considering:

a) How many values the first die can give.
b) How many values the second die can give for each value of the first die.
c) How many values the third die can give for each combination of values for the first and second die.

Clearly the answer in all three cases is 6. We can, therefore, see that for each of 6 values in (a) we obtain 6 values in (b) giving 36 combinations. In (c) we need 6 values for each of the 36 combinations in (a) and (b) giving us 216. In a similar manner for n dice we would have 6^n n-tuples.

The manner in which the n-tuple is formed is said to be sampling *with replacement*. We sample from a set of numbers (six in this example), record the sampled number, and then *replace* it so that it will be available as a possible choice in the next sample.

In general, when sampling is done with replacement the number of outcomes is m^n, where m is the number of different values present and n is the length of the n-tuple.

Example 1-1

What is the probability that in the toss of five dice there will be only odd numbers of dots facing up?

Using the classical approach we must find both the total number of equally likely outcomes and the number of outcomes favorable to the event in question. Imagine tossing the five dice and writing the resulting five values as a 5-tuple. There are $m^n = 6^5 = 7{,}776$ different 5-tuples possible, each of which may be considered equally likely. An outcome of this play is then a 5-tuple. How many of these 5-tuples satisfy the event in question (call it A)? There is a total of three numbers 1, 3, 5 which are eligible for each die. Therefore, there are $3^5 = 243$ different 5-tuples in event A. By the classical definition then

$$P(A) = 243/7776 = 0.03125.$$

WITHOUT REPLACEMENT

Let us say that we are drawing cards from a deck consisting only of the cards labeled 1, 2 and 3, and that we are not replacing drawn cards. We must consider:

a) How many values the first draw can give.
b) How many values the second draw can give for each of the possible values in (a).
c) How many values the third draw can give for each combination of values in (a), and (b).

Clearly the first draw can give any of three values. The second, however, can only give two values since at this draw there are only two cards left. There are, therefore, six combinations for the first two draws. For each of these six combinations, there is only one card left; therefore, there is a total of 6 3-tuples. This type of sampling is said to be *without replacement*; that is, the possible values for the second value are those left after the first has been withdrawn. The value chosen in (a) was not replaced and, therefore, not available for resampling. The value for the third choice is the one left over after the first two have been removed. If there are n different values present, then the first can have any of the n values, the second any of the $(n-1)$ left over, the third any of the $(n-2)$ left over after the second choice, etc. The total number of n-tuples is then $n(n-1)(n-2)(n-3) \ldots 2\cdot 1$. This product is called *n-factorial* and written as $n!(= n(n-1)(n-2)(n-3) \ldots 2\cdot 1)$.

It is often the case that although there are n values we need only an r-tuple. For example, drawing 3 cards from a deck of 10. Clearly the first can have 10 values. The second can have 9 values for each of these 10 giving 90 pairs. The third can have 8 values for each of these 90 pairs, giving 720 3-tuples. In general, the number will be

$$\frac{n!}{(n-r)!} = n(n-1)(n-2) \ldots (n-r+1).$$

For example, the number of ordered 4 card draws from a poker deck containing 52 cards is

$$\frac{52!}{(52-4)!} = 52\cdot 51\cdot 50\cdot 49$$

$$= 6{,}497{,}400$$

Exercise 1-8

Verify that

$$\frac{n!}{(n-r)!} = n(n-1)(n-2) \ldots (n-r+1).$$

Example 1-2

What is the probability that three draws from a poker deck will result in three aces?

Here the draws are without replacement, and there are $n!/(n-r)! = 52!/49! = 52 \cdot 51 \cdot 50 = 132{,}600$ possible 3 card draws. If we consider them equally likely outcomes, then this will be the denominator in the classical definition of probability. Since there are but four aces, the numerator will be the number of 3 card draws from four cards or $4!/(4-3)! = 4 \cdot 3 \cdot 2 = 24$. Therefore, $P(A) = 24/132{,}600 = 0.00018$, where A represents the required event.

COMBINATIONS

In the previous two subsections we considered n-tuples. The order of the entities in an n-tuple was significant. For example, (1, 2, 3) is not the same 3-tuple as (2, 1, 3). Often, however, we use n-tuples merely as a representation of n entities without significance to their order (e.g., the elements in an n element set). To visualize this situation consider choosing a committee of ten men from a group of fifty men. The order in which the men are chosen to the committee is of no significance.

Remark *We are not considering the type of committee in which the first chosen is the chairman.*

A group of distinguishable entities in which no entity appears twice and in which there is no significance to the order in which the entities are displayed is called a *combination*.

We now ask for the number of combinations of r different entities which can be made from n different entities ($r < n$). This will give us, for example, the number of different r-men committees which can be chosen from n men.

Let us say that there are $\binom{n}{r}$ of these combinations. (This can serve as the definition of this symbol; its numerical value will now be determined.) Since there are r different values in this combination, we can form $r!$ different r-tuples with these values. Since no two combinations have the same set of values, and $\binom{n}{r}$ exhausts all the sets of r values, we find that there must be $\binom{n}{r}r!$ different r-tuples; but since there are $n!/(n-r)!$ r-tuples, we obtain that $\binom{n}{r}r! = n!/(n-r)!$

or

$$\binom{n}{r} = \frac{n!}{r!(n-r)!}$$

Remark *We define* 0! *to equal unity, and as a result* $\binom{n}{0} = 1$. *Note also that*

$$\binom{n}{r} = \binom{n}{n-r} = \frac{n!}{r!(n-r)!}$$

by symmetry.

Example 1-3

Let us reconsider the problem in Example 1-2. It seems unreasonable to expect that the manner in which the cards are dealt should influence the probability of being dealt three aces. Consider dealing the cards three at a time. The order within this group of three cards then has no significance and, therefore, we must calculate probabilities using combinations. The total number of three card combinations from the deck (disregarding the order in which they were drawn) $\binom{52}{3} = 52!/49!\cdot 3! = 52\cdot 51\cdot 50/3\cdot 2 = 22{,}100$. If we consider these as our equally likely outcomes, then this will be the denominator in the classical definition. For the numerator we need the number of combinations of three cards taken from four cards (the aces). This number is $4!/3!\cdot 1! = 4$. The desired probability is then $(P(A) = 4/22{,}100 = 0.00018$, which is, as expected, the same result as obtained in Example 1-2.

MULTICOMBINATIONS

There are four suits in a poker deck. If we are interested only in how many of each suit are in an m card draw, then we might ask how many combinations there are with exactly m_1 of the first suit, m_2 of the second, m_3 of the third, and m_4 of the fourth $(m_1+m_2+m_3+m_4 = m)$. In general, of the n card deck there will be n_1 cards of the first suit, n_2 of the second, n_3 of the third, and n_4 of the fourth $(n_1+n_2+n_3+n_4 = n)$. Of course, $m_i \leqq n_i$, $i = 1, 2, 3, 4$, since we cannot draw more cards of a suit than there are cards of that suit in the deck. Since the order of the draws is unimportant, we can consider the draws according to suit. There will be $\binom{n_1}{m_1}$ ways to obtain m_1 of the first suit from the n_1 available. Having done this, and for each of these choices, we can draw m_2 from n_2 in $\binom{n_2}{m_2}$ ways. After this is done, and for each of the choices already made, there are $\binom{n_3}{m_3}$ ways to choose m_3 from n_3, and for each of these $\binom{n_4}{m_4}$ ways to choose m_4 from n_4. Combining all of these we see that there are $\binom{n_1}{m_1}\binom{n_2}{m_2}\binom{n_3}{m_3}\binom{n_4}{m_4}$ ways to obtain exactly

m_1 from the first suit, m_2 from the second, m_3 from the third and m_4 from the fourth.

EXERCISE 1-9

Assuming that each combination is equally likely, find the probability that in m draws from a poker deck of n cards there will be exactly m_1, m_2, m_3, m_4 from the four suits.

1-5 Joint, Marginal and Conditional Probabilities of Events

JOINT PROBABILITIES

Let us say that we have two subdivisions of the sample space $A_1, A_2, A_3, \ldots$ and $B_1, B_2, B_3, \ldots$. The union of all the events $A_1, A_2, A_3, \ldots$ is denoted $\cup_i A_i$, and, by definition $\cup_i A_i = \Omega$. Similarly $\cup_j B_j = \Omega$. Note that the A_i's are disjoint from each other, as are the B_i's, but it is not necessarily the case that $A_i \cap B_j = \phi$ for any i and j. As an example, consider a poker deck subdivided into thirteen card values (ace, deuce, ..., king) and into four suits. The card values are disjoint from each other, as are the suits, but no card value is disjoint from any suit.

Definition The *joint probability* of the events A_i and B_j is given by $P(A_i, B_j) = P(A_i \cap B_j)$ for any i, j.

In order to understand the real-life significance of this definition, we will now give an example using the classical approach.

EXAMPLE 1-4

Let us say that in a given town there are exactly N adults. We can divide the town into two types of disjoint sets by dividing the town into male and female and smokers and nonsmokers. Let us say that $N(m, n)$ is the number of male nonsmokers; $N(m, s)$ is the number of male smokers, $N(f, n)$ is the number of female nonsmokers and $N(f, s)$ is the number of female smokers. There are certain obvious relationships between these numbers, such as the number of males $N(m) = N(m, s)+N(m, n)$, the number of smokers $N(s) = N(m, s) + N(f, s)$, etc. If we assume that each person is equally likely to be chosen, then the probability of choosing a male smoker, for example, is given by $P(m, s) = N(m, s)/N$.

When there are several subdivisions $A_1, A_2, A_3, \ldots, B_1, B_2, B_3, \ldots, C_1, C_2, C_3, \ldots, D_1, D_2, D_3, \ldots$, then the joint probabilities are defined to be $P(A_i, B_j, C_k, D_l) = P(A_i \cap B_j \cap C_k \cap D_l)$ for any values i, j, k, l.

Remark *Since joint probabilities are defined in terms of intersections of events (which are also events), they must satisfy all the properties of probabilities already discussed.*

MARGINAL PROBABILITIES

The situation will arise when we have the complete set of joint probabilities $P(A_i, B_j)$ for all possible combinations of i and j, but we are only interested in the events A_i, $i = 1, 2, 3, \ldots$. In the previous example we may only be interested in the probability of sampling a male or a female, regardless of whether or not the person smokes. The probabilities related to only one of the subdivisions are called the *marginal* probabilities.

Since $A_i = A_i \cap \Omega = A_i \cap (\cup_j B_j) = \cup_j (A_i \cap B_j)$, we have that $P(A_i) = P[\cup_j (A_i \cap B_j)]$. But since $(A_i \cap B_j) \cap (A_i \cap B_k) = \phi$, $j \neq k$, we obtain by Axiom III that $P(A_i) = \Sigma_j P(A_i, B_j)$. This says that in order to obtain the marginal probability of A_i (for any particular i) all we need to do is sum over the joint probabilities containing A_i. In the above example, if we wanted the probability of finding a male in the town, we would divide the number of males by the total number of adults, N. The total number of males is the sum of the smoking males and the nonsmoking males. Thus,

$$P(m) = \frac{N(m, n)+N(m, s)}{N} = \frac{N(m, n)}{N} + \frac{N(m, s)}{N}$$

$$= P(m, n)+P(m, s) \quad \text{which is as it should be.}$$

EXERCISE 1-10

Consider the sample space divided into three subdivisions $A_1, A_2, A_3, \ldots B_1, B_2, B_3, \ldots$ and $C_1, C_2, C_3, \ldots$ Given the joint probabilities $P(A_i, B_j, C_k)$ for all combinations of i, j, k, find the marginal probabilities $P(A_i)$ for i, $P(B_j)$ for any j, and $P(C_k)$ for any k.

CONDITIONAL PROBABILITIES

Consider the situation of Example 1-4. The probability of a male is simply given by $P(m) = N(m)/N$. If, however, we require the probability of a male from among the smokers in the town, then that probability would be $N(m, s)/N(s)$. If we use the notation $P(m \mid s)$ for this probability, then we see that

$$P(m \mid s) = N(m, s)/N(s) = \frac{N(m, s)/N}{N(s)/N} = \frac{P(m, s)}{P(s)}$$

In general, we can make the following definition.

Definition The conditional probability of A_i given B_j (for any i, j) is given by

$$P(A_i \mid B_j) = \frac{P(A_i, B_j)}{P(B_j)}$$

whenever $P(B_j) \neq 0$. [In Section 4-6 we will define certain conditional probabilities when $P(B_j) = 0$.]

A very useful rearrangement of the relation in the definition of conditional probabilities gives $P(A_i, B_j) = P(A_i \mid B_j)P(B_j)$.

EXAMPLE 1-5

What is the probability of drawing an ace on both the first and second draw from a poker deck?

If A represents the event "an ace on the second draw," and B represents the event "an ace on the first draw," then the probability of an ace on both the first and second draw should equal the probability of "an ace on the second draw given one was withdrawn from the deck on the first draw" times the probability of "an ace on the first draw;" that is, $P(A, B) = P(A|B)P(B)$.

Since there are four aces, and 52 cards, the probability of an ace on the first draw is 4/52. The probability of an ace on the second draw, given that an ace was removed on the first draw, is 3/51, since there are 3 aces left and there are now 51 cards.

Therefore, $P(A, B) = 4/52 \cdot 3/51 = .0045$.

INDEPENDENCE

In mathematics, X and Y are said to be independent if there is no functional relation between them. This type of independence is not equivalent to the concept of independence in probability. The deciding criterion for independence in probability is the following. Two events A and B are said to be *independent* of each other if $P(A, B) = P(A)P(B)$.

EXERCISE 1-11

Show that if $P(A) \neq 0$ and $P(B) \neq 0$, then the events A and B are independent if and only if $P(A|B) = P(A)$.

A useful extension to the concept of independence is the following definition.

Definition Two subdivisions of the sample space $A_1, A_2, A_3, \ldots$ and $B_1, B_2, B_3, \ldots$ are said to be independent if $P(A_i, B_j) = P(A_i)P(B_j)$ for every i, j.

EXERCISE 1-12

Find an equivalent definition of independence of subdivisions using conditional probabilities, as was done in Exercise 1-11.

It is quite natural to think that if two sets are disjoint and, therefore, have no elements in common that they are independent. In general, however, the situation is quite the opposite. If two sets A and B are disjoint, then $A \cap B = \phi$, and, therefore, $P(A \cap B) = P(\phi) = 0$. If each event has a nonzero probability of occurring, then $P(A)P(B) \neq 0$. In fact, it is clear that disjoint events cannot be independent unless one of them is the impossible event. It is interesting to note that this shows that the impossible event and any other event are independent.

BAYES THEOREM

EXAMPLE 1-6

Consider the situation in which there are two urns, A and B, each containing one white ball and one black ball. A ball is chosen randomly from urn A and placed in urn B. A ball is then chosen randomly from urn B, and it is noted to be black. What is the probability that the first ball chosen was white?

Without information as to what the second ball was, the probability that the first ball was white is 1/2. Knowledge of the second ball, however, alters the situation drastically. Consider the frequency interpretation. We are not going to find the frequency of the ball being white for all picks of the first ball. We are only allowed to use the results of the picks when the second ball was black. Clearly, it is less likely that the first ball was white if the second ball was black, than if it had been white. Therefore, the frequency, if we were to sample in situations identical to this, should approach something less than 1/2. To be more precise, we will use the properties of conditional probabilities: where W_1, W_2, B_1, B_2 represent a white ball on the first and second pick and a black ball on the first and second pick, in that order.

We can equate the joint probability $P(W_1, B_2)$ with a conditional probability multiplied by a marginal probability in two ways. They are

1) $P(W_1, B_2) = P(W_1 \mid B_2)P(B_2)$

and

2) $P(W_1, B_2) = P(B_2 \mid W_1)P(W_1)$.

From these we obtain

$$P(W_1 \mid B_2) = \frac{P(W_1, B_2)}{P(B_2)} = \frac{P(B_2 \mid W_1)P(W_1)}{P(B_2)}.$$

Noting that $P(B_2) = P(W_1, B_2)+P(B_1, B_2)$, by the definition of marginal probabilities, and replacing the joint probabilities with their equivalents, we obtain $P(B_2) = P(B_2 \mid W_1)P(W_1) + P(B_2 \mid B_1)P(B_1)$, giving us

$$P(W_1 \mid B_2) = \frac{P(B_2 \mid W_1)P(W_1)}{P(B_2 \mid W_1)P(W_1)+P(B_2 \mid B_1)P(B_1)}.$$

There is a very simple reason for representing $P(W_1 \mid B_2)$ in this seemingly complicated form. All the probabilities on the right-hand side of the equation are very simple to calculate.

We already know that $P(W_1) = P(B_1) = 1/2$. $P(B_2 \mid W_1)$ is the probability that the second pick is a black ball given a white was added to the one white and one black ball already there. Since there is but one black ball to choose from out of three, $P(B_2 \mid W_1) = 1/3$. Similarly for $P(B_2 \mid B_1)$ we choose from two black balls out of three giving $P(B_2 \mid B_1) = 2/3$. Thus,

$$P(W_1 \mid B_2) = \frac{1/3 \cdot 1/2}{1/3 \cdot 1/2 + 2/3 \cdot 1/2} = 1/3$$

The above example illustrates a very useful relation between probabilities. If $A_1, A_2, A_3, \ldots$ and $B_1, B_2, B_3, \ldots$ are subdivisions of the sample space, then for any i, j

$$P(A_i, B_j) = P(A_i \mid B_j)P(B_j)$$

and

$$P(A_i, B_j) = P(B_j \mid A_i)P(A_i).$$

But

$$\begin{aligned} P(A_i) &= \sum_j P(A_i, B_j) \\ &= \sum_j P(A_i \mid B_j)P(B_j) \end{aligned}$$

and, therefore,

$$P(A_i, B_j) = P(B_j \mid A_i) \sum_k P(A_i \mid B_k)P(B_k).$$

Equating this with the first equation, we obtain

$$P(B_j \mid A_i) = \frac{P(A_i \mid B_j)P(B_j)}{\Sigma_k P(A_i \mid B_k)P(B_k)}$$

for any i, j. This relation is known as *Bayes Theorem*.

1-6 Problems

1. Let Ω be the set whose elements are the numbers 1, 2, 3, 4, 5, 6, 7, 8, 9, 10.
   a. Draw a picture of the certain event.
   b. Draw a picture of the event "an even number larger than 6."
   c. Give an example of an event such that its outcomes add up to seven.
   d. How many events are there which satisfy the conditions in part c?
2. Three billiard balls numbered 1, 2, and 3 lie on a table. An outcome is the order in which the balls are chosen. What is the certain event? How many outcomes are there in this event? How many events are there which have the ball numbered 1 chosen second?
3. Consider the tossing of a die. There are six outcomes 1, 2, 3, 4, 5, and 6. Give a function $P(\cdot)$ which could represent a probability function on the subsets of the certain event. Be sure to check that the properties of $P(\cdot)$ satisfy Axioms I–III. After you have completed this, find another such function $P'(\cdot)$.
4. Consider the sample space as given in Problem 1. If each outcome is considered to be equally likely, then give the probability of each of the events described in the answers to the questions of Problem 1.
5. There are 52 possible outcomes of a draw from a poker deck. Considering each of these outcomes equally likely find:
   a. The probability of drawing an ace.

d. The probability of drawing a spade.
c. The probability of drawing the ace of spades. (Are the events "drawing an ace" and "drawing a spade" independent?)
d. The probability of drawing a spade or a heart.
e. The probability of drawing either the ace of spades or the ace of hearts.

6. There are 10 members of a particular club. Committees are chosen by choosing five names, with replacement, and listing them on the bulletin board. How many different ordered lists can there be on the bulletin board?

7. How many different committees are represented by the lists in Problem 6?

8. How many of the lists on the bulletin board of Problem 6 have five different names on them?

9. If in the club of Problem 6 there are five men and five women, how many five-member committees are there with exactly three women on them? How many with exactly two women on them?

10. If in the club of Problem 6 there was but one woman, what would be the probability that she was on a randomly selected five-member committee? What is the probability that she is not on the committee?

11. Consider the sample space of Problem 1. Let the event A be made up of all the even numbers less than seven. Let B be all the odd numbers, and let C be all the remaining numbers. How many combinations of six numbers will have two from each event A, B, and C? If all combinations of six numbers are equally likely, what is the probability that there will be two from each event, A, B, and C?

12. Consider the following table. The rows represent the sex and marital status for all the entering freshmen at a particular college. The columns represent the age of these freshmen.

	Below 18	18–19	19–20	20–21	Above 21
Married male	0	12	34	79	74
Married female	1	5	8	2	50
Unmarried male	10	104	362	26	6
Unmarried female	15	203	775	7	1

a. What is the joint probability that a student chosen at random will be a married female, with age between 19 and 20 years old?
b. What is the marginal probability that a student chosen at random will be a married female?
c. What is the conditional probability that a married female student chosen at random will be between 19 and 20 years old?

13. The following questions do not refer to the marital status of the students in Problem 12. Using the data given there, construct a new table which neglects this characteristic.
a. What is the joint probability that a student chosen at random will be a female between 19 and 20 years old?
b. What is the probability that a student chosen at random will be a female?
c. What is the conditional probability that a female student chosen at random will be between 19 and 20 years old?
d. If you had not constructed a new table but used the table in Problem 12, would your answers be the same?

14. A particular office is polled to see how many people smoke, and if so, whether they are light or heavy smokers. The results are given in the following table.

	Nonsmoker	Light smoker	Heavy smoker
Male	20	30	20
Female	10	15	10

Considering only people in this office:

a. What is the joint probability of a person chosen at random being a male smoker?
b. What is the marginal probability that a person chosen at random will be a non-nonsmoker?
c. What is the probability that a person chosen at random will be a male?
d. Are the events "being male" and "being a nonsmoker" independent?

15. Considering the table in Problem 14.
a. List at least 10 different events.
b. Test each pair of these for independence.

16. Consider the 10 events given in answer to Problem 15.
a. Which of these pairs are disjoint?
b. Are any that are disjoint independent?

17. Assume that you have three urns, A, B, and C. Urn A contains one white ball and one black ball. Urn B contains two white balls, and urn C contains two black balls. Someone chooses an urn at random and you pick from it one ball at random which turns out to be black. What is the probability that the urn was urn A?

18. Let us assume that you have three urns, labeled A, B, and C. Each urn has five balls in it. Urn A has one white ball and four black balls; urn B has two white balls and three black balls, and urn C has three white balls and two black balls. You roll a die, and if it turns up *one* you pick a ball from urn A, if it rolls an *even number* you pick a ball from urn B, otherwise you pick a ball from urn C.
a. What is the probability of obtaining a white ball?
b. What is the probability that you picked from urn B, given that the ball picked was white?

19. Consider the following simplification: It will either rain or be fair on any day, each with probability 1/2. When it is going to rain, the weatherman predicts it 1/3 of the time. When it is going to be fair, he predicts it 1/2 of the time. Given that the weatherman predicts rain tomorrow, what is the probability that it will be fair?

20. Consider the situation in Problem 14. Are sex and smoking habit mutually independent?

CHAPTER 2

Random Variables and their Distributions

Consider the situation in which we will perform an experiment or play a game, the outcome of which cannot be completely determined in advance. Let X represent a quantity whose value will be determined by the outcome. Before the experiment is performed, or the game played, the value of X is unknown. As a result of performing the experiment or playing the game the value, x (of X), can be determined from the outcome of the experiment or game. We say that X is a *random variable* until its value x becomes known. When we perform the experiment or play the game, we say that we are observing the value of the random variable X. We will use an upper case letter to represent a random variable, and the corresponding lower case letter to represent the observed value of the random variable. We will, wherever possible, also use the lower case letter of the random variable as its dummy variable of summation or integration.

2-1 Discrete Random Variables

When there are a countable number of possible values for X we say that X is a *discrete* random variable. For example, in the tossing of a coin, the possible outcomes are heads and tails. In counting the number of nonsmokers in a town, the possible values are 0, 1, 2, 3, ..., N, where N is the total population of the town. For this latter example X can represent the number of nonsmokers. Its numerical value becomes known only after we survey the entire town. We can, however, speculate about the value of X before making the survey.

PROBABILITY MASS FUNCTIONS

It is often useful to think of x, the value of X, as an outcome of an associated experiment. For example, if X is the number of heads in ten flips of a coin, we can think of the possible outcomes of the ten flips as the numbers

0, 1, 2, ..., 10, rather than as sequence of heads and tails. When this is the case, we can ask for the probability of the event consisting of the one outcome x. This should be written $P(\{x\})$, but for simplicity will be written $P(x)$ or occasionally $P(X = x)$. We will similarly simplify the probability notation of more general sets. For example, the probability that X will prove to be some number between 0 and 5 inclusive should be written $P(\{x: 0 \leqq x \leqq 5\})$, but will be written $P(0 \leqq X \leqq 5)$.

Consider now the simple situation in which there are N equally likely values for X, $x = 1, 2, 3, \ldots, N$. We can, in this case write

$$P(x) = \frac{1}{N} \qquad x = 1, 2, 3, \ldots, N.$$

Often the possible values for X are not equally likely, but are themselves made up of equally likely outcomes. For example, if we flip a coin twice and let X represent the number of heads ($x = 0, 1, 2$), then clearly one head is more likely (will come up more often) than either no heads or two heads. However, if we consider the outcome of the two flips to be one of the ordered pairs (H, H), (H, T), (T, H), (T, T) (where H represents a head and T represents a tail), then we might consider each of these outcomes equally likely. The event $\{0\}$, no heads is the same as the event $\{(T, T)\}$ and contains only one outcome. The event $\{1\}$ is the same as the event $\{(H, T), (T, H)\}$ and contains two outcomes. The event $\{2\}$ is the same as the event $\{(H, H)\}$ and contains only one outcome. From this we obtain that

$$P(X = 0) = 1/4$$
$$P(X = 1) = 2/4 = 1/2$$
$$P(X = 2) = 1/4.$$

This function $P(x)$ of the possible values of X is called the *probability mass function* of the discrete random variable X.

The random variable X subdivides the sample space of its possible values into disjoint events, the events containing one value each. Since we are considering each value an outcome of an associated experiment, we see that the mass function is actually a set of probabilities, each of which is the probability of an event containing one outcome. Therefore, the probability mass function must satisfy all the properties discussed in Chapter One for events. For example, $0 \leqq P(x) \leqq 1$ for each x. It should also be evident that since the events containing one outcome each are disjoint and their union (all the outcomes in one event) is the certain event, the probability mass function must be such that

$$\sum_{\text{all } x} P(x) = 1.$$

For example, if X represents the number of dots which will appear on the upper face of a die after it is tossed, then the possible values for X are $x = 1, 2, 3, 4, 5, 6$. If we assume equally likely outcomes, then $P(x) = 1/6$, $x = 1, 2, 3, 4, 5, 6$.

EXERCISE 2-1

Demonstrate that the probability mass function described above for the tossing of a die satisfies Axioms I–III of Section 1-3.

CUMULATIVE DISTRIBUTION FUNCTIONS

When the values x can be ordered as in the case where they are real numbers we can define the *cumulative distribution function*

$$F(x) = \sum_{z<x} P(z) \qquad \text{for all } x.$$

(Note that it was convenient here not to use x as the dummy variable for X.)

If the values of x are indeed real numbers, then the cumulative distribution function is defined for all values of x on the whole real line even though the probability mass function $P(x)$ is defined only at a countable number of values. It is interesting to note that $F(x)$ is the probability that X will take on some value less than or equal to x.

Some properties of the cumulative distribution function should be clear from its definition. For example, $F(x)$, being the sum of nonnegative numbers cannot itself be negative. Since all $x > -\infty$, $F(-\infty) = 0$. On the other hand, all $x < \infty$ and so $F(\infty) = 1$. The function $F(x)$ changes value only at those points for which $P(x) > 0$ and at those points by an amount $P(x)$.

In Figure 2-1 we see a plot of the cumulative distribution function for the die tossing situation with equally likely outcomes. It should be noted that in a plot of this type the value of $F(x)$ at a point of discontinuity is the higher of the two values.

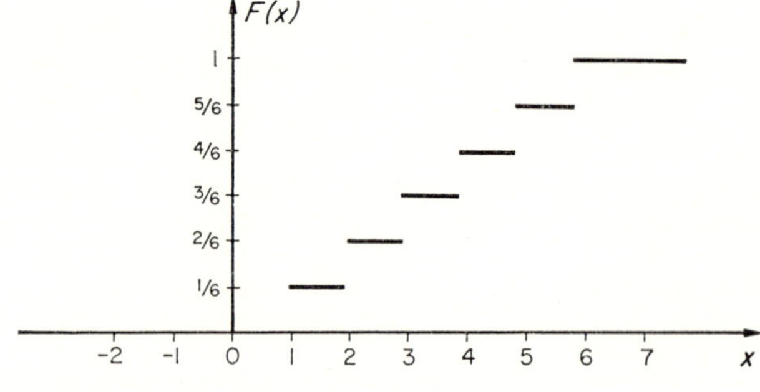

Figure 2-1

2-2 Some Probability Mass Functions

BERNOULLI TRIALS

Let us say that we have a coin which has probability p of turning up heads on any flip. If we let X equal the number of heads in one flip of the coin then the possible values for x are 0 and 1. Clearly the probability mass function is $P(1) = p$ and $P(0) = 1-p$. This can be written in a single functional form as

$$P(x) = p^x(1-p)^{1-x} \qquad x = 0, 1.$$

It is interesting to note that this mass function is given in terms of an unspecified constant p. For any particular value of p between zero and unity, this is called a *bernoulli* mass function. An unspecified constant, such as p, is called a *parameter* of the distribution.

THE BINOMIAL DISTRIBUTION

Let us assume that a coin is flipped n times. Let X equal the number of heads in those flips. Clearly $x = 0, 1, 2, \ldots, n$. The problem is to find $P(x)$.

Consider a possible n-tuple which could result from these n flips. As an example, let n be three, and consider the particular triple (H, T, H) where H represents a head, and T a tail. Since the result of one flip in no way influences the relative frequency (or probability) of any other flip, we find that $P(H, T, H) = P(H)P(T)P(H) = p(1-p)p = p^2(1-p)$. In general, the order of the heads and tails is immaterial, and only the actual numbers of heads and tails are needed to determine the joint probability. We now notice that if there are n flips resulting in heads at any x specified positions in the sequence of flips, then the probability is $p^x(1-p)^{n-x}$.

Since we are interested in the probability of having exactly x heads without regard to the order in which they appear, we must group all those n-tuples with exactly x heads into the same event. The events are disjoint, and therefore by Axiom III the probabilities will add, and since each has probability $p^x(1-p)^{n-x}$, all we need do is count how many there are. This is a combination problem. Out of n places we must choose x places in which to place an H. This can be done in $\binom{n}{x}$ different ways. Therefore there are exactly $\binom{n}{x}$ n-tuples with exactly x heads. The mass function is therefore given by

$$P(x) = \binom{n}{x} p^x(1-p)^{n-x} \qquad x = 0, 1, 2, \ldots, n.$$

Generally speaking, this is the probability of obtaining exactly x successes in n independent trials when the probability for success at each trial is p. It is of course important to check to see that

$$\sum_{x=0}^{n} P(x) = 1.$$

In order to do this we will use the Binomial Theorem given by the relation

$$\sum_{v=0}^{v} \binom{n}{v} x^v y^{n-v} = (x+y)^n.$$

Using this we easily see that

$$\sum_{\text{all } x} P(x) = \sum_{x=0}^{n} \binom{n}{x} p^x (1-p)^{n-x} = [p+(1-p)]^n = 1^n = 1.$$

EXERCISE 2-2

Show that a. $\sum_{v=0}^{n} \binom{n}{v} = 2^n$

b. $\sum_{v=0}^{n} (-1)^v \binom{n}{v} = 0$

for any $n = 1, 2, 3, \ldots$

THE POISSON DISTRIBUTION

The poisson distribution is very useful in the physical sciences. For example, the number of particles decaying in a given time from a mass of radioactive material follows this distribution. There are several ways of deriving the poisson probability mass function. We will obtain it by a limiting process on a binomial probability mass function.

Consider the binomial probability mass function

$$P(x) = \binom{n}{x} p^x (1-p)^{n-x}, \qquad x = 0, 1, 2, \ldots, n.$$

We ask: What is the limiting value of $P(x)$ with fixed x as $n \to \infty$ and $p \to 0$ in such a way that $np = \lambda$ (a constant)?

Note that for any n and p

$$P(x) = \binom{n}{x} p^x (1-p)^{n-x}$$

$$= \frac{n(n-1) \ldots (n-x+1)}{x!} p^x \frac{(1-p)^n}{(1-p)^x}.$$

Dividing and multiplying by n^x and noting that there are exactly x terms in the product $n(n-1) \dots (n-x+1)$, we obtain

$$P(x) = \frac{n}{n} \cdot \frac{n-1}{n} \cdot \frac{n-2}{n} \dots \frac{n-x+1}{n} \frac{(np)^x}{x!} \frac{(1-p)^n}{(1-p)^x}.$$

Now as $n \to \infty$, $p \to 0$ with $np = \lambda$ remaining constant, we obtain

$$\frac{n}{n} = 1, \frac{n-1}{n} \to 1, \frac{n-x+1}{n} \to 1, (1-p)^x \to 1, (np)^x = \lambda^x$$

and $(1-p)^n = (1-\lambda/n)^n \to e^{-\lambda}$ by the definition of the exponential function.

Putting all the pieces back together we find that in the limit

$$P(x) = \frac{e^{-\lambda}\lambda^x}{x!} \qquad x = 0, 1, 2, \dots$$

which is known as the *poisson probability mass function.* Notice that this function is defined for any value of $\lambda > 0$ which is called the parameter of the poisson distribution. We must now check to see that

$$\sum_{x=0}^{\infty} P(x) = \sum_{x=0}^{\infty} \frac{e^{-\lambda}\lambda^x}{x!} = 1.$$

$e^{-\lambda}$ comes out of the summation and

$$\sum_{x=0}^{\infty} \frac{\lambda^x}{x!} = e^{\lambda}.$$

Therefore the above equation is checked.

Example 2-1

The human body contains approximately 6×10^{10} white blood cells randomly dispersed in about 6×10^6 cubic millimeters of blood. If we were to take a quantity of 10^{-3} of a cubic millimeter of blood and place it under a microscope, what would be the probability of observing x white blood cells?

If we assume that the blood cells are too small to get in the way of each other, then each cell has a probability $p = 1/6 \times 10^9$ of appearing in the drop. (There are 6×10^9 such drops in the body.) Since there are 6×10^{10} cells which either are, or are not, there, x is the number of successes in 6×10^{10} bernoulli trials with probability p for success; i.e.

$$P(x) = \binom{n}{x} p^x(1-p)^{n-x} \qquad x = 0, 1, 2, 3, \dots, n$$

where $n = 6 \times 10^{10}$ and $p = 1/6 \times 10^9$. The arithmetic involved becomes hopelessly difficult without the poisson approximation. We now note that n is very large, and p is very small, but that $np = \lambda = 10$. Therefore, we can set

$$P(x) = \frac{e^{-\lambda}\lambda^x}{x!} \qquad x = 0, 1, 2, \dots.$$

These values are easily obtained; see, for example, Table B-2 of the Appendix.

NEGATIVE BINOMIAL DISTRIBUTION

Let us say that we have a coin with probability p that heads will turn up on any flip. If we were to flip the coin until we obtained exactly r heads, what would be the probability of obtaining exactly x tails? In order to have exactly r heads and x tails, there must have been exactly $r+x$ flips of the coin. A particular $(r+x)$-tuple outcome which would be in the event "exactly x tails before the rth head" would have its last value a head and $r-1$ others somewhere in the remaining $r+x-1$ places. The probability of each of these $(r+x)$-tuples is $p^r(1-p)^x$. How many of them are there? The last must be a head, but there are $r+x-1$ places left out of which we choose $r-1$ for the remaining heads. This can be done in $\binom{r+x-1}{r-1}$ ways. Therefore

$$P(x) = \binom{r+x-1}{r-1} p^r(1-p)^x \qquad x = 0, 1, 2, \ldots.$$

Generally speaking, this is the probability of obtaining exactly x failures before the rth success when there is probability p for success. This is called the *negative binomial probability mass junction.* In order to check that

$$\sum_{x=0}^{\infty} P(x) = \sum_{x=0}^{\infty} \binom{x+r-1}{r-1} p^r(1-p)^x = 1$$

we expand the series,

$$\sum_{x=0}^{\infty} \binom{x+r-1}{r-1} q^x = 1+rq+\frac{(r+1)r}{2} q^2+ \ldots \frac{(x+r-1)\ldots r}{x!} q^x+ \ldots.$$

We now make a Taylor series expansion about zero of $1/(1-q)^r$, giving

$$\frac{1}{(1-q)^r} = 1+rq+\frac{(r+1)r}{2!} q^2+ \ldots \frac{(r+x-1)\ldots r}{x!} q^x+ \ldots.$$

Therefore, if $q = 1-p$ we obtain

$$\sum_{x=0}^{\infty} P(x) = p^r \frac{1}{[1-(1-p)]^r} = \frac{p^r}{p^r} = 1.$$

THE HYPERGEOMETRIC DISTRIBUTION

Let us say that there are n objects, m of which have a property A, while the remaining $n-m$ do not. If r of the n objects were selected at random, without replacement, what would be the probability that exactly x of these r would have the property A?

There is a total of $\binom{n}{r}$ ways to choose r objects from n (without regard to order). How many of these have exactly x objects with property A? Using the properties of multicombinations we see that we must choose x from m and $r-x$ from $n-m$, giving $\binom{m}{x}\binom{n-m}{r-x}$ combinations. Therefore

$$P(x) = \frac{\binom{m}{x}\binom{n-m}{r-x}}{\binom{n}{r}} \qquad x = \max(0, r-n+m), \max(0, r-n+m)+1, \ldots, \min(m, r).$$

The limits come about since x cannot be less than $r-n+m$ nor greater than m or r. This is called the *hypergeometric probability mass function.* In order to check that

$$\sum_{x=\max(0,\, r-n+m)}^{\min(m,\, r)} P(x) = \sum_{x=\max(0,\, r-n+m)}^{\min(m,\, r)} \frac{\binom{m}{x}\binom{n-m}{r-x}}{\binom{n}{r}} = 1$$

we consider the problem of multiplying out $(U+V)^m(W+V)^{n-m}$. The coefficient of $U^xW^{r-x}V^{n-r}$ can be obtained by choosing x number of U's from m number of $(U+V)$'s and $(r-x)$ number of W's from $(n-m)$ number of $(W+V)$'s giving the coefficient as $\binom{m}{x}\binom{n-m}{r-x}$. The term is therefore $\binom{m}{x}\binom{n-m}{r-x} U^xW^{r-x}V^{n-r}$. If we now let $U = W = 1$, then each term for any x is simply V^{n-r}. There are

$$\sum_{x=\max(0,\, r-n+m)}^{\min(m,\, r)} \binom{m}{x}\binom{n-m}{r-x}$$

such terms. Therefore

$$(1+V)^m(1+V)^{n-m} = \sum_{r=0}^{n} \sum_{x=\max(0,\, r-n+m)}^{\min(m,\, r)} \binom{m}{x}\binom{n-m}{r-x} V^{n-r}.$$

But

$$(1+V)^m(1+V)^{n-m} = (1+V)^n = \sum_{r=0}^{n} \binom{n}{r} V^{n-r}$$

and by equating coefficients,

$$\binom{n}{r} = \sum_{x=\max(0,\, r-n+m)}^{\min(m,\, r)} \binom{m}{x}\binom{n-m}{r-x}.$$

Using this result we obtain

$$\sum_{x=\max(0,\, r-n+m)}^{\min(m,\, r)} P(x) = \sum_{x=\max(0,\, r-n+m)}^{\min(m,\, r)} \frac{\binom{m}{x}\binom{n-m}{r-x}}{\binom{n}{r}} = \frac{\binom{n}{r}}{\binom{n}{r}} = 1.$$

Remark *Usually $r < m$ and $n > r+m$, so that $x = 0, 1, 2, \ldots, r$.*

Example 2-2

Let us say that you have a batch of 1000 bolts from which you will choose 20 at random to put in a box. If there are five defective bolts in the batch, what is the probability that there will be no defectives in the box?

This is an example of the situation above with $n = 1000$, $m = 5$, and $r = 20$. We are asking for the probability that $X = 0$. We obtain this from

$$P(X=0) = \frac{\binom{5}{0}\binom{995}{20}}{\binom{1000}{20}} = \frac{(995)!/(20)!(975)!}{(1000)!/(20)!(980)!}$$

$$= \frac{(995)(994)(993)\ldots(976)}{(1000)(999)(998)\ldots(981)}$$

$$= \frac{(980)(979)(978)(977)(976)}{(1000)(999)(998)(997)(996)} = 0.904$$

THE DISCRETE UNIFORM DISTRIBUTION

If there are n equally likely outcomes, then $P(x) = (1/n)$ for all n values of x. If $x = 1, 2, 3, \ldots, n$, then

$$P(x) = \frac{1}{n} \qquad x = 1, 2, 3, \ldots, n.$$

Clearly

$$\sum_{x=1}^{n} P(x) = \sum_{x=1}^{n} \frac{1}{n} = 1.$$

This is the *uniform probability mass function* which we have already used in the example of the tossing of a die.

2-3 Expected Values

Let us say that we have a die which is not quite balanced. The underlying mass function will be written as $P(x)$, $x = 1, 2, 3, 4, 5, 6$. Imagine that we have tossed this die ten times and obtained the outcomes 1, 2, 5, 6, 3, 2, 5, 4,2,3. What is the average value of the outcomes?

$$\text{Av.} = \frac{1+2+5+6+3+2+5+4+2+3}{10} = 3.3.$$

Note that 2 appears three times, 5 and 3 appear twice and 1, 4, and 6 appear once. We can therefore write

$$\text{Av.} = 1\cdot\tfrac{1}{10}+2\cdot\tfrac{3}{10}+3\cdot\tfrac{2}{10}+4\cdot\tfrac{1}{10}+5\cdot\tfrac{2}{10}+6\cdot\tfrac{1}{10} = 3.3.$$

That is, $\frac{1}{10}$ of the time a 1 appeared, $\frac{3}{10}$ of the time a 2 appeared, $\frac{2}{10}$ of the time a 3 appeared, etc. What we are saying is that

$$\text{Av.} = \sum_{all\ x} x\cdot f_n(x)$$

where $f_n(x)$ is the relative frequency of x for a total of n outcomes. In the long run, as the number of tosses becomes larger and larger, the relative frequency of x should approach $P(x)$ as a limit. If n is large enough, one would expect that the average should approach $E(X)$ where

$$\begin{aligned} E(X) &= 1P(1)+2P(2)+3P(3)+4P(4)+5P(5)+6P(6) \\ &= \sum_x xP(x) \end{aligned}$$

where we use the x under the summation sign to signify the sum over all x. $E(X)$ is called the *expectation* of the random variable X. Using the same observations we may ask for the average value of some function $g(x)$. We would obtain,

$$\begin{aligned} \text{Av.}[g(X)] &= \frac{g(1)+g(2)+g(5)+g(6)+g(3)+g(2)+g(5)+g(4)+g(2)+g(3)}{10} \\ &= g(1)\tfrac{1}{10}+g(2)\tfrac{3}{10}+g(3)\tfrac{2}{10}+g(4)\tfrac{1}{10}+g(5)\tfrac{2}{10}+g(6)\tfrac{1}{10}. \end{aligned}$$

For a very large number of tosses one would expect the relative frequencies to approach the corresponding probabilities, and the average would then approach $Eg(X)$ where

$$\begin{aligned} Eg(X) &= g(1)P(1)+g(2)P(2)+g(3)P(3)+g(4)P(4)+g(5)P(5)+g(6)P(6) \\ &= \sum_x g(x)P(x). \end{aligned}$$

Definition *If $P(x)$ is the probability that $X = x$, then* the *expected value* (also called the *expectation*) of any function $g(x)$ is given by

$$Eg(X) = \sum_x g(x)P(x)$$

provided that the sum converges. (When the sum does not converge we say that the expectation does not exist.)

It is important to note that even though $g(x)$ and $P(x)$ are both functions of x, the expectation $Eg(X)$ is not a function of x.

MOMENTS

A class of important expectations is known as the *moments* of the distribution. These moments are defined by

$$\mu_r' = EX^r = \sum_x x^r P(x) \qquad r = 1, 2, 3, \ldots .$$

The prime and subscript on μ_1' are usually dropped, and μ is called *the mean.* The moments about the mean defined by

$$\mu_r = E(X-\mu)^r = \sum_x (x-\mu)^r P(x) \qquad r = 2, 3, 4, \ldots$$

are known as the *central moments.* The first central moment (where $r = 1$) has been omitted since

$$\sum_x (x-\mu)P(x) = \sum_x xP(x) - \sum_x \mu P(x) = \mu - \mu$$
$$= 0$$

and does not depend on the distribution.

The second central moment μ_2 is called the *variance*, and is usually written Var X and in some special cases σ^2.

PROPERTIES OF THE VARIANCE

The variance has widespread application in statistics. It will benefit us to investigate some of its properties.

1) Since Var X is the expected value of a squared quantity it can never be negative. That is, Var $X \geqq 0$.
2) $\text{Var } X = E(X-\mu)^2 = \sum_x (x-\mu)^2 P(x) = \sum_x (x^2 - 2\mu x + \mu^2)P(x)$

$$= \sum_x x^2 P(x) - 2\mu \sum_x xP(x) + \mu^2 \sum_x P(x).$$

Now

$$\sum_x x^2 P(x) = \mu_2', \qquad \sum_x xP(x) = \mu, \qquad \text{and} \qquad \sum_x P(x) = 1$$

giving us

$$\text{Var } X = \mu_2' - 2\mu^2 + \mu^2 = \mu_2' - \mu^2$$

or

$$\text{Var } X = EX^2 - (EX)^2.$$

Since Var $X \geqq 0$ it follows that

$$EX^2 \geqq (EX)^2.$$

EXERCISE 2-3

Derive the following very useful properties of expectations.

1) $E(aX+b) = aEX+b$
2) $\mathrm{Var}(aX+b) = a^2 \mathrm{Var}\, X$

where a and b are constants.

2-4 The Mean and Variance of Some Discrete Distributions

It will be quite worthwhile for us to devote a section to finding the mean and variance, and in some cases, the higher moments of the distributions we have already obtained.

THE BERNOULLI DISTRIBUTION

The mass function for the bernoulli distribution is given by

$$P(x) = p^x(1-p)^{1-x} \qquad x = 0, 1$$
$$0 < p < 1.$$

Therefore

$$\mu = EX = \sum_{x=0}^{1} xP(x) = 0\cdot(1-p)+1\cdot p = p.$$

Similarly

$$\mu_r' = EX^r = \sum_{x=0}^{1} x^r P(x) = 0(1-p)+1\cdot p = p.$$

Therefore

$$\mathrm{Var}\, X = \mu_2' - \mu^2 = p - p^2 = p(1-p).$$

THE BINOMIAL DISTRIBUTION

The mass function of the binomial distribution is given by

$$P(x) = \binom{n}{x} p^x(1-p)^{n-x} \qquad x = 0, 1, 2, \ldots, n$$
$$0 < p < 1$$

and so

$$\mu = EX = \sum_{x=0}^{n} x\binom{n}{x} p^x(1-p)^{n-x} = \sum_{x=1}^{n} x\binom{n}{x} p^x(1-p)^{n-x}$$

(Note that since we are multiplying by x we can eliminate the term in the sum where $x = 0$.)

$$= \sum_{x=1}^{n} \frac{xn!}{x!(n-x)!} p^x(1-p)^{n-x} = \sum_{x=1}^{n} \frac{n!}{(x-1)!(n-x)!} p^x(1-p)^{n-x}$$

$$= np\sum_{x=1}^{n} \frac{(n-1)!}{(x-1)!(n-x)!} p^{x-1}(1-p)^{n-x} = np\sum_{x=1}^{n} \binom{n-1}{x-1} p^{x-1}(1-p)^{n-x}.$$

If we now let $y = x-1$ and $m = n-1$ we obtain

$$\mu = np\sum_{y=0}^{m} \binom{m}{y} p^y(1-p)^{m-y}.$$

However we already know that

$$\sum_{y=0}^{m} \binom{m}{y} p^y(1-p)^{m-y} = 1$$

giving

$$\mu = np.$$

The method used here is a very important and useful procedure in probability manipulations. The idea is to reduce a complicated formula to one which we have already solved. We will use this technique again in finding μ_2'. Here

$$\mu_2' = EX^2 = \sum_{x=0}^{n} x^2\binom{n}{x} p^x(1-p)^{n-x} = \sum_{x=1}^{n} x^2\binom{n}{x} p^x(1-p)^{n-x}$$

$$= \sum_{x=1}^{n} [x(x-1)+x]\binom{n}{x} p^x(1-p)^{n-x}$$

$$= \sum_{x=2}^{n} x(x-1)\binom{n}{x} p^x(1-p)^{n-x} + \sum_{x=1}^{n} x\binom{n}{x} p^x(1-p)^{n-x}$$

(Note the increase of subscript again. Why is this possible?)

$$= \sum_{x=2}^{n} \frac{x(x-1)n!}{x!(n-x)!} p^x(1-p)^{n-x} + \mu$$

$$= \sum_{x=2}^{n} \frac{n!}{(x-2)!(n-x)!} p^x(1-p)^{n-x} + \mu$$

$$= n(n-1)p^2 \sum_{x=2}^{n} \binom{n-2}{x-2} p^{x-2}(1-p)^{n-x} + \mu$$

$$= n(n-1)p^2 \sum_{z=0}^{r} \binom{r}{z} p^z(1-p)^{r-z} + \mu \text{ where } z = x-2 \text{ and } r = n-2$$

$$= n(n-1)p^2 + np = n^2p^2 + np(1-p).$$

Therefore

$$\text{Var } X = \mu_2' - \mu^2 = np(1-p).$$

THE POISSON DISTRIBUTION

The mass function for the poisson distribution is given by

$$P(x) = \frac{e^{-\lambda}\lambda^x}{x!} \qquad x = 0, 1, 2, \dots$$

$$\lambda > 0$$

and therefore

$$\mu = EX - \sum_{x=0}^{\infty} \frac{xe^{-\lambda}\lambda^x}{x!} = \sum_{x=1}^{\infty} \frac{e^{-\lambda}\lambda^x}{(x-1)!}$$

$$= \lambda \sum_{x=1}^{\infty} \frac{e^{-\lambda}\lambda^{x-1}}{(x-1)!} = \lambda \sum_{y=0}^{\infty} \frac{e^{-\lambda}\lambda^y}{y!}$$

where $y = x-1$ giving

$$\mu = \lambda.$$

We can arrive at the same result by another method,

$$e^{\lambda} = \sum_{x=0}^{\infty} \frac{\lambda^x}{x!}$$

and so

$$e^{\lambda} = \frac{de^{\lambda}}{d\lambda} = \frac{d}{d\lambda} \sum_{x=0}^{\infty} \frac{\lambda^x}{x!} \sum_{x=1}^{\infty} \frac{\lambda^{x-1}}{(x-1)!} \cdot$$

giving

$$\mu = \sum_{x=1}^{\infty} \frac{e^{-\lambda}\lambda^x}{(x-1)!} = \lambda e^{-\lambda} \sum_{x=1}^{\infty} \frac{\lambda^{x-1}}{(x-1)!} = \lambda.$$

We can find μ_2' by proceeding in either way. For example

$$\mu_2' = EX^2 = \sum_{x=0}^{\infty} x^2 \frac{e^{-\lambda}\lambda^x}{x!} = \sum_{x=1}^{\infty} [x(x-1)+x] \frac{e^{-\lambda}\lambda^x}{x!}$$

$$= \sum_{x=2}^{\infty} \frac{x(x-1)e^{-\lambda}\lambda^x}{x!} + \sum_{x=1}^{\infty} x \frac{e^{-\lambda}\lambda^x}{x!}$$

$$= \lambda^2 \sum_{x=2}^{\infty} \frac{e^{-\lambda}\lambda^{x-2}}{(x-2)!} + \lambda = \lambda^2 \sum_{z=0}^{\infty} \frac{e^{-\lambda}\lambda^{z}}{z!} + \lambda \quad \text{where} \quad z = x-2$$

giving

$$\mu_2' = \lambda^2 + \lambda.$$

It is then clear that

$$\text{Var } X = \mu_2' - \mu^2 = \lambda^2 + \lambda - \lambda^2 = \lambda.$$

Note that in the poisson distribution the mean and variance are equal.

Exercise 2-4

Find Var X by using the equality

$$e^{\lambda} = \frac{d^2}{d\lambda^2} e^{\lambda} = \frac{d^2}{d\lambda^2} \sum_{x=0}^{\infty} \frac{\lambda^x}{x!} = \sum_{x=2}^{\infty} \frac{\lambda^{x-2}}{(x-2)!}.$$

THE NEGATIVE BINOMIAL DISTRIBUTION

The mass function for the negative binomial distribution is given by

$$P(x) = \binom{x+r-1}{r-1} p^r (1-p)^x \qquad x = 0, 1, 2, \ldots \quad 0 < p < 1$$

giving

$$\mu = EX = \sum_{x=0}^{\infty} xP(x) = \sum_{x=0}^{\infty} x \binom{x+r-1}{r-1} p^r (1-p)^x$$

$$= \sum_{x=1}^{\infty} \frac{x(x+r-1)!}{x!(r-1)!} p^r (1-p)^x$$

$$= \sum_{x=1}^{\infty} \frac{(x+r-1)!}{(x-1)!(r-1)!} p^r (1-p)^x.$$

If we now let $y = x-1$ and $s = r+1$, we have

$$\frac{(x+r-1)!}{(x-1)!(r-1)!} = \frac{(y+s-1)!}{y!(s-1)!} (s-1)$$

and therefore

$$\mu = (s-1) \sum_{y=0}^{\infty} \binom{y+s-1}{s-1} p^{s-1} (1-p)^{y+1}$$

$$= \frac{(s-1)(1-p)}{p} \sum_{y=0}^{\infty} \binom{y+s-1}{s-1} p^s (1-p)^y$$

$$= \frac{(s-1)(1-p)}{p} = \frac{r(1-p)}{p}.$$

EXERCISE 2-5

In a manner similar to that above, show that

$$\mu_2' = \frac{r^2(1-p)^2}{p^2} + \frac{r(1-p)}{p^2}$$

and

$$\operatorname{Var} X = \frac{r(1-p)}{p^2} .$$

THE HYPERGEOMETRIC DISTRIBUTION

The mass function for the hypergeometric distribution is given by

$$P(x) = \frac{\binom{m}{x}\binom{n-m}{r-x}}{\binom{n}{r}} \quad x = \max(0, r-n+m), \max(0, r-n+m)+1, \ldots, \min(m, r)$$

where m and r are any fixed integers no larger than n. For simplicity, however, we will work only with the special case where $r < m$ and $n > r+m$. For this case

$$\mu = EX = \sum_{x=0}^{r} \frac{x\binom{m}{x}\binom{n-m}{r-x}}{\binom{n}{r}} = \frac{\sum_{x=1}^{r} \frac{m!}{(x-1)!(m-x)!}\binom{n-m}{r-x}}{\binom{n}{r}} .$$

If we let $y = x-1$, $s = r-1$, $M = m-1$, and $N = n-1$, we obtain

$$\mu\binom{n}{r} = \sum_{y=0}^{s} \frac{(M+1)M!}{y!(M-y)!}\binom{N-M}{s-y} = (M+1)\sum_{y=0}^{s}\binom{M}{y}\binom{N-M}{s-y}$$

$$= (M+1)\binom{N}{s} = m\binom{n-1}{r-1}$$

giving

$$\mu = \frac{m\binom{n-1}{r-1}}{\binom{n}{r}} = \frac{m(n-1)!r!(n-r)!}{(r-1)!(n-r)!n!} = \frac{mr}{n} .$$

EXERCISE 2-6

In a similar manner, show that

$$\operatorname{Var} X = \frac{m(n-m)r(n-r)}{n^2(n-1)} .$$

THE DISCRETE UNIFORM DISTRIBUTION

For the discrete uniform distribution the mass function is

$$P(x) = \frac{1}{n} \qquad x = 1, 2, 3, \ldots, n$$

giving

$$\mu = \sum_{x=1}^{n} xP(x) = \frac{1}{n}\sum_{x=1}^{n} x.$$

In order to calculate this sum consider stacking squares of unit size on an axis, as in Figure 2-2. The total area will be $\Sigma_{x=1}^{n} x$. This area can easily

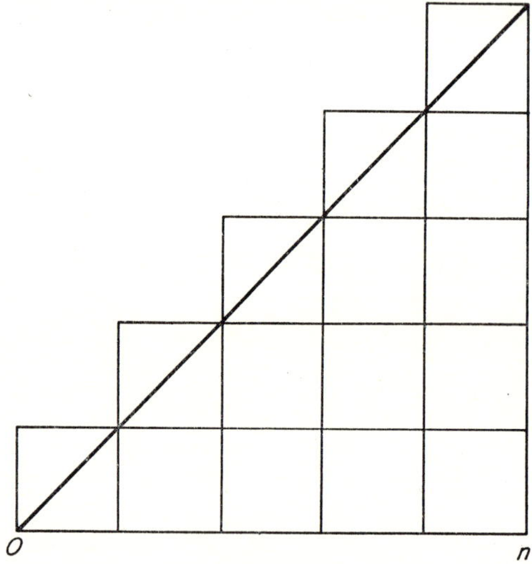

Figure 2-2

be seen to be the area of a right triangle of base n and height n plus n half squares. Therefore

$$\sum_{x=1}^{n} x = \frac{n^2}{2} + \frac{n}{2} = \frac{n(n+1)}{2}.$$

Using this we see that

$$\mu = \frac{n+1}{2}.$$

Now

$$\mu_2' = \sum_{x=1}^{n} x^2 P(x) = \frac{1}{n}\sum_{x=1}^{n} x^2.$$

In order to evaluate this sum, consider the identity $(x+1)^3 - x^3 = 3x^2+3x+1$ and sum it.

$$\sum_{x=1}^{n} (x+1)^3 - \sum_{x=1}^{n} x^3 = 3\sum_{x=1}^{n} x^2 + 3\sum_{x=1}^{n} x + \sum_{x=1}^{n} 1$$

which becomes

$$(n+1)^3 - 1 = 3\sum_{x=1}^{n} x^2 + \frac{3n(n+1)}{2} + n$$

giving

$$\sum_{x=1}^{n} x^2 = \frac{(n+1)^3}{3} - \frac{n(n+1)}{2} - \frac{n}{3} - \frac{1}{3}$$

or

$$\sum_{x=1}^{n} x^2 = \frac{n^3}{3} + \frac{n^2}{2} + \frac{n}{6}$$

and therefore

$$\mu_2' = \frac{n^2}{3} + \frac{n}{2} + \frac{1}{6}$$

and

$$\text{Var } X = \frac{n^2}{3} + \frac{n}{2} + \frac{1}{6} - \frac{(n+1)^2}{4} = \frac{n^2}{12} - \frac{1}{12}.$$

2-5 Continuous Random Variables and Their Distributions

When the random variable X can take on an uncountable number of values no probability mass function can be defined. It is well known that if there are an uncountable number of values for which $P(x) > 0$, then it could not be the case that $\Sigma_{\text{all } x} P(x) = 1$; in fact, this sum must be infinite. In this case it is often appropriate to take intervals on the real line as the events for which we define probabilities. For example, it may make sense to speak of the probability that X will take on some value between a and b, i.e., $P(a \leqq X \leqq b)$.

If we construct a subdivision of the real line $I_1, I_2, I_3, \ldots$, where the I_i are disjoint intervals such that $\cup_i I_i = (-\infty, \infty)$, then Axioms I–III must hold for these I_i. In particular, for each i, $P(X \in I_i) \geqq 0$ and $\Sigma_i P(X \in I_i) = 1$. When X takes on an uncountable number of values and there is no one value of x such that $P(x) > 0$, then we say that X is a *continuous random variable.*

PROBABILITY DENSITY FUNCTIONS

If there exists a non-negative function $f(x)$ such that

$$P(a \leqq X \leqq b) = \int_a^b f(x)\, dx$$

for every a and b then $f(x)$ is said to be the *probability density function* of X. A convenient interpretation of $f(x)$ is to consider $f(x)\, dx$ as the probability that X will take on a value in the small interval $(x, x+dx)$. The integral can then be interpreted as a sum over intervals.

Some useful properties of the probability density function $f(x)$ to which we will later refer, will now be listed:

1) $f(x)$ is single valued.
2) $f(x) \geqq 0$.
3) $f(x)$ is continuous, or at least piecewise continuous, i.e., it has at most a countable number of discontinuities.
4) $P(X \in I) = \int_I f(x)\, dx$, for any interval I.
5) If $I = \cup_i I_i$ with $I_i \cap I_j = \phi, \quad i \neq j$, then $P(X \in I) = \sum_i \int_{I_i} f(x)\, dx$.
6) $\int_{-\infty}^{\infty} f(x)\, dx = 1$.

CUMULATIVE DISTRIBUTION FUNCTIONS

We can now define the *cumulative distribution function* for a continuous random variable by

$$F(x) = \int_{-\infty}^{x} f(t)\, dt.$$

From this definition we can find another relation between the cumulative distribution function and the probability density function by differentiation. Namely,

$$\frac{\partial}{\partial x} F(x) = f(x).$$

Since an indefinite integral of a piecewise continuous function is continuous, $F(x)$ is a continuous function of x. Since $f(x) \geqq 0$, it is easily seen that $F(x)$ is a monotonic nondecreasing function which is zero at $x = -\infty$ and unity at $x = +\infty$. In general it looks something like the curve in Figure 2-3. Another very useful property of distribution functions is that

$$P(a \leqq X \leqq b) = \int_a^b f(x)\,dx = \int_{-\infty}^{b} f(x)\,dx - \int_{-\infty}^{a} f(x)\,dx$$
$$= F(b) - F(a).$$

Cumulative distribution functions will play a much greater role in the development when X is continuous than when it is discrete.

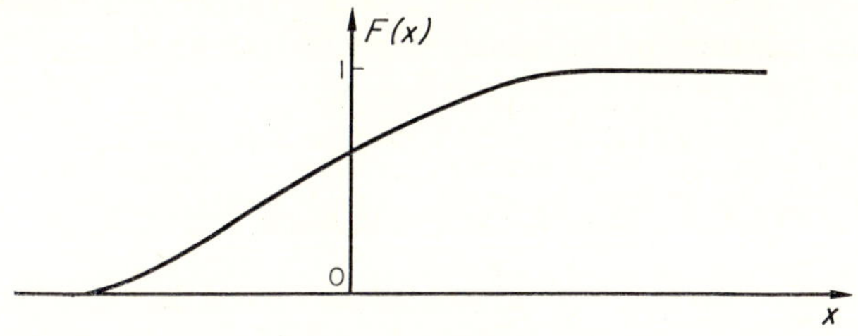

Figure 2-3

EXPECTATIONS

Using integrals rather than sums, we can define expectations by

$$Eg(X) = \int_{-\infty}^{\infty} g(x)\,f(x)\,dx$$

when the integral is finite. When it is not finite, we say that the expectation does not exist. In particular

$$\mu = \int x f(x)\,dx$$

(where the omission of limits of integration signifies limits from $-\infty$ to $+\infty$) and

$$\begin{aligned}\text{Var } X &= \int (x-\mu)^2 f(x)\,dx = \int (x^2 - 2\mu x + \mu^2) f(x)\,dx \\ &= \int x^2 f(x)\,dx - 2\mu \int x f(x)\,dx + \mu^2 \int f(x)\,dx \\ &= EX^2 - 2\mu\cdot\mu + \mu^2 = \mu_2{}' - \mu^2 \\ &= EX^2 - (EX)^2\end{aligned}$$

as in the discrete case.

EXERCISE 2-7

Show that

1) $E(aX+b) = aEX+b$
2) $\text{Var}(aX+b) = a^2 \text{ Var } X$

(where a and b are constants) hold for continuous random variables.

2-6 Some Distributions of Continuous Random Variables

THE UNIFORM DISTRIBUTION

Let us say that the probability of X having a value in any subinterval of the interval (α, β) is proportional to the length of that subinterval. The whole interval has length $\beta-\alpha$ and if we assume that $P(\alpha \leqq X \leqq \beta) = 1$, then we find that the factor of proportionality is $1/(\beta-\alpha)$ for any subinterval (γ,δ) where $a \leqq \gamma \leqq \delta \leqq \beta$. Therefore, $P(\gamma \leqq X \leqq \delta) = (\delta-\gamma)/(\beta-\alpha)$. The probability density function for this situation can be seen to be

$$f(x) = \frac{1}{\beta-\alpha} \qquad \alpha \leqq x \leqq \beta$$
$$= 0 \qquad \text{otherwise.}$$

Clearly

$$\int_{-\infty}^{\infty} f(x)\,dx = \int_{\alpha}^{\beta} \frac{1}{\beta-\alpha}\,dx = \frac{\beta-\alpha}{\beta-\alpha} = 1.$$

Now

$$\mu = \frac{1}{\beta-\alpha}\int_{\alpha}^{\beta} x\,dx = \frac{x^2}{2(\beta-\alpha)}\Bigg|_{\alpha}^{\beta} = \frac{\beta^2-\alpha^2}{2(\beta-\alpha)} = \frac{(\beta-\alpha)(\beta+\alpha)}{2(\beta-\alpha)} = \frac{\beta+\alpha}{2}$$

and

$$\mu_2' = \frac{1}{\beta-\alpha}\int_{\alpha}^{\beta} x^2\,dx = \frac{x^3}{3(\beta-\alpha)}\Bigg|_{\alpha}^{\beta} = \frac{\beta^3-\alpha^3}{3(\beta-\alpha)}$$

and therefore

$$\text{Var } X = \frac{\beta^3-\alpha^3}{3(\beta-\alpha)} - \left(\frac{\beta+\alpha}{2}\right)^2 = \frac{4(\beta^3-\alpha^3)}{12(\beta-\alpha)} - \frac{3(\beta-\alpha)(\beta+\alpha)^2}{12(\beta-\alpha)}$$
$$= \frac{4\beta^3-4\alpha^3-3\beta^3-3\beta^2\alpha+3\beta\alpha^2+3\alpha^3}{12(\beta-\alpha)}$$
$$= \frac{\beta^3-\alpha^3-3\beta^2\alpha+3\alpha^2\beta}{12(\beta-\alpha)} = \frac{(\beta-\alpha)^3}{12(\beta-\alpha)} = \frac{(\beta-\alpha)^2}{12}.$$

THE NORMAL DISTRIBUTION

The *normal probability density function* given by

$$f(x) = \frac{1}{\sqrt{(2\pi)}\sigma} e^{-(x-\mu)^2/2\sigma^2} \quad -\infty < x < \infty$$

is one of the most useful probability density functions in statistics. The parameter μ can be any finite real number, while the parameter σ can be any positive real number. The notation indicates that μ is the mean and σ^2 the variance for this distribution; we will, indeed, derive these results in this section.

We will derive this density function by a limiting process on the binomial distribution in Section 7-4, but concentrate here on demonstrating that it is indeed a density function, and on finding its properties.

It is immediate that this $f(x)$ satisfies properties 1-5 of the last section. We will now demonstrate that it satisfies property 6. We will show that

$$\int_{-\infty}^{\infty} f(x)\,dx = 1$$

by showing that its square is equal to unity. Thus, we find

$$\int_{-\infty}^{\infty} f(x)\,dx \int_{-\infty}^{\infty} f(y)\,dy = \frac{1}{2\pi\sigma^2} \iint_{-\infty}^{\infty} e^{-[(x-\mu)^2+(y-\mu)^2]/2\sigma^2} dx\,dy$$

$$= \frac{1}{\pi} \iint_{-\infty}^{\infty} e^{-(u^2+v^2)} du\,dv$$

where $u = (x-\mu)/\sqrt{(2)}\sigma$ and $v = (y-\mu)/\sqrt{(2)}\sigma$.

If we now transform u and v to polar coordinates with $u = r\cos\theta$ and $v = r\sin\theta$, we obtain

$$\int_{-\infty}^{\infty} f(x)\,dx \int_{-\infty}^{\infty} f(y)\,dy = \frac{1}{\pi} \int_{0}^{\infty}\int_{0}^{2\pi} e^{-r^2} r\,dr\,d\theta$$

$$= 2\int_{0}^{\infty} e^{-r^2} r\,dr = \int_{0}^{\infty} e^{-z} dz = 1$$

where $z = r^2$. Note that $du\,dv = r\,dr\,d\theta$ for this transformation. The variable r is called the Jacobian of the transformation. Although the theory of Jacobians will not be included in this text, we will, in Section 6-3 discuss methods for finding them.

We now demonstrate that μ and σ^2 actually are the mean and variance for this distribution.

$$EX = \frac{1}{\sqrt{(2\pi)}\sigma} \int_{-\infty}^{\infty} x\, e^{-(x-\mu)^2/2\sigma^2} dx$$

$$= \frac{1}{\sqrt{(2\pi)}\sigma}\int_{-\infty}^{\infty}(x-\mu)\,e^{-(x-\mu)^2/2\sigma^2}dx+\frac{\mu}{\sqrt{(2\pi)}\sigma}\int_{-\infty}^{\infty}e^{-(x-\mu)^2/2\sigma^2}dx$$

$$= \sqrt{\left(\frac{2}{\pi}\right)}\sigma\int_{-\infty}^{\infty}y\,e^{-y^2}dy+\mu$$

[where $y = (x-\mu)/\sqrt{(2)}\sigma$]

$$= \sqrt{\left(\frac{2}{\pi}\right)}\sigma\int_{0}^{\infty}y\,e^{-y^2}dy-\sqrt{\left(\frac{2}{\pi}\right)}\sigma\int_{0}^{\infty}y\,e^{-y^2}dy+\mu$$

[where $z = y^2$]

$$= \frac{\sigma}{\sqrt{(2\pi)}}\int_{0}^{\infty}e^{-z}dz-\frac{\sigma}{\sqrt{(2\pi)}}\int_{0}^{\infty}e^{-z}dz+\mu$$

$$= \mu.$$

Remark *The integral of an odd function integrated over a finite symmetric range about zero is always zero. This is not necessarily the case when the limits of integration are* $-\infty, +\infty$. *In this case both half integrals could be infinite, and we would find that the value of the integral was* $\infty-\infty$, *which is undefined. We will be taking great care with this type of integral in the following chapters. As will be seen in Section 3-2, the effort is necessary.*

Now

$$\text{Var } X = E(X-\mu)^2 = \frac{1}{\sqrt{(2\pi)}\sigma}\int_{-\infty}^{\infty}(x-\mu)^2e^{-(x-\mu)^2/2\sigma^2}dx$$

$$= \frac{1}{\sqrt{(2\pi)}\sigma}\int_{-\infty}^{\infty}y^2e^{-y^2/2\sigma^2}dy$$

where $y = x-\mu$.

We now integrate by parts, i.e.

$$\int U\,dV = UV| - \int V\,dU$$

with

$$U = \frac{y}{\sqrt{(2\pi)}\sigma} \quad \text{and} \quad dV = d(-\sigma^2\,e^{-y^2/2\sigma^2})$$

giving

$$\text{Var } X = \frac{y\sigma^2}{\sqrt{(2\pi)}\sigma}\,e^{-y^2/2\sigma^2}\Bigg|_{-\infty}^{\infty}+\frac{\sigma^2}{\sqrt{(2\pi)}\sigma}\int_{-\infty}^{\infty}e^{-y^2/2\sigma^2}dy$$

The first term is zero since $y\,e^{-y^2} \to 0$ as $y \to \infty$ which can be seen by using l'Hopital's rule. Therefore, Var $X = \sigma^2$.

PROPERTIES OF THE NORMAL DENSITY

Since the maximum value of e^{-z}, namely unity, occurs when $z = 0$, the normal density function

$$f(x) = \frac{1}{\sqrt{(2\pi)}\sigma}\, e^{-(x-\mu)^2/2\sigma^2} \quad -\infty < x < \infty$$

attains its maximum at $x = \mu$. Here $f(\mu) = 1/\sqrt{(2\pi)}\sigma$. The function is symmetric about $x = \mu$ decreasing exponentially to zero when $x = -\infty$ or $x = +\infty$. (It drops to approximately 60% of its maximum when $|x-\mu| = \sigma$ and to approximately 1% of its maximum when $|x-\mu| = 3\sigma$.) The density has a characteristic bell shape which is depicted in Figure 2-4.

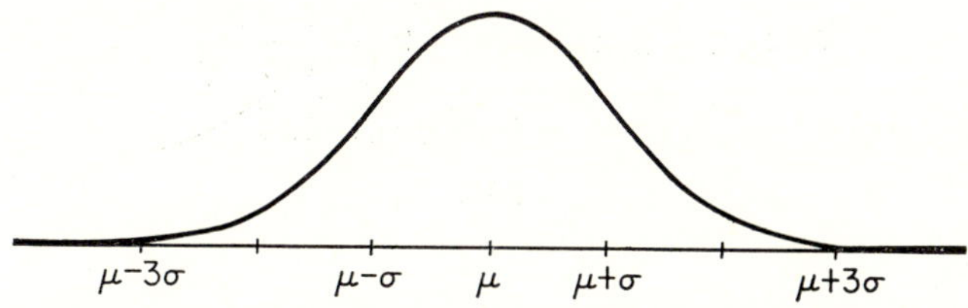

Figure 2-4

One very important property of the normal distribution, which is not shared by some other distributions, is the following. We can tabulate the cumulative distribution function for the particular set of parameters $\mu = 0$ and $\sigma^2 = 1$, and use this table to obtain the properties of the normal distribution with any other set of parameters μ, σ^2.

Let us define the function

$$\Phi(x) = \frac{1}{\sqrt{(2\pi)}} \int_{-\infty}^{x} e^{-t^2/2} dt.$$

$\Phi(x)$ is then the cumulative distribution function of the *standard normal distribution* ($\mu = 0$, $\sigma^2 = 1$).

We will now derive probability statements about a random variable which has a nonstandard normal distribution in terms of the function

$\Phi(x)$. For any such random variable the probability that X will be in the interval (a, b)

$$P(a \leqq X \leqq b) = \frac{1}{\sqrt{(2\pi)}\sigma}\int_a^b e^{-(x-\mu)^2/2\sigma^2}dx = \frac{1}{\sqrt{(2\pi)}}\int_{a-\mu/\sigma}^{b-\mu/\sigma} e^{-y^2/2}dy$$

$$= \Phi\left(\frac{b-\mu}{\sigma}\right) - \Phi\left(\frac{a-\mu}{\sigma}\right) \qquad [\text{where } y = (x-\mu)/\sigma].$$

This shows that one table of $\Phi(x)$ values will be sufficient to find the values of probability statements about any normal random variable.

EXAMPLE 2-3

The diameter of a particular cylinder is 3.50 cm. If the micrometer reading to be taken on this cylinder has a normal distribution with mean 3.50 and variance 0.0001 cm^2, what is the probability that the reading will be greater than 3.51 cm? The question simply asks for $P(3.51 \leqq X \leqq \infty)$ where X has a mean $\mu = 3.50$ and variance $\sigma^2 = 0.0001$. The answer is given by

$$P(3.51 \leqq X \leqq \infty) = \Phi(\infty) - \Phi\left(\frac{3.51-3.50}{0.01}\right)$$
$$= 1-\Phi(1) = 1-0.84$$
$$= 0.16.$$

EXERCISE 2-8

For the normal distribution with mean μ and variance σ^2 show that:

a) $\mu_{2n-1} = 0 \qquad n = 1, 2, 3, \ldots$

b) $\mu_{2n} = 1\cdot3\cdot5 \ldots (2n-1)\sigma^{2n} \qquad n = 1, 2, 3, \ldots.$

2-7 Problems

1. Consider the binomial distribution with $n = 10$. For values of $p = 0.2, 0.5, 0.8$ graph both the probability mass function and the cumulative distribution.
2. Assume that we have a box containing 1000 ball point pens of which 100 are defective.
   a. What is the probability that a pen chosen at random will be defective?
   b. If ten pens are chosen with replacement, what is the probability that there will be exactly one defective?
   c. What is the probability that there will be at most one defective in the ten pens chosen in part b?
3. Consider poisson distributions with $\lambda = 1, 5, 10$. For each distribution, plot both the mass function and the cumulative function for values of x up to fifteen.
4. A particular bakery makes chocolate chip cookies by mixing enough batter for 1000 cookies, dumping in 5000 chocolate chips, mixing, and then making the cookies.

a. Find a rationale to justify using a poisson distribution with $\lambda = 5$ on the number of chocolate chips per cookie.
b. What is the probability of finding exactly five chips on a given cookie?
c. What is the probability that there are at least five in the cookie?

5. Consider negative binomial distributions with $r = 2$ and $p = 0.2, 0.5, 0.8$. For each distribution, plot the probability mass function and the cumulative distribution for x values up to ten.

6. For the situation in problem 2 (with replacement), find:
   a. The probability of obtaining exactly two defectives before three good pens.
   b. The probability of obtaining exactly two good pens before three defectives.

7. Consider the hypergeometric distribution with $n = 100$, $m = 10$ and $r = 5$. Plot the probability mass function and the cumulative distribution.

8. If in Problem 2 the pens were chosen without replacement (which is the only sensible manner), then:
   a. What is the probability that a pen chosen at random will be defective?
   b. If ten pens are chosen at random, what is the probability that exactly one will be defective?
   c. What is the probability that there will be at most one defective among the ten pens chosen in part b?
   d. What is the probability that there will be two or more defectives among the ten pens?

9. The third central moment can be given in terms of the first three non-central moments. Find this relation.

10. The third central moment can be given in terms of the third non-central moment, the variance, and the mean. Find this relation.

11. Find μ_3' for the binomial distribution.

12. Find μ_3 for the binomial distribution.

13. Check the relationship between the moments found in Problems 9 and 10 for the binomial distribution.

14. Find μ_3' for the poisson distribution.

15. Find μ_3 for the poisson distribution.

16. Check the relationship between the moments found in Problems 9 and 10 for the poisson distribution.

17. Find μ_3' for the discrete uniform distribution.

18. Find μ_3 for the discrete uniform distribution.

19. Check the relationship between the moments found in Problems 9 and 10 for the discrete uniform distribution.

20. Show that if $f(x)$ is symmetric about $x = 0$, then for $a > 0$:
    a. $F(0) = 0.5$.
    b. $F(-a) = 1-F(a)$.
    c. $F(a) = 1-F(-a)$.
    d. $P(|X| < a) = 2F(a)-1$.
    e. $P(|X| > a) = 2F(-a)$.
    f. $\mu = 0$, when it exists.

21. Show that the relationship between the moments in Problems 9 and 10 hold for distributions of continuous random variables.
22. Find μ_3' for the uniform distribution.
23. Find μ_3 for the uniform distribution.
24. Check the relationship between the moments found in Problems 9 and 10 for the uniform distribution.
25. Plot the standard normal density and cumulative distribution functions. (Use Table B-3, Appendix B.)
26. Trace the curves in previous problem on plain white paper and by relabeling the axes, but have them represent the density and cumulative functions for a normal distribution with mean ten and variance twenty-five.
27. Find $P(1 \leqq X \leqq 3)$ when X is normally distributed with:
    a. $\mu = 0, \sigma^2 = 1$.
    b. $\mu = 1, \sigma^2 = 1$.
    c. $\mu = 0, \sigma^2 = 2$.
    d. $\mu = 2, \sigma^2 = 2$.
28. Find μ_3' for the normal distribution.
29. Find μ_3 for the normal distribution.
30. We have a box containing 10,000 pens 1,000 of which are defective. We choose 100 pens from this box at random. What is the probability that the number of defectives is some number between eight and twelve exclusive (9, 10, 11)? Answer this problem in each of the following ways:
    a. Exactly, considering no replacement.
    b. Exactly, considering sampling with replacement.
    c. Approximately, by using the poisson approximation.
31. Repeat Problem 30 but now find the probability of the number of defectives being between zero and four exclusive (1, 2, 3). Does this result imply something about the poisson approximation?

CHAPTER 3

The Gamma and Beta Distributions and Distributions Related to Them

3-1 The Gamma Distribution

THE GAMMA FUNCTION

The gamma function is defined for any $\alpha > 0$ by

$$\Gamma(\alpha) = \int_0^\infty x^{\alpha-1}e^{-x}dx.$$

This function has some rather interesting and useful properties. When $\alpha = 1$ for example

$$\Gamma(\alpha) = \Gamma(1) = \int_0^\infty e^{-x}dx = 1.$$

For any $\alpha > 1$ we can integrate by parts with $U = x^{\alpha-1}$ and $V = -e^{-x}$ obtaining

$$\begin{aligned}\Gamma(\alpha) &= -x^{\alpha-1}e^{-x}\Big|_0^\infty + (\alpha-1)\int_0^\infty x^{\alpha-2}e^{-x}dx \\ &= (\alpha-1)\Gamma(\alpha-1).\end{aligned}$$

Therefore, if α is an integer (n) greater than unity, we obtain

$$\Gamma(n) = (n-1)\Gamma(n-1).$$

Using the same relation on $\Gamma(n-1)$ when $n > 2$, we obtain

$$\Gamma(n) = (n-1)(n-2)\Gamma(n-2).$$

This can be continued until we obtain

$$\Gamma(n) = (n-1)(n-2)(n-3)\ldots 1\cdot\Gamma(1) = (n-1)!$$

Another interesting special case occurs when α is an odd integer divided by two. Then we obtain

$$\Gamma\left(\frac{n}{2}\right) = \left(\frac{n}{2}-1\right)\left(\frac{n}{2}-2\right)\left(\frac{n}{2}-3\right)\ldots(1/2)\Gamma(1/2).$$

Now

$$\Gamma(1/2) = \int_0^\infty x^{-1/2} e^{-x} dx = \sqrt{2} \int_0^\infty e^{-z^2/2} dz$$

where $z = \sqrt{(2x)}$.
Noting that the integrand is an even function, we obtain

$$\Gamma(1/2) = \frac{\sqrt{2}}{2} \int_{-\infty}^{\infty} e^{-z^2/2} dz = \frac{1}{\sqrt{2}} \int_{-\infty}^{\infty} e^{-z^2/2} dz.$$

But

$$\frac{1}{\sqrt{(2\pi)}} \int_{-\infty}^{\infty} e^{-z^2/2} dz = 1,$$

and therefore,

$$\Gamma(1/2) = \sqrt{\pi} \left(\frac{1}{\sqrt{(2\pi)}} \int_{-\infty}^{\infty} e^{-z^2/2}\, dz \right) = \sqrt{\pi}.$$

When α is neither integer nor half integer, then we can use the reduction formula

$$\Gamma(\alpha) = (\alpha-1)(\alpha-2)(\alpha-3) \ldots \beta\Gamma(\beta)$$

where β is between zero and unity. From this it is seen that the gamma function need only be tabulated for values between zero and unity. Often, however, the tables are listed for β values between one and two. See for example Table B-4 in the Appendix.

Another useful form of the gamma function is a slight modification of the transformation used above. Let $y = \sqrt{x}$, then

$$\Gamma(\alpha) = 2 \int_0^\infty y^{2\alpha-1} e^{-y^2} dy.$$

THE GAMMA DISTRIBUTION

The gamma distribution is defined by the density function

$$f(x) = \frac{1}{\Gamma(\alpha)\beta^\alpha} x^{\alpha-1} e^{-x/\beta} \qquad 0 < x < \infty$$

$$\alpha, \beta > 0$$

$$= 0 \qquad \text{otherwise.}$$

For fixed α and β this clearly satisfies Properties 1-5 of Section 2-5. We now check to see that it integrates to unity.

$$\int_0^\infty f(x)\,dx = \frac{1}{\Gamma(\alpha)\beta^\alpha}\int_0^\infty x^{\alpha-1}e^{-x/\beta}dx = \frac{\beta^\alpha}{\Gamma(\alpha)\beta^\alpha}\int_0^\infty y^{\alpha-1}e^{-y}dy \qquad \text{(where } y = x/\beta\text{)}$$

$$= \frac{\Gamma(\alpha)}{\Gamma(\alpha)} = 1.$$

We will now find the moments of the gamma distribution.

$$EX = \frac{1}{\Gamma(\alpha)\beta^\alpha}\int_0^\infty x^{\alpha}e^{-x/\beta}dx = \frac{\beta^{\alpha+1}}{\Gamma(\alpha)\beta^\alpha}\int_0^\infty y^{\alpha}e^{-y}dy \qquad \text{(where } y = x/\beta\text{)}$$

$$= \frac{\beta\Gamma(\alpha+1)}{\Gamma(\alpha)} = \alpha\beta.$$

This procedure can be generalized to yield the rth moment.

$$EX^r = \frac{1}{\Gamma(\alpha)\beta^\alpha}\int_0^\infty x^{(\alpha+r)-1}e^{-x/\beta}dx = \frac{\beta^{\alpha+r}}{\Gamma(\alpha)\beta^\alpha}\int_0^\infty y^{(\alpha+r)-1}e^{-y}dy \qquad \text{(where } y = x/\beta\text{)}$$

$$= \beta^r\frac{\Gamma(\alpha+r)}{\Gamma(\alpha)} = \beta^r(\alpha+r-1)(\alpha+r-2)\ldots\alpha.$$

In particular

$$EX^2 = \beta^2\alpha(\alpha+1)$$

and therefore

$$\text{Var } X = EX^2-(EX)^2 = \beta^2\alpha(\alpha+1)-\alpha^2\beta^2 = \alpha\beta^2.$$

THE EXPONENTIAL DISTRIBUTION

When $\alpha = 1$ the gamma distribution is called the *exponential* distribution. If we reparameterize so that $\beta = 1/\lambda$, we obtain the density of the exponential distribution as

$$f(x) = \lambda e^{-\lambda x} \qquad 0 < x < \infty, \quad \lambda > 0$$

$$= 0 \qquad \text{otherwise.}$$

From the general properties of the gamma distribution it is immediate that

$$EX^r = \frac{r!}{\lambda^r}$$

giving

$$EX = 1/\lambda, \quad EX^2 = 2/\lambda^2, \quad \text{Var } X = 1/\lambda^2.$$

EXAMPLE 3-1

The time to burnout of a light bulb has been found to follow an exponential distribution. If it is known that the mean burnout time for a particular type of bulb is 100 hours, then what is the probability that the bulb will burn out in less than 80 hours?

Since the mean burnout time is 100

$$\lambda = \frac{1}{EX} = \frac{1}{100} = 0.01.$$

Therefore

$$P(X \leqq 80) = .01 \int_0^{80} e^{-.01x} dx = 1 - e^{-.8} = 0.55.$$

THE CHI-SQUARED DISTRIBUTION

Another important special case of the gamma distribution occurs when $\alpha = n/2$ with n any positive integer. We reparameterize so that $\beta = 2\sigma^2$ and write the density of the chi-squared (χ^2) distribution as

$$\begin{aligned} f(x) &= \frac{1}{\Gamma(n/2)2^{n/2}\sigma^n} x^{n/2-1} e^{-x/2\sigma^2} && 0 < x < \infty \\ & && n = 1, 2, \ldots \\ & && \sigma > 0 \\ &= 0 && \text{otherwise.} \end{aligned}$$

The use of σ as a parameter here is misleading. As we will see immediately σ^2 is not the variance of this distribution. However, the chi-squared distribution arises most often in practice as the distribution of a sum of squared independent normal random variables with zero means which themselves have variance σ^2. (We will define independent random variables in Section 4-1 and demonstrate the relation between the chi-squared distribution and the normal distribution in Section 9-3.)

The moments for this distribution follow immediately from those of the gamma distribution, namely

$$EX^r = 2^r\sigma^{2r}\Gamma(r+n/2)/\Gamma(n/2),$$

and in particular

$$EX = n\sigma^2, \quad EX^2 = n(n+2)\sigma^4, \quad \text{Var } X = 2n\sigma^4.$$

THE CHI DISTRIBUTION

The chi distribution is not just a special case of the gamma distribution, as was the chi-squared distribution, but is derived from the gamma distribution, and in particular, from the chi-squared distribution. In general, if $0 \leqq X$ and $Y = \sqrt{X}$, then

$$P(a \leqq Y \leqq b) = P(a^2 \leqq X \leqq b^2).$$

If in particular, X was a chi-squared random variable, then Y would be the square root of a chi squared, and is called a *chi random variable*. We will find its density by noting that

$$P(a \leqq Y \leqq b) = P(a^2 \leqq X \leqq b^2) = \frac{1}{\Gamma(n/2)2^{n/2}\sigma^n} \int_{a^2}^{b^2} x^{n/2-1} e^{-x/2\sigma^2} dx$$

(and by substituting $y = \sqrt{x}$)

$$= \frac{1}{\Gamma(n/2)2^{n/2-1}\sigma^n} \int_a^b y^{n-1} e^{-y^2/2\sigma^2} dy$$

giving

$$f(x) = \frac{1}{\Gamma(n/2)2^{n/2-1}\sigma^n} x^{n-1} e^{-x^2/2\sigma^2} \qquad 0 < x < \infty$$

$$\sigma < 0$$

$$= 0 \qquad \text{otherwise.}$$

as the probability density function of the chi random variable.

We now find the moments of this distribution by transforming back to the chi-squared distribution. Notice that for any r

$$EX^r = \frac{1}{\Gamma(n/2)2^{n/2-1}\sigma^n} \int_0^\infty x^{n+r-1} e^{-x^2/2\sigma^2} dx$$

$$= \frac{1}{\Gamma(n/2)2^{n/2}\sigma^n} \int_0^\infty y^{(n+r)/2-1} e^{-y/2\sigma^2} dy$$

(where $y = x^2$)

$$= \frac{\Gamma[(n+r)/2]2^{(n+r)/2}\sigma^{n+r}}{\Gamma(n/2)2^{n/2}\sigma^n}$$

$$[(1/\{\Gamma[(n+r)/2]2^{(n+r)/2}\sigma^{n+r}\}) \int_0^\infty y^{(n+r)/2-1} e^{-y/2\sigma^2} dy]$$

and since the term in large brackets integrates to unity

$$= \Gamma[(n+r)/2]2^{r/2}\sigma^r/\Gamma(n/2).$$

In particular when $r = 0$ we have the required property

$$\int_0^\infty f(x)\,dx = \frac{1}{\Gamma(n/2)2^{n/2-1}\sigma^n}\int_0^\infty x^{n-1}e^{-x^2/2\sigma^2}dx = \frac{\Gamma(n/2)}{\Gamma(n/2)} = 1.$$

We can also easily obtain

$$EX = \frac{\sqrt{(2)}\sigma\Gamma[(n+1)/2]}{\Gamma(n/2)}$$

$$EX^2 = \frac{2\sigma^2\Gamma[(n+2)/2]}{\Gamma(n/2)} = n\sigma^2$$

and

$$\text{Var } X = n\sigma^2 - 2\sigma^2\left\{\frac{\Gamma[(n+1)/2]}{\Gamma(n/2)}\right\}^2.$$

Exercise 3-1

The *Rayleigh distribution* is given by the density function

$$\begin{aligned} f(x) &= \beta^{-2}xe^{-x^2/2\beta^2} && 0 < x < \infty \\ &= 0 && \beta > 0 \\ & && \text{otherwise.} \end{aligned}$$

Find EX^r for any r, and then find the mean and variance of this distribution.

Exercise 3-2

The *Maxwell distribution* is given by the density function

$$\begin{aligned} f(x) &= \frac{4}{\sqrt{\pi}(kT)^3}x^2e^{-x^2/(kT)^2} && 0 < x < \infty \\ & && kT > 0 \\ & && \text{otherwise.} \end{aligned}$$

Find EX^r for any r, and then find the mean and variance of this distribution.

3-2 The Beta Distribution

THE BETA FUNCTION

The beta function is defined by

$$B(\alpha, \beta) = \int_0^1 x^{\alpha-1}(1-x)^{\beta-1}dx$$

with $\alpha, \beta > 0$.

We will evaluate this integral by toying with the product of two gamma functions.

$$\Gamma(\alpha)\Gamma(\beta) = \int_{\infty}^{0} x^{\alpha-1}e^{-x}dx \int_{\infty}^{0} y^{\beta-1}e^{-y}\,dy$$

$$= \int_0^\infty\int_0^\infty x^{\alpha-1}y^{\beta-1}e^{-(x+y)}dx\,dy$$

$$= 4\int_0^\infty\int_0^\infty w^{2\alpha-1}z^{2\beta-1}e^{-(w^2+z^2)}dw\,dz$$

where $w = \sqrt{x}$ and $z = \sqrt{y}$. If we now let $w = r\cos\theta$ and $z = r\sin\theta$, then $dw\,dz = r\,dr\,d\theta$, giving

$$\Gamma(\alpha)\Gamma(\beta) = 4\int_0^\infty\int_0^{\pi/2} r^{2(\alpha+\beta)-2}\cos^{2\alpha-1}\theta\,\sin^{2\beta-1}\theta\,e^{-r^2}r\,dr\,d\theta$$

$$= [2\int_0^\infty r^{2(\alpha+\beta)-1}e^{-r^2}dr][2\int_0^{\pi/2}\cos^{2\alpha-1}\theta\,\sin^{2\beta-1}\theta\,d\theta].$$

Letting $u = r^2$ in the first integral gives us

$$2\int_0^\infty r^{2(\alpha+\beta)-1}e^{-r^2}dr = \int_0^\infty u^{(\alpha+\beta)-1}e^{-u}du = \Gamma(\alpha+\beta).$$

By letting $v = \cos^2\theta$ in the second integral and noting that $dv = -2\cos\theta\,\sin\theta\,d\theta$, we obtain

$$2\int_0^{\pi/2}\cos^{2\alpha-1}\theta\,\sin^{2\beta-1}\theta\,d\theta = -\int_1^0 v^{\alpha-1}(1-v)^{\beta-1}dv = \int_0^1 v^{\alpha-1}(1-v)^{\beta-1}dv.$$

Recombining terms, we find that

$$\Gamma(\alpha)\Gamma(\beta) = \Gamma(\alpha+\beta)\int_0^1 x^{\alpha-1}(1-x)^{\beta-1}dx$$

or

$$B(\alpha,\beta) = \frac{\Gamma(\alpha)\Gamma(\beta)}{\Gamma(\alpha+\beta)}.$$

THE BETA DISTRIBUTION

The *beta distribution* is defined by the density function

$$f(x) = \frac{\Gamma(\alpha+\beta)}{\Gamma(\alpha)\Gamma(\beta)}x^{\alpha-1}(1-x)^{\beta-1} \qquad 0 < x < 1 \quad \alpha,\beta > 0.$$

$$= 0 \qquad \text{otherwise.}$$

From the derivation of $B(\alpha, \beta)$ it is clear that $\int_0^1 f(x)\,dx = 1$. The moments of the beta distribution are rather simple to obtain. Note that

$$EX^r = \frac{\Gamma(\alpha+\beta)}{\Gamma(\alpha)\Gamma(\beta)} \int_0^1 x^{(\alpha+r)-1}(1-x)^{\beta-1}dx$$

$$= \frac{\Gamma(\alpha+\beta)}{\Gamma(\alpha)\Gamma(\beta)} \frac{\Gamma(\alpha+r)\Gamma(\beta)}{\Gamma(\alpha+\beta+r)} = \frac{\Gamma(\alpha+\beta)\Gamma(\alpha+r)}{\Gamma(\alpha)\Gamma(\alpha+\beta+r)}$$

$$= \frac{(\alpha+r-1)(\alpha+r-2)\ldots\alpha}{(\alpha+\beta+r-1)(\alpha+\beta+r-2)\ldots(\alpha+\beta)}.$$

In particular

$$EX = \frac{\Gamma(\alpha+\beta)\Gamma(\alpha+1)}{\Gamma(\alpha)\Gamma(\alpha+\beta+1)} = \frac{\alpha}{\alpha+\beta}$$

$$EX^2 = \frac{\Gamma(\alpha+\beta)\Gamma(\alpha+2)}{\Gamma(\alpha)\Gamma(\alpha+\beta+2)} = \frac{\alpha(\alpha+1)}{(\alpha+\beta)(\alpha+\beta+1)}$$

from which we can obtain

$$\text{Var } X = \frac{\alpha\beta}{(\alpha+\beta)^2(\alpha+\beta+1)}.$$

THE F DISTRIBUTION

The F distribution is very useful in statistics, as will be seen in Section 9-3. We obtain this distribution here as a transformation from the beta distribution. If X is a beta variable and $Y = \beta X/\alpha(1-X)$, then $X = \alpha Y/(\beta+\alpha Y)$, and

$$P(a \leqq Y \leqq b) = P[\alpha a/(\beta+\alpha a) < X < \alpha b/(\beta+\alpha b]$$

$$= \frac{\Gamma(\alpha+\beta)}{\Gamma(\alpha)\Gamma(\beta)} \int_{\alpha a/(\beta+\alpha a)}^{\alpha b/(\beta+\alpha b)} x^{\alpha-1}(1-x)^{\beta-1}dx$$

$$= \frac{\Gamma(\alpha+\beta)}{\Gamma(\alpha)\Gamma(\beta)} \int_a^b \left(\frac{\alpha}{\beta}\right)^{\alpha} y^{\alpha-1}\left(1+\frac{\alpha}{\beta}y\right)^{-(\alpha+\beta)} dy$$

where $y = \beta x/\alpha(1-x)$, giving

$$f(y) = \frac{\Gamma(\alpha+\beta)}{\Gamma(\alpha)\Gamma(\beta)}\left(\frac{\alpha}{\beta}\right)^{\alpha} y^{\alpha-1}\left(1+\frac{\alpha}{\beta}y\right)^{-(\alpha+\beta)} \qquad 0 < y < \infty$$

$$\alpha, \beta > 0$$

$$= 0 \qquad \text{otherwise.}$$

If we now take $\alpha = m/2$ and $\beta = n/2$, and restrict m, n to be integers, we obtain the density of the F distribution as

$$f(x) = \frac{\Gamma[(m+n)/2]}{\Gamma(m/2)\Gamma(n/2)}\left(\frac{m}{n}\right)^{m/2} x^{m/2-1}\left(1+\frac{m}{n}x\right)^{-[(m+n)/2]} \qquad 0 < x < \infty$$

$$= 0 \qquad \text{otherwise.}$$

The parameters m, n are called the *degrees of freedom* of the F distribution. One often sees the notation $F_{m,n}$ for an F variable with m and n degrees of freedom. The reason for this will become evident in Section 9-3.

EXERCISE 3-3

The above derivation does not contain the details of the transformation. Put in these details.

We will now find the moments of this distribution. For any r we have

$$EX^r = \frac{\Gamma[(m+n)/2]}{\Gamma(m/2)\Gamma(n/2)}\int_0^\infty \left(\frac{m}{n}\right)^{m/2} x^{m/2+r-1}\left(1+\frac{m}{n}x\right)^{-[(m+n)/2]} dx.$$

Reversing the manner in which the F density was derived we let $x = ny/m(1-y)$ or $y = mx/(n+mx)$ and obtain

$$EX^r = \frac{\Gamma[(m+n)/2]}{\Gamma(m/2)\Gamma(n/2)}\int_0^1 \left(\frac{n}{m}\right)^r y^{m/2+r-1}(1-y)^{(n/2)-r-1} dx$$

$$= \frac{\Gamma[(m+n)/2]}{\Gamma(m/2)\Gamma(n/2)}\left(\frac{n}{m}\right)^r \frac{\Gamma(m/2+r)\Gamma(n/2-r)}{\Gamma[(m+n)/2]} = \frac{\Gamma(m/2+r)\Gamma(n/2-r)}{\Gamma(m/2)\Gamma(n/2)}\left(\frac{n}{m}\right)^r$$

$$= \frac{(m/2+r-1)(m/2+r-2)\ldots(m/2)}{(n/2-1)(n/2-2)\ldots(n/2-r)}\left(\frac{n}{m}\right)^r.$$

Remark *Since the beta function was defined only when α, $\beta > 0$, we must restrict this solution to the case where $n/2-r > 0$ or $r < n/2$.*

In particular, for $r = 0$, we find that

$$\int_0^\infty f(x)\,dx = 1.$$

When $n \geqq 3$ we obtain

$$EX = \frac{(m/2)}{(n/2-1)}\frac{n}{m} = \frac{n}{n-2}$$

regardless of the value of m. When $n \geqq 5$, we obtain

$$EX^2 = \frac{(m/2+1)(m/2)}{(n/2-2)(n/2-1)}\left(\frac{n}{m}\right)^2 = \frac{n^2(m+2)}{m(n-2)(n-4)}$$

giving

$$\text{Var } X = \frac{2n^2(m+n-2)}{m(n-2)^2(n-4)}.$$

Exercise 3-4

The details of the transformation were not included. Put in these details.

Exercise 3-5

Show that when $r \geqq n/2$, $EX^r = \infty$.

THE t DISTRIBUTION

We now look for a symmetric density function for Y such that if X has an F distribution, then $Y = \sqrt{X}$ has this distribution. This can be most easily done by finding the density function for $Y = \sqrt{X}$ defined only on the non-negative real axis, and then extending it symmetrically to the negative real axis. If one is restricted to the positive axis, then

$$P(a < Y < b) = P(a^2 < X < b^2).$$

But

$$P(a^2 < X < b^2) = \frac{[\Gamma(m+n)/2]}{\Gamma(m/2)\Gamma(n/2)} \int_{a^2}^{b^2} \left(\frac{m}{n}\right)^{m/2} x^{m/2-1}\left(1+\frac{m}{n}x\right)^{-[(m+n)/2]} dx$$

$$= \frac{2\Gamma[(m+n)/2]}{\Gamma(m/2)\Gamma(n/2)} \int_{a}^{b} \left(\frac{m}{n}\right)^{m/2} y^{m-1}\left(1+\frac{m}{n}y^2\right)^{-[(m+n)/2]} dy$$

where $y = \sqrt{z}$, giving the density defined on the positive axis as

$$f(y) = \frac{2\Gamma[(m+n)/2]}{\Gamma(m/2)\Gamma(n/2)}\left(\frac{m}{n}\right)^{m/2} y^{m-1}\left(1+\frac{m}{n}y^2\right)^{-[(m+n)/2]} \qquad 0 < y < \infty$$

$$m, n = 1, 2, 3, \ldots$$

$$= 0 \qquad \text{otherwise.}$$

In order to extend this to the negative axis we will have to multiply by 1/2 so that the total integral remains unity. The density function is then written

$$f(x) = \frac{\Gamma[(m+n)/2]}{\Gamma(m/2)\Gamma(n/2)}\left(\frac{m}{n}\right)^{m/2} |x|^{m-1}\left(1+\frac{m}{n}x^2\right)^{-[(m+n)/2]} \qquad -\infty < x < \infty$$

$$m, n = 1, 2, 3, \ldots$$

An important special case of the above occurs when $m = 1$; $f(x)$ is called the density of the t distribution, which is given by

$$f(x) = \frac{\Gamma[(n+1)/2]}{\sqrt{(2\pi)}\Gamma(n/2)}\left(1+\frac{x^2}{n}\right)^{-[(n+1)/2]} \qquad -\infty < x < \infty$$

$$n = 1, 2, 3, \ldots.$$

By the symmetry of this density it is clear that if an odd moment exists, then it must be equal to zero. On the other hand, as we have already seen

$$EX^2 \geqq (EX)^2 \text{ which implies}$$

$$EX^{2r} \geqq (EX^r)^2 \text{ for any } r$$

which further implies that if EX^{2r} exists, then EX^r exists, and if r is odd then $EX^r = 0$. We will now find a relation for the $2r$th moment of the t distribution:

$$EX^{2r} = \frac{\Gamma[(n+1)/2]}{\sqrt{(n\pi)}\Gamma(n/2)}\int_{-\infty}^{\infty} x^{2r}\left(1+\frac{x^2}{n}\right)^{-[(n+1)/2]} dx$$

$$= \frac{\Gamma[(n+1)/2]}{2\sqrt{(n\pi)}\Gamma(n/2)}\int_{\infty}^{\infty} y^{[(2r-1)/2]}\left(1+\frac{y}{n}\right)^{-[(n+1)/2]} dy$$

(where $y = x^2$)

$$= \frac{\Gamma[(n+1)/2]}{\sqrt{(n\pi)}\Gamma(n/2)}\int_{0}^{\infty} y^r y^{-1/2}\left(1+\frac{y}{n}\right)^{-[(n+1)/2]} dy = EY^r$$

where Y has an F distribution with one, and n degrees of freedom. We have already solved this problem, giving

$$EX^{2r} = \frac{\Gamma(r+1/2)\Gamma(n/2-r)}{\sqrt{(\pi)}\Gamma(n/2)}\, n^r = \frac{(r+1/2-1)(r+1/2-2)\ldots(1/2)}{(n/2-1)(n/2-2)\ldots(n/2-r)}\, n^r$$

when $r < n/2$.

In particular, for $r = 0$, we have that

$$\int_{-\infty}^{\infty} f(x)\, dx = 1,$$

and when $n \geqq 3$ we have

$$EX = 0$$

and

$$EX^2 = n/(n-2)$$

giving

$$\text{Var } X = n/(n-2).$$

THE CAUCHY DISTRIBUTION

An interesting special case of the t distribution occurs when $n = 1$, in which case we obtain the density of the cauchy distribution

$$f(x) = \frac{1}{\pi(1+x^2)} \qquad -\infty < x < \infty.$$

Before investigating an interesting property of this distribution, we will derive it from another point of view. This will emphasize the fact that such distributions can exist in nature (in so far as the distribution of any continuous variable can exist in nature). Consider the two-headed arrow in Figure 3-1. We spin a well balanced pointer (the probability of falling in any

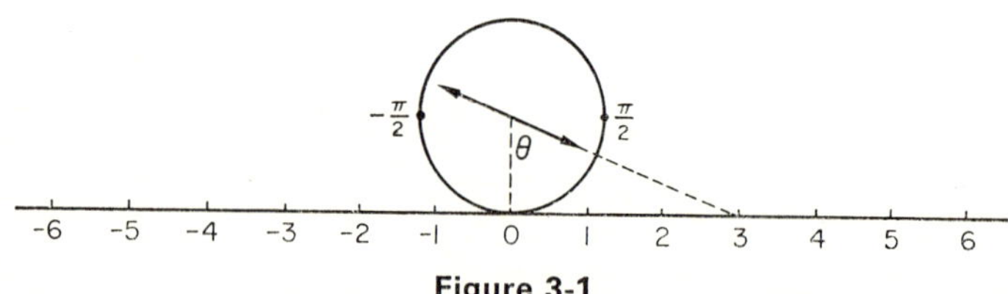

Figure 3-1

interval on the circle is proportional to the length of that interval). The unit circle touches the coordinate axis at the origin. We ask for the probability density of the random variable whose outcome is the coordinate to which the pointer points. Since the probability of being in any range of θ values $-\pi/2$ and $\pi/2$ is proportional to the length of the range. We therefore obtain the density for θ as

$$\begin{aligned} f(\theta) &= \frac{1}{\pi} \qquad -\frac{\pi}{2} < \theta < \frac{\pi}{2} \\ &= 0 \qquad \text{otherwise.} \end{aligned}$$

Now

$$P(a \leqq X \leqq b) = P(\tan^{-1}a \leqq \theta \leqq \tan^{-1}b) = \frac{\tan^{-1}b - \tan^{-1}a}{\pi}.$$

However

$$\frac{1}{\pi}\int_a^b \frac{dx}{1+x^2} = \frac{\tan^{-1}b - \tan^{-1}a}{\pi}$$

and therefore, by definition of a density

$$f(x) = \frac{1}{\pi(1+x^2)} \qquad -\infty < x < \infty.$$

We will now attempt to find the mean of this distribution.

$$EX = \frac{1}{\pi}\int_{-\infty}^{\infty}\frac{x\,dx}{1+x^2} = \frac{1}{\pi}\int_{0}^{\infty}\frac{x}{1+x^2}\,dx+\frac{1}{\pi}\int_{-\infty}^{0}\frac{x}{1+x^2}\,dx$$

$$= \frac{1}{2\pi}\int_{0}^{\infty}\frac{dy}{1+y}+\frac{1}{2\pi}\int_{\infty}^{0}\frac{dy}{1+y}$$

(where $y = x^2$)

$$= \frac{1}{2\pi}\ln(1+y)\Big|_{0}^{\infty}+\frac{1}{2\pi}\ln(1+y)\Big|_{\infty}^{0}$$

$$= \infty-\infty$$

which is undefined. It should be clear that if the mean is undefined, then the higher moments cannot exist. To see this we will try to find the second moment.

$$EX^2 = \frac{1}{\pi}\int_{-\infty}^{\infty}\frac{x^2\,dx}{1+x^2} = \frac{2}{\pi}\int_{0}^{\infty}\frac{x^2\,dx}{1+x^2} = \frac{2}{\pi}\int_{0}^{\infty}\left[1-\frac{1}{1+x^2}\right]dx$$

$$= \frac{2}{\pi}\int_{0}^{\infty}dx-\frac{2}{\pi}\int_{0}^{\infty}\frac{dx}{1+x^2} = \infty-1 = \infty$$

as predicted.

Remark *It is often said that the mean of the cauchy distribution "does not exist". Here this phrase means that the mean is undefined. This should not be interpreted to mean that the mean of a cauchy distribution is infinite for it is not.*

3-3 Problems

1. If $\Gamma(1.33)$ is known to be 0.8934, what is $\Gamma(5.33)$? What is $\Gamma(0.33)$?
2. Using Table B-4, find
   a. $\Gamma(5.23)$
   b. $\Gamma(10.27)$
   c. $\Gamma(0.25)$.
3. Plot the density of the gamma distribution function with parameters
   a. $\alpha = 1, \beta = 1$
   b. $\alpha = 3, \beta = 1$
   c. $\alpha = 1, \beta = 3$.

4. Give the first three moments and the variance of the distribution with density

$$f(x) = (1/54)\,x^2 e^{-x/3} \qquad 0 < x < \infty$$
$$= 0 \qquad \text{otherwise.}$$

5. Give the first three moments and the variance of the distribution with density

$$f(x) = (1/16)\,x^2 e^{-x/2} \qquad 0 < x < \infty$$
$$= 0 \qquad \text{otherwise.}$$

6. Give the first three moments and the variance of the distribution with density

$$f(x) = (1/9)\,x e^{-x/3} \qquad 0 < x < \infty$$
$$= 0 \qquad \text{otherwise.}$$

7. Plot the density of the exponential distribution with
   a. $\lambda = 1$
   b. $\lambda = 10$.

8. The lifetime of an incandescent bulb is reputed to follow the exponential law. If $1/\lambda = 10$ days, what is the probability that the bulb will burn out within 500 days? What is the probability that it will last at least 100 days?

9. The probability of a radioactive decay in nuclear physics follows an exponential law with λ related to the half life. With the half life defined as the time at which the probability of decay is 0.5, find the relation between λ and T (the half life).

10. Plot the density of the chi-squared distribution with parameters
    a. $n = 2$, $\sigma^2 = 1$
    b. $n = 10$, $\sigma^2 = 1$
    c. $n = 2$, $\sigma^2 = 25$
    d. $n = 10$, $\sigma^2 = 25$.

11. Give the first three moments and the variance of the chi-squared distribution with
    a. $n = 2$, $\sigma^2 = 1$
    b. $n = 10$, $\sigma^2 = 1$
    c. $n = 2$, $\sigma^2 = 25$
    d. $n = 10$, $\sigma^2 = 25$.

12. Plot the density of the chi distribution with the same parameters as in Problem 10.

13. Give the first three moments and the variance of the distribution with density

$$f(x) = x e^{-x^2/2} \qquad 0 < x < \infty$$
$$= 0 \qquad \text{otherwise.}$$

14. Give the first three moments and the variance of the distribution with density

$$f(x) = \sqrt{\frac{2}{\pi}} \left(\frac{1}{27}\right) x^2 e^{-x^2/18} \qquad 0 < x < \infty$$
$$= 0 \qquad \text{otherwise.}$$

15. Plot the density of the beta distribution with parameters
    a. $\alpha = 1$, $\beta = 1$
    b. $\alpha = 10$, $\beta = 1$
    c. $\alpha = 1$, $\beta = 10$
    d. $\alpha = 10$, $\beta = 10$.

16. Give the first three moments and the variance of the distribution with density

$$f(x) = 12(x^2 - x^3) \qquad 0 < x < 1$$
$$= 0 \qquad \text{otherwise.}$$

17. Give the first three moments and the variance of the distribution with density

$$f(x) = 20(x^3 - x^4) \qquad 0 < x < 1$$
$$= 0 \qquad \text{otherwise.}$$

18. Plot the density of the F distribution for parameters
   a. $m = 1, n = 1$
   b. $m = 2, n = 2$
   c. $m = 2, n = 10$
   d. $m = 10, n = 2$
   e. $m = 10, n = 10.$

19. Give the first three moments and the variance of the distribution with density

$$f(x) = (1 + x/5)^{-6} \qquad 0 < x < \infty$$
$$= 0 \qquad \text{otherwise.}$$

20. Give the first three moments and the variance of the distribution with density

$$f(x) = \frac{128}{3\pi} x^{3/2}(1+x)^{-5} \qquad 0 < x < \infty$$
$$= 0 \qquad \text{otherwise.}$$

21. Plot the density of the t distribution with parameter (degrees of freedom) given by
   a. $n = 2$
   b. $n = 4$
   c. $n = 10.$

22. Give the first four moments and the variance of the distribution with density

$$f(x) = \frac{128}{105\pi}(1 \times x^2/9)^{-5} \qquad -\infty < x < \infty.$$

23. Give the first four moments and the variance of the distribution with density

$$f(x) = \frac{315}{256\sqrt{10}}(1 \times x^2/10)^{-11/2} \qquad -\infty < x < \infty.$$

24. Plot the density of the cauchy distribution.

CHAPTER 4

Multivariate Distributions

4-1 Discrete Random Variables

When there are at most three random variables under consideration, it is convenient to use the letters X, Y, and Z. However, the situation is quite different when there is a larger number of random variables, say n of them. For this case we will denote by $X_1, X_2, X_3, \ldots, X_n$, the n different random variables. It must be emphasized here that this is not a sequence of values for one variable. It is quite possible that the range of values X_1 takes on is negative, while the range for X_2 is positive. Each random variable, such as X_1, represents a partition of the sure set which we labeled $A_1, A_2, A_3, \ldots$ in Chapter One.

With this notational convention, we cannot now use $x_1, x_2, x_3, \ldots$ to represent values for X_1 since these might just as well be used for X_2. In order to avoid a double subscript, we will use x_1 to represent a generic element of the values which X_1 can have, and x_2 a generic element for the values that X_2 can have, and so on until x_n. Therefore, by the set $x_1, x_2, x_3, \ldots, x_n$ we mean $X_1 = x_1, X_2 = x_2, X_3 = x_3, \ldots, X_n = x_n$. By the statement "all $x_1, x_2, x_3, \ldots, x_n$," for example, we mean all possible n-tuples with one value from each of the n ranges. Keeping these notational conventions in mind, we can now make the following definition.

Definition If $X_1, X_2, X_3, \ldots, X_n$ are discrete random variables (*i.e. they take on only a countable number of values each*) then any function $P(x_1, x_2, x_3, \ldots, x_n)$ satisfying

a) $0 \leqq P(x_1, x_2, x_3, \ldots, x_n) \leqq 1 \qquad \text{all } x_1, x_2, x_3, \ldots, x_n.$

and

b) $\sum_{x_1} \sum_{x_2} \sum_{x_3} \cdots \sum_{x_n} P(x_1, x_2, x_3, \ldots, x_n) = 1$

is called a *joint probability mass function* for these random variables. The *joint cumulative distribution function* can then be defined by

$$F(x_1, x_2, x_3, \ldots, x_n) = \sum_{y_1 \leqq x_1} \sum_{y_2 \leqq x_2} \sum_{y_3 \leqq x_3} \cdots \sum_{y_n \leqq x_n} P(y_1, y_2, y_3, \ldots, y_n)$$

where we have used $y_1, y_2, y_3, \ldots, y_n$ as the generic element of $X_1, X_2, X_3, \ldots, X_n$.

The properties of the cumulative distribution function are easily seen. Since it is the sum of non-negative terms, it is itself non-negative and is a non-decreasing function in each of its variables. If all $x_i = \infty$, $i = 1, 2, 3, \ldots, n$, its value is unity, and if any $x_i = -\infty$, $i = 1, 2, 3, \ldots, n$, its value is zero.

We can also ask for the *marginal probability mass function* of any subset of the n random variables. In Chapter One this marginal probability was obtained by summing out all the unwanted partitions. Similarly we can obtain the marginal probability mass function of $X_i, X_j, \ldots, X_k$ by summing out all the remaining X's; That is

$$P(x_i, x_j, \ldots, x_k) = \underset{\substack{\text{all} \\ x_1, x_2, x_3, \ldots, x_n \\ \text{except} \\ x_i, x_j, \ldots x_k}}{\sum \sum \cdots \sum} P(x_1, x_2, x_3, \ldots, x_n).$$

In particular

$$P(x_i) = \underset{\substack{\text{all} \\ x_1, x_2, x_3, \ldots, x_n \\ \text{except } x_i}}{\sum \sum \cdots \sum} P(x_1, x_2, x_3, \ldots, x_n).$$

Remark *We will use the letter P to signify a mass function whether it be joint or marginal.*

EXERCISE 4-1

a) Show that the marginal probability of the subset $X_i, X_j, \ldots, X_k$ satisfies the conditions of the definition of a joint probability mass function.

b) Show that the marginal probability of X_i satisfies the conditions for a univariate probability mass function.

EXPECTATIONS

The expectation of any function of the n random variables $g(X_1, X_2, X_3, \ldots, X_n)$ is defined by

$$Eg(X_1, X_2, X_3, \ldots, X_n) = \sum_{x_1} \sum_{x_2} \sum_{x_3} \cdots \sum_{x_n} g(x_1, x_2, x_3, \ldots, x_n) \, P(x_1, x_2, x_3, \ldots, x_n).$$

The expectation of a function of a subset of the n random variables has a rather interesting property. Note that

$$Eg(X_i, X_j, \ldots, X_k) = \sum_{x_1} \sum_{x_2} \sum_{x_3} \cdots \sum_{x_n} g(x_i, x_j, \ldots, x_k) P(x_1, x_2, x_3, \ldots, x_n)$$

$$= \sum_{x_i}\sum_{x_j}\cdots\sum_{x_k} g(x_i, x_j, \ldots, x_k) \sum\sum_{\substack{\text{all} \\ x_1, x_2, x_3, \ldots, x_n \\ \text{except} \\ x_i, x_j, \ldots, x_k}}\cdots\sum P(x_1, x_2, x_3, \ldots, x_n)$$

$$= \sum_{x_i}\sum_{x_j}\cdots\sum_{x_k} g(x_i, x_j, \ldots, x_k)P(x_i, x_j, \quad , x_k).$$

This says that the expectation of a function of a subset of the random variables is the same in the joint distribution of the subset as it is in the joint distribution of all the variables. In particular, then

$$Eg(X_i) = \sum_{x_i} g(x_i)P(x_i).$$

Another interesting property of joint expectations is seen when the function in question is a sum of other functions. Note that

$$E[g_1(X_1, \ldots, X_j)+g_2(X_k, \ldots, X_l)+g_3(X_r, \ldots, X_s)]$$
$$= \sum_{x_1}\sum_{x_2}\sum_{x_3}\cdots\sum_{x_n}[g_1(x_i, \ldots, x_j)+g_2(x_k, \ldots, x_l)+g_3(x_r, \ldots, x_s)] \cdot P(x_1, x_2, x_3, \ldots, x_n)$$

(where there may be overlapping in the three subsets $X_i, \ldots, X_j$ and $X_k, \ldots, X_l$ and $X_r, \ldots, X_s$)

$$= \sum_{x_1}\sum_{x_2}\sum_{x_3}\cdots\sum_{x_n} g_1(x_i, \ldots, x_j)P(x_1, x_2, x_3, \ldots, x_n)$$
$$+ \sum_{x_1}\sum_{x_2}\sum_{x_3}\cdots\sum_{x_n} g_2(x_k, \ldots, x_l)P(x_1, x_2, x_3, \ldots, x_n)$$
$$+ \sum_{x_1}\sum_{x_2}\sum_{x_3}\cdots\sum_{x_n} g_3(x_r, \ldots, x_s)P(x_1, x_2, x_3, \ldots, x_n)$$
$$= Eg_1(x_i, \ldots, x_j)+Eg_2(x_k, \ldots, x_l)+Eg_3(x_r, \ldots, x_s).$$

Simply stated, the expectation of a sum is the sum of the expectations. In particular

$$E\sum_i g_i(X_i) = \sum_i Eg_i(X_i).$$

Combining this with the previous result we obtain

$$E\sum_i g_i(X_i) = \sum_{x_1} g_1(x_1)P(x_1)+ \sum_{x_2} g_2(x_2)P(x_2) \ldots + \sum_{x_n} g_n(x_n)P(x_n).$$

It is interesting to note that

$$E\sum_i X_i = \sum_i EX_i$$

and

$$E\sum_i X_i^2 = \sum_i EX_i^2.$$

It is, however, *incorrect* to write

$$\operatorname{Var} \sum_i X_i = \sum_i \operatorname{Var} X_i.$$

This can be seen from the following:

$$\begin{aligned}\operatorname{Var} \sum_i X_i &= E(\sum_i X_i - \sum_i EX_i)^2 = E(\sum_i X_i)^2 - 2E(\sum_i EX_i \sum_j X_j) + (\sum_i EX_i)^2 \\ &= E(\sum_i X_i)^2 - (\sum_i EX_i)^2 = \sum_i EX_i^2 + \sum_{i \neq j}\sum E(X_i X_j) \\ &\qquad - \sum_i (EX_i)^2 - \sum_{i \neq j}\sum EX_i EX_j \\ &= \sum_i EX_i^2 - \sum_i (EX_i)^2 + \sum_{i \neq j}\sum [E(X_i X_j) - EX_i EX_j] \\ &= \sum_i \operatorname{Var} X_i + \sum_{i \neq j}\sum [E(X_i X_j) - EX_i EX_j].\end{aligned}$$

The term $E(X_i X_j) - EX_i EX_j$ is called the covariance of X_i and X_j and is written

$$\operatorname{Cov}(X_i, X_j) = E(X_i X_j) - EX_i EX_j.$$

Hence we may write

$$\operatorname{Var} \sum_i X_i = \sum_i \operatorname{Var} X_i$$

if and only if

$$\sum_{i \neq j}\sum \operatorname{Cov}(X_i, X_j) = 0.$$

Remark *It is also interesting to note that*

$$\operatorname{Cov}(X_i, X_j) = E\{[x_i - E(x_i)][x_j - E(x_j)]\}.$$

In order to analyze this covariance term further we will have to define independence for random variables.

Definition The discrete random variables $X_1, X_2, X_3, \ldots, X_n$ are said to be statistically independent if

$$P(x_1, x_2, x_3, \ldots, x_n) = P(x_1)P(x_2)P(x_3) \ldots P(x_n) \quad \text{all } x_1, x_2, x_3, \ldots, x_n.$$

EXERCISE 4-2
Show that the equation in the preceding definition is equivalent to

$$F(x_1, x_2, x_3, \ldots, x_n) = F(x_1)\, F(x_2)\, F(x_3) \ldots F(x_n) \quad \text{all } x_1, x_2, x_3, \ldots, x_n.$$

If in the previous definition we sum over any subset of the variables, we find for the remaining set that

$$P(x_i, x_j, \ldots, x_k) = P(x_i)P(x_j) \ldots P(x_k) \quad \text{all } x_i, x_j, \ldots, x_k$$

which says that if $X_1, X_2, X_3, \ldots, X_n$ are statistically independent (for simplicity we say *independent*), then any subset of them is also independent.

Now if X_i and X_j are independent, then

$$E(X_iX_j) = \sum_{x_i}\sum_{x_j} x_ix_jP(x_ix_j) = \sum_{x_i}\sum_{x_j} x_ix_jP(x_i)P(x_j)$$
$$= \sum_{x_i} x_iP(x_i)\sum_{x_j} x_jP(x_j) = EX_iEX_j.$$

Since $\text{Cov}(X_i, X_j) = E(X_iX_j) - EX_iEX_j$, we have that if X_i and X_j are independent, then

$$\text{Cov}(X_i, X_j) = 0.$$

Therefore if $X_1, X_2, X_3, \ldots, X_n$ are independent, then each pair is independent and each covariance term is zero. From this we conclude that if $X_1, X_2, X_3, \ldots, X_n$ are independent, then

$$\text{Var} \sum_i X_i = \sum_i \text{Var}\, X_i.$$

4-2 The Trinomial Distribution

A very interesting example of a discrete multivariate distribution is afforded by the following. Let us say that we spin a three sided top which has probability p of landing on its red side, probability q of landing on its green side, and probability $1-p-q$ of landing on its blue side. We spin the top n times and define the random variables X and Y by

X = number of times the top landed on its red side

Y = number of times the top landed on its green side.

Each possible outcome can be considered an n-tuple with entries consisting of "red," "green," and "blue." All outcomes with x reds, y greens, and of course $n-x-y$ blues are *equally likely* to occur, having probability $p^xq^y(1-p-q)^{n-x-y}$. There are $\binom{n}{x}$ ways to obtain x reds out of n spins and $\binom{n-x}{y}$ ways to obtain y greens, given x reds have already been chosen. All the others must be blue, and there is only one way that can happen $\left(\binom{n-x-y}{n-x-y} = 1\right)$. From this we see that

$$P(x, y) = \binom{n}{x}\binom{n-x}{y} p^xq^y(1-p-q)^{n-x-y} \qquad 0 \leqq x+y \leqq n.$$

We will now verify that this is a probability mass function.

$$\sum_{x=0}^{n}\sum_{y=0}^{n-x} P(x, y) = \sum_{x=0}^{n}\sum_{y=0}^{n-x}\binom{n}{x}\binom{n-x}{y} p^x q^y (1-p-q)^{n-x-y}$$

$$= \sum_{x=0}^{n}\binom{n}{x} p^x \sum_{y=0}^{m}\binom{m}{y} q^y (1-p-q)^{m-y}$$

(where $m = n-x$)

$$= \sum_{x=0}^{n}\binom{n}{x} p^x (1-p)^m$$

(by the Binomial Theorem)

$$= \sum_{x=0}^{n}\binom{n}{x} p^x (1-p)^{n-x} = 1$$

and since each term is non-negative, $p(x, y)$ satisfies the conditions of the definition. Let us find the marginal distribution of X.

$$P(x) = \sum_{y} P(x, y) = \sum_{y=0}^{n-x}\binom{n}{x}\binom{n-x}{y} p^x q^y (1-p-q)^{n-x-y}$$

which, proceeding as above

$$= \binom{n}{x} p^x (1-p)^{n-x} \qquad 0 \leqq x \leqq n.$$

In order to find the marginal distribution of y, we find that the simplest way is to return to the original problem and count the green sides first, giving us an equivalent

$$P(x, y) = \binom{n}{n}\binom{n-y}{x} p^x q^y (1-p-q)^{n-x-y} \qquad 0 \leqq x+y \leqq n.$$

Then

$$P(y) = \sum_{x=0}^{n-y}\binom{n}{y}\binom{n-y}{x} p^x q^y (1-p-q)^{n-x-y}$$

$$= \binom{n}{y} q^y \sum_{x=0}^{m}\binom{m}{x} P^x (1-p-q)^{m-x}$$

(where $m = n-y$)

$$= \binom{n}{y} q^y (1-q)^{n-y} \qquad 0 \leqq y \leqq n$$

as we could have predicted from the symmetry between X and Y.

EXERCISE 4-3

a) Find EX in the joint distribution, and show that it equals EX in the marginal distribution, i.e., $EX = np$.

b) Find EY in the joint distribution, and show that it equals EY in the marginal distribution, i.e., $EY = nq$.

Having obtained the expectations of X and Y, we can obtain the covariance of X and Y by first finding the expectation of their product. We now do this:

$$E(XY) = \sum_{x=0}^{n}\sum_{y=0}^{n-x} xy\binom{n}{x}\binom{n-x}{y} p^x q^y (1-p-q)^{n-x-y}$$

$$= \sum_{x=1}^{n} x\binom{n}{x} p^x \sum_{y=1}^{n-x} y\binom{n-x}{y} q^y(1-p-q)^{n-x-y}$$

$$= \sum_{x=1}^{n} x\binom{n}{x} p^x (n-x)q \sum_{y=1}^{n-x} \frac{(n-x-1)!}{(y-1)!(n-x-y)!} q^{y-1}(1-q-p)^{n-x-y}$$

$$= \sum_{x=1}^{n} x\binom{n}{x} p^x (n-x)q \sum_{z=0}^{m} \binom{m}{z} q^z(1-p-q)^{m-z}$$

(where $m = n-x-1$ and $z = y-1$)

$$= \sum_{x=1}^{n} x\binom{n}{x} p^x (n-x)q(1-p)^{n-x-1}$$

$$= q\sum_{x=1}^{n-1} \frac{n!}{(x-1)!(n-x-1)!} p^x(1-p)^{n-x-1}$$

$$= n(n-1)pq\sum_{x=1}^{n-1} \frac{(n-2)!}{(x-1)!(n-x-1)!} p^{x-1}(1-p)^{n-x-1}$$

$$= n(n-1)pq\sum_{z=0}^{m} \frac{m!}{z!(m-z)!} p^z(1-p)^{m-z}$$

where $m = n-2$ and $z = x-1$

$$= n(n-1)pq.$$

Therefore, we find

$$\text{Cov}(X, Y) = E(XY) - EXEY = n(n-1)pq - n^2pq = -npq.$$

Clearly X and Y cannot be independent random variables unless one of them cannot occur (p or q equals zero). That they are not independent can also be seen in the fact that their joint probability mass function is not the product of their marginal probability mass functions.

4-3 Continuous Random Variables

If $X_1, X_2, X_3, \ldots, X_n$ are random variables each of which takes on values in an interval, then we can speak of a joint probability density function. If for any two sets of values for $X_1, X_2, X_3, \ldots, X_n$, say $a_1, a_2, a_3, \ldots, a_n$ and $b_1, b_2, b_3, \ldots, b_n$, we have

$$P(a_1 \leqq X_1 \leqq b_1, a_2 \leqq X_2 \leqq b_2, \ldots, a_n \leqq X_n \leqq b_n)$$

$$= \int_{a_n}^{b_n} \int_{a_{n-1}}^{b_{n-1}} \cdots \int_{a_1}^{b_1} f(x_1, x_2, x_3, \ldots, x_n)\, dx_1\, dx_2\, dx_3 \ldots dx_n$$

then $f(x_1, x_2, x_3, \ldots, x_n)$ will be called the *joint probability density function* for $X_1, X_2, X_3, \ldots, X_n$. As in the univariate case, we will restrict $f(x_1, x_2, x_3, \ldots, x_n)$ to be a non-negative piecewise continuous function in each of the variables. From its definition it is clear that

$$\int \int \cdots \int f(x_1, x_2, x_3, \ldots, x_n)\, dx_1\, dx_2\, dx_3 \ldots dx_n = 1$$

(where the omission of limits of integration implies limits from $-\infty$ to $+\infty$).

The *marginal density function* of the random variables $X_i, X_j, \ldots, X_k$ can now be defined as

$$f(x_i, x_j, \ldots, x_k) = \underbrace{\int \int \cdots \int}_{\substack{\text{omitting} \\ \text{integration over} \\ x_i, x_j, \ldots, x_k}} f(x_1, x_2, x_3, \ldots, x_n) \underbrace{dx_1\, dx_2\, dx_3 \ldots dx_n}_{\substack{\text{omitting} \\ dx_i, dx_j \ldots dx_k}}$$

Remark *The letter f will represent a density, whether it be joint or marginal.*

The marginal density of a single random variable X_i is then

$$f(x_i) = \underbrace{\int \int \cdots \int}_{\substack{\text{omitting} \\ \text{integration over} \\ x_i}} f(x_1, x_2, x_3, \ldots, x_n) \underbrace{dx_1\, dx_2\, dx_3 \ldots dx_n}_{\substack{\text{omitting} \\ dx_i}}.$$

Clearly since $f(x_i)$ is the integral of a non-negative function it is itself non-negative. Besides this, we see that

$$\int f(x_i)\, dx_i = \int \int \cdots \int f(x_1, x_2, x_3, \ldots, x_n)\, dx_1\, dx_2\, dx_3 \ldots dx_n = 1$$

which indicates that $f(x_i)$ satisfies the requirements for a probability density function.

The *joint cumulative distribution* function for $X_1, X_2, X_3, \ldots, X_n$ is defined by

$$F(x_1, x_2, x_3, \ldots, x_n) = \int_{-\infty}^{x_n} \cdots \int_{-\infty}^{x_1} f(t_1, t_2, t_3, \ldots, t_n)\, dt_1\, dt_2\, dt_3 \ldots dt_n.$$

The *marginal cumulative distributions* can now be found in two ways. The first method eliminates the unwanted variables by setting them at $+\infty$, giving

$$F(x_i, x_j, \ldots, x_k) = F(\infty, \infty, \ldots, x_i, x_j, \ldots, x_k, \infty, \ldots)$$

with all $x_1, x_2, x_3, \ldots, x_n$ set to ∞ except $x_i, x_j, \ldots, x_k$.

The second method utilizes the relation between a density function and its cumulative distribution function giving

$$F(x_i, x_j, \ldots, x_k) = \int_{-\infty}^{x_k} \cdots \int_{-\infty}^{x_j} \int_{-\infty}^{x_i} f(t_i, t_j, \ldots, t_k)\, dt_i dt_j \ldots dt_k.$$

EXERCISE 4-4

Show that the two definitions of the cumulative distribution function are equivalent.

EXPECTATIONS

We can define the *expectation* of a function of the n random variables $g(X_1, X_2, X_3, \ldots, X_n)$ by

$$Eg(X_1, X_2, X_3, \ldots, X_n) = \int\int \cdots \int g(x_1, x_2, x_3, \ldots, x_n) f(x_1, x_2, x_3, \ldots, x_n)\, dx_1 dx_2 dx_3 \ldots dx_n.$$

When $g(\cdot)$ is a function of fewer than n random variables, then we obtain

$$Eg(X_i, X_j, \ldots, X_k) = \int\int \cdots \int g(x_i, x_j, \ldots, x_k) f(x_1, x_2, x_3, \ldots, x_n)\, dx_1 dx_2 dx_3 \ldots dx_n$$

$$= \int\int \cdots \int g(x_i, x_j, \ldots, x_k) f(x_i, x_j, \ldots, x_k)\, dx_i dx_j \ldots dx_k.$$

As in the discrete case, this says that the joint expectation of a function of fewer than n random variables is the same as it is in the marginal distribution of those variables. The sum property can be seen to hold here, too. That is, the integral of a sum (a finite sum) is the sum of the integrals. This leads almost immediately to the fact that

$$E\sum g_i(X_i) = \sum_i Eg_i(X_i).$$

as in the discrete case. Therefore, we see that even for continuous random variables

$$E\sum_i X_i = \sum_i EX_i$$

and

$$E\sum_i X_i^2 = \sum_i EX_i^2$$

but

$$\operatorname{Var}\sum_i X_i = \sum_i \operatorname{Var} X_i + \sum_{i \neq j}\sum \operatorname{Cov}(X_i, X_j)$$

where $\operatorname{Cov}(X_i, X_j) = E(X_iX_j) - EX_iEX_j$.

The term *independence* is also used with continuous random variables. Its meaning, however, is not as intuitive as in the discrete case since a probability density is not a probability.

Definition The continuous random variables $X_1, X_2, X_3, \ldots, X_n$ are said to be statistically independent if

$$f(x_1, x_2, x_3, \ldots, x_n) = f(x_1)f(x_2)f(x_3)\ldots f(x_n) \qquad \text{for all } x_1, x_2, x_3, \ldots, x_n.$$

Since the number of n-tuples $x_1, x_2, x_3, \ldots, x_n$ for which this must be satisfied is uncountable, this is a stronger definition than in the discrete case. That it is equivalent to the discrete case when we are restricted to interval events is easily seen by the following:

$$\begin{aligned}
&P(a_1 \leqq X_1 \leqq b_1, a_2 \leqq X_2 \leqq b_2, \ldots a_n \leqq X_n \leqq b_n) \\
&\quad = \int_{a_1}^{b_1}\int_{a_2}^{b_2} \cdots \int_{a_n}^{b_n} f(x_1, x_2, \ldots, x_n)\, dx_1 dx_2 \ldots dx_n \\
&\quad = \int_{a_1}^{b_1}\int_{a_2}^{b_2} \cdots \int_{a_n}^{b_n} f(x_1)f(x_2)\ldots f(x_n)\, dx_1 dx_2 \ldots dx_n \\
&\quad = \int_{a_1}^{b_1} f(x_1)\, dx_1 \int_{a_2}^{b_2} f(x_2)\, dx_2 \ldots \int_{a_n}^{b_n} f(x_n)\, dx_n \\
&\quad = P(a_1 \leqq X_1 \leqq b_1)P(a_2 \leqq X_2 \leqq b_2) \ldots P(a_n \leqq X_n \leqq b_n).
\end{aligned}$$

EXERCISE 4-5

a) Show that if $X_1, X_2, X_3, \ldots, X_n$ are statistically independent, then any two of them, X_i and X_j, are independent.

b) Show that if $X_1, X_2, X_3, \ldots, X_n$ are independent, then

$$\operatorname{Var} \Sigma_i X_i = \Sigma_i \operatorname{Var} X_i$$

4-4 The Bivariate Normal Distribution

As an example of a joint probability density function, we will consider the bivariate normal density function of the random variables X and Y given by

$$f(x, y) = \frac{1}{2\pi\sigma_x\sigma_y\sqrt{(1-\rho^2)}} \exp\left\{\frac{1}{-2(1-\rho^2)}\left[\left(\frac{x-\mu_x}{\sigma_x}\right)^2 - 2\rho\left(\frac{x-\mu_x}{\sigma_x}\right)\left(\frac{y-\mu_y}{\sigma_y}\right) + \left(\frac{y-\mu_y}{\sigma_y}\right)^2\right]\right\}$$

$$-\infty < x, y < \infty$$

$$-\infty < \mu_x, \mu_y < \infty$$

$$\sigma_x, \sigma_y > 0$$

$$-1 < \rho < +1.$$

When $\mu_x = \mu_y = 0$ and $\sigma_x = \sigma_y = 1$, the resulting bivariate normal density function

$$f(x, y) = \frac{1}{2\pi\sqrt{(1-\rho^2)}} \exp\left\{-\frac{1}{2(1-\rho^2)}(x^2 - 2\rho xy + y^2)\right\} \qquad -\infty < x, y < \infty$$

$$-1 < \rho < +1$$

is known as the standard bivariate normal density function with correlation ρ.

The bivariate normal density function is everywhere positive and continuous in both variables.

The marginal distribution of Y can now be found by

$$f(y) = \frac{1}{2\pi\sigma_x\sigma_y\sqrt{(1-\rho^2)}} \int_{-\infty}^{\infty} \exp\left\{-\frac{1}{2(1-\rho^2)}\left[\left(\frac{x-\mu_x}{\sigma_x}\right)^2 - 2p\left(\frac{x-\mu_x}{\sigma_x}\right)\left(\frac{y-\mu_y}{\sigma_y}\right) + \left(\frac{y-\mu_y}{\sigma_y}\right)^2\right]\right\} dx$$

$$= \frac{1}{2\pi\sigma_x\sigma_y\sqrt{(1-\rho^2)}} \int_{-\infty}^{\infty} \exp\left\{-\frac{1}{2(1-\rho^2)}[u^2 - 2\rho uv + v^2]\right\} \sigma_x du$$

$$\left(\text{where} \quad u = \frac{x-\mu_x}{\sigma_x} \quad \text{and} \quad v = \frac{y-\mu_y}{\sigma_y}\right)$$

$$= \frac{1}{2\pi\sigma_y} \int e^{-(w^2+v^2)/2} dw$$

$$\left(\text{where} \quad w = \frac{u-\rho v}{\sqrt{(1-\rho^2)}}\right)$$

$$= \frac{1}{\sqrt{(2\pi)}\sigma_y} e^{-v^2/2} \frac{1}{\sqrt{(2\pi)}} \int_{-\infty}^{\infty} e^{-w^2/2} dw = \frac{1}{\sqrt{(2\pi)}\sigma_y} e^{-v^2/2}$$

$$= \frac{1}{\sqrt{(2\pi)}\sigma_y} e^{-(y-\mu_y)^2/2\sigma^2{}_y} \qquad -\infty < y < \infty.$$

Similarly, by integrating out y, we obtain the marginal density for X as

$$f(x) = \frac{1}{\sqrt{(2\pi)}\sigma_x} e^{-(x-\mu_x)^2/2\sigma_x{}^2} \qquad -\infty\; x < \infty.$$

Clearly then

$$\int\int f(x, y)\, dx\, dy = 1.$$

Since the marginal density functions are density functions for a univariate normal random variable, we easily obtain

$$EX = \mu_x, \quad EY = \mu_y, \quad \text{Var } X = \sigma_x{}^2, \quad \text{and} \quad \text{Var } Y = \sigma_y{}^2.$$

In order to find the covariance between X and Y we recall for any random variables X and Y

$$\text{Cov}(X, Y) = E(XY) - EXEY = E[(X-\mu_x)(Y-\mu_y)].$$

We now find this in the following manner.

$$E[(X-\mu_x)(Y-\mu)]$$

$$= \frac{1}{2\pi\sigma_x\sigma_y\sqrt{(1-\rho^2)}} \iint (x-\mu_y)(y-\mu_y)$$

$$\cdot \exp\left\{-\frac{1}{2(1-\rho^2)}\left[\left(\frac{x-\mu_x}{\sigma_x}\right)^2 - 2\rho\left(\frac{x-\mu_y}{\sigma_x}\right)\left(\frac{y-\mu_y}{\sigma_y}\right) + \left(\frac{y-\mu_y}{\sigma_y}\right)^2\right]\right\} dx\, dy$$

$$= \frac{\sigma_x\sigma_y}{2\pi\sqrt{(1-\rho^2)}} \iint uv \exp\left\{-\frac{1}{2(1-\rho^2)}(u^2 - 2\rho uv + v^2)\right\} du\, dv$$

(where $u = (x-\mu_x)/\sigma_x$ and $v = (y-\mu_y)/\sigma_y$)

$$= \frac{\sigma_x\sigma_y}{2\pi\sqrt{(1-\rho^2)}} \iint_{-\infty}^{\infty} uv \exp\left\{-\frac{1}{2(1-\rho^2)}[(u-\rho v)^2 + v^2(1-\rho^2)]\right\} du\, dv$$

$$= \frac{\sigma_x\sigma_y}{2\pi\sqrt{(1-\rho^2)}} \int_{-\infty}^{\infty} v\, e^{-v^2/2} \int_{-\infty}^{\infty} u \exp\left\{-\frac{1}{2(1-\rho^2)}[(u-\rho v)^2]\right\} du\, dv$$

$$= \frac{\sigma_x\sigma_y}{2\pi\sqrt{(1-\rho^2)}} \int_{-\infty}^{\infty} v\, e^{-v^2/2} \int_{-\infty}^{\infty} (u-\rho v) \exp\left\{-\frac{1}{2(1-\rho^2)}[(u-\rho v)^2]\right\} du\, dv$$

$$+ \frac{\sigma_x\sigma_y}{2\pi\sqrt{(1-\rho^2)}} \int_{-\infty}^{\infty} v\, e^{-v^2/2} \int_{-\infty}^{\infty} \rho v \exp\left\{-\frac{1}{2(1-\rho^2)}[(u-\rho v)^2]\right\} du\, dv.$$

By setting $w = u-\rho v$ in the inner integral of the first term on the right-hand side, we obtain an integration similar to that used in finding the mean of a normal variable with mean zero and variance $(1-\rho^2)$. This first term is therefore zero. Thus

$$\mathrm{Cov}(X, Y) = \frac{\sigma_x\sigma_y}{2\pi\sqrt{(1-\rho^2)}} \int_{-\infty}^{\infty} \rho v^2\, e^{-v^2/2} \int_{-\infty}^{\infty} \exp\left\{-\frac{1}{2(1-\rho^2)}(u-\rho v)^2\right\} du\, dv$$

$$= \frac{\sigma_x\sigma_y}{\sqrt{(2\pi)}} \int_{-\infty}^{\infty} \rho v^2\, e^{-v^2/2} \frac{1}{\sqrt{(2\pi(1-\rho^2))}} \int_{-\infty}^{\infty} \exp\left\{-\frac{1}{2(1-\rho^2)} w^2\right\} dw\, dv$$

(where $w = u-\rho v$)

$$= \frac{\sigma_x\sigma_y}{\sqrt{(2\pi)}} \int_{-\infty}^{\infty} \rho v^2 e^{-v^2/2} dv = \rho\sigma_x\sigma_y \left[\frac{2}{\sqrt{(2\pi)}} \int_{0}^{\infty} v^2 e^{-v^2/2} dv\right]$$

(where the bracketed term is the integral of the chi density function with $n = 3$ and $\sigma = 1$)

$$= \rho\sigma_x\sigma_y.$$

Since neither σ_x nor σ_y can be zero, this covariance term is zero if and only if $\rho = 0$. However, when $\rho = 0$ the bivariate density becomes

$$f(x, y) = \frac{1}{2\pi\sigma_x\sigma_y} \exp\left\{-\frac{1}{2}\left[\left(\frac{x-\mu_x}{\sigma_x}\right)^2 + \left(\frac{y-\mu_y}{\sigma_y}\right)^2\right]\right\}$$

$$= \frac{1}{\sqrt{(2\pi)}\sigma_x} e^{-(x-\mu_x)^2/2\sigma^2{}_x} \frac{1}{\sqrt{(2\pi)}\sigma_y} e^{-(y-\mu_y)^2/2\sigma^2{}_y} = f(x) f(y)$$

which says that X and Y are independent. Thus in the special case of a bivariate normal distribution we see that X and Y are independent if and only if the covariance between X and Y is zero. This is however not a general result.

EXERCISE 4-6

a) If $U = aX$ and $V = bY$, show that $\text{Cov}(U, V) = ab\,\text{Cov}(X, Y)$.
b) If $U = X+a$ and $V = Y+b$, show that $\text{Cov}(U, V) = \text{Cov}(X, Y)$.

4-5 Conditional Distributions: The Discrete Case

To be consistent with Chapter One, the probability of the event that the random variable X will be observed to have value x, given the event that the random variable Y has value y, must be given by

$$P(x \mid Y = y) = \frac{P(x, y)}{P(y)} \qquad \text{when } P(y) \neq 0.$$

This is a non-negative function such that

$$\sum_x P(x \mid Y = y) = \sum_x \frac{P(x, y)}{P(y)} = \frac{P(y)}{P(y)} = 1$$

and is therefore a probability mass function. This concept can be extended to any number of variables $X_1, X_2, X_3, \ldots, X_n$ by defining the conditional probability of the random variables $X_i, X_j, \ldots, X_k$ given another set of random variables $X_\mu, X_\nu, \ldots, X_\rho$ by

$$P(x_i, x_j, \ldots, x_k \mid x_\mu, x_\nu, \ldots, x_\rho) = \frac{P(x_i, x_j, \ldots, x_k, x_\mu, x_\nu, \ldots x_\rho)}{P(x_\mu, x_\nu, \ldots, x_\rho)}$$

$$\text{when } P(x_\mu, x_\nu, \ldots, x_\rho) \neq 0.$$

Since both probabilities on the right-hand side are non-negative this is a non-negative function. It is not quite so clear that this ratio is less than unity. However

$$P(x_\mu, x_\nu, \ldots, x_\rho) = \sum_i \sum_j \cdots \sum_k P(x_i, x_j, \ldots, x_k, x_\mu, x_\nu, \ldots, x_\rho)$$

and since a sum of non-negative quantities is never smaller than one of its parts, we have that the ratio is not greater than unity. As a matter of fact

$$\begin{aligned} &\sum_i \sum_j \cdots \sum_k P(x_i, x_j, \ldots, x_k \mid x_\mu, x_\nu, \ldots, x_\rho) \\ &= \sum_i \sum_j \cdots \sum_k \frac{P(x_i, x_j, \ldots, x_k, x_\mu, x_\nu, \ldots, x_\rho)}{P(x_\mu, x_\nu, \ldots, x_\rho)} \\ &= \frac{P(x_\mu, x_\nu, \ldots, x_\rho)}{P(x_\mu, x_\nu, \ldots, x_\rho)} = 1 \end{aligned}$$

as it should.

Remark *The conditional probability* $P(x_i, x_j, \ldots, x_k \mid x_\mu, x_\nu, \ldots, x_\rho)$ *is interpreted as a function of* $x_i, x_j, \ldots, x_k$ *with* $X_\mu = x_\mu$, $X_\nu = x_\nu, \ldots X_\rho = x_\rho$ *fixed.*

At this point it is useful to extend the concept of independence to independence between two sets of the X's—say $X_i, X_j, \ldots, X_k$ and $X_\mu, X_\nu, \ldots, X_\rho$. This can be done in two ways. First, one can say that these sets are independent if and only if

$$P(x_i, x_j, \ldots, x_k, x_\mu, x_\nu, \ldots, x_\rho) = P(x_i, x_j, \ldots, x_k)P(x_\mu, x_\nu, \ldots, x_\rho) \qquad \text{all } x_i, x_j, \ldots, x_k \text{ and } x_\mu, x_\nu, \ldots, x_\rho.$$

The second way is to say that these sets are independent if and only if

$$P(x_i, x_j, \ldots, x_k \mid x_\mu, x_\nu, \ldots, x_\rho) = P(x_i, x_j, \ldots, x_k) \qquad \text{all } x_i, x_j, \ldots, x_k \text{ and } x_\mu, x_\nu, \ldots, x_\rho.$$

That these two definitions are equivalent can be seen as follows. Assuming the first form, we obtain

$$\begin{aligned} P(x_i, x_j, \ldots, x_k \mid x_\mu, x_\nu, \ldots, x_\rho) &= \frac{P(x_i, x_j, \ldots, x_k, x_\mu, x_\nu, \ldots, x_\rho)}{P(x_\mu, x_\nu, \ldots, x_\rho)} \\ &= \frac{P(x_i, x_j, \ldots, x_k)P(x_\mu, x_\nu, \ldots, x_\rho)}{P(x_\mu, x_\nu, \ldots, x_\rho)} \\ &= P(x_i, x_j, \ldots, x_k). \end{aligned}$$

If we now assume the second form we obtain

$$\begin{aligned} P(x_i, x_j, \ldots, x_k, x_\mu, x_\nu, \ldots, x_\rho) &= P(x_i x_j, \ldots, x_k \mid x_\mu, x_\nu, \ldots, x_\rho) \\ &\qquad P(x_\mu, x_\nu, \ldots, x_\rho) \\ &= P(x_i, x_j, \ldots, x_k)P(x_\mu, x_\nu, \ldots, x_\rho) \end{aligned}$$

demonstrating the equivalence of the two definitions. If there are just two random variables, then this shows that one can define independence of X and Y by

$$P(x \mid Y = y) = P(x) \qquad \text{all } x \text{ for each } y.$$

EXPECTATIONS

The expectation of a function of $X_i, X_j, \ldots, X_k$, given $X_\mu = x_\mu$, $X_\nu = x_\nu, \ldots, X_\rho = x_\rho$ is defined simply as the expectation of the function in the conditional distribution of $X_i, X_j, \ldots, X_k$ given $x_\mu, x_\nu, \ldots, x_\rho$, and is written

$$\begin{aligned} &E[g(X_i, X_j, \ldots, X_k) \mid x_\mu, x_\nu, \ldots, x_\rho] \\ &\qquad = \sum_{x_i}\sum_{x_j} \cdots \sum_{x_k} g(x_i, x_j, \ldots, x_k)P(x_i, x_j, \ldots, x_k \mid x_\mu, x_\nu, \ldots, x_\rho). \end{aligned}$$

For just two variables X and Y this becomes

$$E(g(x) \mid y) = \sum_x g(x)P(x \mid y) = \sum_x g(x)\frac{P(x, y)}{P(y)}.$$

For fixed y this is simply a number, but it is often useful to leave y as an unspecified parameter, making this conditional expectation a function of y. That is $E[g(X) \mid y] = h(y)$, where $h(y)$ is some function of y. In general, one writes

$$E[g(X_i, X_j, \ldots, X_k) \mid x_\mu, x_\nu, \ldots, x_\rho] = h(x_\mu, x_\nu, \ldots, x_\rho)$$

where $h(x_\mu, x_\nu, \ldots, x_\rho)$ is some function of $x_\mu, x_\nu, \ldots, x_\rho$.

If $X_i, X_j, \ldots, X_k$ and $X_\mu, X_\nu, \ldots, X_\rho$ are independent sets of random variables, then

$$\begin{aligned} &E[g(X_i, X_j, \ldots, X_k) \mid x_\mu, x_\nu, \ldots, x_\rho] \\ &\quad = \sum_{x_i}\sum_{x_j}\cdots\sum_{x_k} g(x_i, x_j, \ldots, x_k)P(x_i, x_j, \ldots, x_k \mid x_\mu, x_\nu, \ldots, x_\rho) \\ &\quad = \sum_{x_i}\sum_{x_j}\cdots\sum_{x_k} g(x_i, x_j, \ldots, x_k)P(x_i, x_j, \ldots, x_k) \\ &\quad = Eg(X_i, X_j, \ldots, X_k). \end{aligned}$$

Therefore, we see that in the special case of independence the function $h(x_\mu, x_\nu, \ldots, x_\rho)$ must be a constant for all values of $x_\mu, x_\nu, \ldots, x_\rho$, this constant being the expectation of the function in the joint distribution of $X_i, X_j, \ldots, X_k$. When the sets are not independent this function may not be constant at all, but

$$\begin{aligned} &E\{E[g(X_i, X_j, \ldots, X_k) \mid X_\mu, X_\nu, \ldots, X_\rho] \\ &\quad = E\,h\,(X_\mu, X_\nu, \ldots, X_\rho) \\ &\quad = \sum_{x_\mu}\sum_{x_\nu}\cdots\sum_{x_\rho}\Bigg[\sum_{x_i}\sum_{x_j}\cdots\sum_{x_k} g(x_i, x_j, \ldots, x_k) \\ &\qquad\qquad P(x_i, x_j, \ldots, x_k \mid x_\mu, x_\nu, \ldots, x_\rho)\Bigg]P(x_\mu, x_\nu, \ldots, x_\rho) \\ &\quad = \sum_{x_\mu}\sum_{x_\nu}\cdots\sum_{x_\rho}\sum_{x_i}\sum_{x_j}\cdots\sum_{x_k} g(x_i, x_j, \ldots, x_k) \\ &\qquad\qquad \frac{P(x_i, x_j, \ldots, x_k, x_\mu, x_\nu, x_\rho)}{P(x_\mu, x_\nu, \ldots, x_\rho)}P(x_\mu, x_\nu, \ldots, x_\rho) \\ &\quad = \sum_{x_\mu}\sum_{x_\nu}\cdots\sum_{x_\rho}\sum_{x_i}\sum_{x_j}\cdots\sum_{x_k} g(x_i, x_j, \ldots, x_k) \\ &\qquad\qquad P(x_i, x_j, \ldots, x_k, x_\mu, x_\nu, \ldots, x_\rho) \\ &\quad = Eg(X_i, X_j, \ldots, X_k). \end{aligned}$$

This says that the expectation of this conditional expectation is the expectation of the function in the joint distribution of $X_i, X_j, \ldots, X_k$. In particular

$$E\{E[g(X) \mid Y]\} = E\{g(X)\}.$$

Remark *It is a notational convention that when we are taking an expectation of a conditional expectation we use the random variable Y rather than its observed value y in the inner expectation.*

An interesting relation comes about when we ask for $E[\operatorname{Var}(X \mid Y)]$. An immediate but incorrect impression might be that it equals Var X. However

$$\begin{aligned} E[\operatorname{Var}(X \mid Y)] &= E\{E(X^2 \mid Y) - [E(X \mid Y)]^2\} \\ &= E\{E(X^2 \mid Y) - (EX)^2 + (EX)^2 - [E(X \mid Y)]^2\} \\ &= EX^2 - (EX)^2 - \{E[E(X \mid Y)]^2 - (EX)^2\} \\ &= \operatorname{Var} X - \{E[E(X \mid Y)]^2 - (E[E(X \mid Y)])^2\} \\ &= \operatorname{Var} X - \operatorname{Var}[E(X \mid Y)]. \end{aligned}$$

This is usually stated as

$$\operatorname{Var} X = E[\operatorname{Var}(X \mid Y)] + \operatorname{Var}[E(X \mid Y)].$$

This equation is often useful. In particular, since a variance cannot be negative, it shows that

$$\operatorname{Var} X \geqq E[\operatorname{Var}(X \mid Y)].$$

THE TRINOMIAL DISTRIBUTION

We will now obtain conditional probabilities and conditional expectations for the trinomial distribution. Recall that

$$P(x, y) = \binom{n}{y}\binom{n-y}{x} p^x q^y (1-p-q)^{n-x-y} \qquad \begin{array}{l} 0 < p+q < 1 \\ 0 \leqq x+y \leqq n \end{array}$$

and

$$P(y) = \binom{n}{y} q^y (1-q)^{n-y} \qquad \begin{array}{l} 0 < q < 1 \\ 0 \leqq y \leqq n \end{array}$$

from which we obtain

$$P(x \mid y) = \frac{P(x, y)}{P(y)} = \frac{\binom{n}{y}\binom{n-y}{x} p^x q^y (1-p-q)^{n-x-y}}{\binom{n}{y} q^y (1-q)^{n-y}}$$

$$= \binom{n-y}{x} p^x \frac{(1-p-q)^{n-x-y}}{(1-q)^{n-y}} \qquad 0 < p+q < 1$$

$$0 \leqq x \leqq n-y$$

for any $0 \leqq y \leqq n$.

We see that this is a mass function since

$$\sum_x P(x \mid y) = \sum_{x=0}^{n-y} \binom{n-y}{x} p^x \frac{(1-p-q)^{n-y-x}}{(1-q)^{n-y}} = \frac{(p+1-p-q)^{n-y}}{(1-q)^{n-y}} = 1.$$

It is also interesting to note that the mass function can be written

$$P(x \mid y) = \binom{n-y}{x} \left(\frac{p}{1-q}\right)^x \left(\frac{1-p-q}{1-q}\right)^{n-y-x}$$

$$0 \leqq x \leqq n-y$$

which is the probability of x successes in $n-y$ trials with probability $p/(1-q)$ for success at each trial. This is therefore a binomial distribution. Similarly, since

$$P(x) = \binom{n}{x} p^x (1-p)^{n-x} \qquad 0 < p < 1$$

$$0 \leqq x \leqq n$$

and

$$P(x, y) = \binom{n-y}{x} \binom{n}{x} p^x q^y (1-p-q)^{n-x-y} \qquad 0 < p+q < 1$$

$$0 \leqq x+y \leqq n$$

we obtain

$$P(y \mid x) = \frac{P(x, y)}{p(y)} = \frac{\binom{n-x}{y} \binom{n}{x} p^x q^y (1-p-q)^{n-x-y}}{\binom{n}{x} p^x (1-p)^{n-x}}$$

$$= \binom{n-x}{y} q^y \frac{(1-p-q)^{n-x-y}}{(1-p)^{n-x}} \qquad 0 < p+q < 1$$

$$0 \leqq y \leqq n-x$$

for any $0 \leqq y \leqq n$.

$$= \binom{n-x}{y} \left(\frac{q}{1-p}\right)^y \left(\frac{1-p-q}{1-p}\right)^{n-x-y} \qquad 0 \leqq y \leqq n-x.$$

From these we obtain

$$E(X \mid y) = \sum_{x=0}^{n-y} x\binom{n-y}{x} p^x \frac{(1-p-q)^{n-y-x}}{(1-q)^{n-y}}$$

$$= \frac{1}{(1-q)^N} \sum_{x=1}^{N} \frac{N!}{(x-1)!(N-x)!} p^x(1-p-q)^{N-x}$$

$$\text{(where } N = n-y\text{)}$$

$$= \frac{Np}{(1-q)^N} \sum_{z=0}^{M} \frac{M!}{(M-z)!z!} p^z(1-p-q)^{M-z}$$

$$\text{(where } M = N-1 \quad \text{and} \quad z = x-1\text{)}$$

$$= \frac{Np}{(1-q)^N}(1-q)^M = (n-y)\frac{p}{1-q} \quad \text{for any } 0 \leqq y \leqq n.$$

From the similarity of the mass functions one can easily see that

$$E(Y \mid x) = (n-x)\frac{q}{1-p} \qquad \text{for any } 0 \leqq x \leqq n.$$

These results could have been obtained without the tedious algebra simply by noting that the conditional distributions are themselves binomial distributions. The mean of any binomial distribution with R trials and probability P for success is given by $EZ = RP$ which in this case implies

$$E(X \mid y) = RP = (n-y)\frac{p}{1-q}$$

and

$$E(Y \mid x) = (n-x)\frac{q}{1-p}.$$

Similarly, the variance of the binomial distribution is $\operatorname{Var} Z = RP(1-P)$ which implies

$$\operatorname{Var}(X \mid y) = (n-y)\left(\frac{p}{1-q}\right)\left(1-\frac{p}{1-q}\right) = (n-y)p(1-p-q)/(1-q)^2$$

and

$$\operatorname{Var}(Y \mid x) = (n-x)\left(\frac{q}{1-p}\right)\left(1-\frac{q}{1-p}\right) = (n-x)q(1-p-q)/(1-p)^2.$$

These conditional expectations are not constants for all values of the conditioned variable. This could have been anticipated since we already know that X and Y are not independent. Let us now find the expected values of these conditional expectations.

$$E[E(X \mid Y)] = E\left[(n-Y)\frac{p}{1-q}\right] = E\left[\frac{np}{1-q}-\frac{Yp}{1-q}\right] = \frac{np}{1-q}-\frac{p}{1-q}EY$$

$$= \frac{np}{1-q}-\frac{p}{1-q}nq = np(1-q)/(1-q) = np = EX$$

as it should.

$$E[E(Y \mid X)] = E\left[(n-X)\frac{q}{1-p}\right] = \frac{nq}{1-p}-\frac{q}{1-p}EX$$

$$= \frac{nq}{1-p}-\frac{q}{1-p}np = nq = EY$$

as it should. Similarly

$$E[\mathrm{Var}(X \mid Y)] = (n-EY)p(1-p-q)/(1-q)^2$$
$$= n(1-q)p(1-p-q)/(1-q)^2 = np(1-p-q)/(1-q)$$

and

$$E[\mathrm{Var}(Y \mid X)] = (n-EX)q(1-p-q)/(1-p)^2 = nq(1-p-q)/(1-p).$$

It will now be of interest to find

$$\mathrm{Var}[E(X \mid Y)] = \mathrm{Var}[(n-Y)p/(1-q)] = \frac{p^2}{(1-q)^2}\mathrm{Var}\ Y = \frac{np^2q}{(1-q)}$$

and similarly

$$\mathrm{Var}[E(Y \mid X)] = \mathrm{Var}[(n-X)q/(1-p) = \frac{q^2}{(1-p)^2}\mathrm{Var}\ X = \frac{npq^2}{(1-p)}.$$

From these we can check the equation

$$\mathrm{Var}\ X = E[\mathrm{Var}(X \mid Y)]+\mathrm{Var}[E(X \mid Y)] = \frac{np(1-p-q)}{(1-q)}+\frac{np^2q}{(1-q)}$$

$$= \frac{np(1-p)-np(1-p)q}{1-q} = np(1-p)$$

as it should; similarly

$$\mathrm{Var}\ Y = E[\mathrm{Var}(Y \mid X)]+\mathrm{Var}[E(Y \mid X)] = \frac{nq(1-p-q)}{(1-p)}+\frac{npq^2}{(1-p)}$$

$$= \frac{nq(1-q)-nq(1-q)p}{(1-p)} = nq(1-q)$$

as it should.

4-6 Conditional Distributions: The Continuous Case

The concepts of conditional probabilities can be extended to continuous random variables. But we must be cautious here since the denominator which was always assumed to be non-zero in the discrete case will always be zero in the continuous case. Let us first define the conditional probability in the continuous case and then analyze it.

Definition If $X_i, X_j, \ldots, X_k$ and $X_\mu, X_\nu, \ldots, X_\rho$ are all continuous random variables then by the conditional probability of $X_i, X_j, \ldots, X_k$ being simultaneously between the limits $a_i, a_j, \ldots, a_k$ and $b_i, b_j, \ldots, b_k$, given that $X_\mu = x_\mu, X_\nu = x_\nu, \ldots, X_\rho = x_\rho$ we mean

$$P(a_i \leqq X_i \leqq b_i, a_j \leqq X_j \leqq b_j, \ldots, a_k \leqq X_k \leqq b_k \mid X_\mu = x_\mu, X_\nu = x_\nu, \ldots, X_\rho = x_\rho)$$

$$= \lim_{\varepsilon \to 0} P(a_i \leqq X_i \leqq b_i, a_j \leqq X_j \leqq b_j, \ldots, a_k \leqq X_k \leqq b_k \mid x_\mu - \varepsilon \leqq X_\mu \leqq x_\mu + \varepsilon, x_\upsilon - \varepsilon \leqq X_\nu \leqq x_\nu + \varepsilon, \ldots, x_\rho - \varepsilon \leqq X_\rho \leqq x_\rho + \varepsilon)$$

assuming the limit exists, and where the probability on the right-hand side is the conditional probability of events as defined previously.

We see that this defines probabilities in the continuous case in terms of a limit of probabilities in the discrete case which we have already discussed. The essential features can be seen without notational confusion in the two dimensional case, which we will now discuss.

Let us assume that the joint probability density for X and Y is given by $f(x, y)$ and is at least piecewise continuous in each variable. In order for a limit to exist at a point $Y = y$, we must restrict the points y to be the points of continuity (in y) for every x. It will help to assume that $f(x, y)$ is continuous in both variables in what follows, but this will not be necessary. If we use $f(y)$ to represent the marginal distribution of Y, then

$$\lim_{\varepsilon \to 0} P(a \leqq X \leqq b \mid y-\varepsilon \leqq Y \leqq y+\varepsilon) = \lim_{\varepsilon \to 0} \frac{P(a \leqq X \leqq b, y-\varepsilon \leqq Y \leqq y+\varepsilon)}{P(y-\varepsilon \leqq Y \leqq y+\varepsilon)}$$

$$= \lim_{\varepsilon \to 0} \frac{\int_a^b \int_{y-\varepsilon}^{y+\varepsilon} f(x, t)\, dt\, dx}{\int_{y-\varepsilon}^{y+\varepsilon} f(t)\, dt}$$

$$= \lim_{\varepsilon \to 0} \frac{2\varepsilon \int_a^b f(x, y)\, dx}{2\varepsilon f(y)}$$

(by the mean value theorem)

$$= \int_a^b \frac{f(x, y)}{f(y)}\, dx.$$

Therefore, the probability density in the conditional distribution is given by

$$f(x \mid y) = \frac{f(x, y)}{f(y)}$$

at the points of continuity in y. Thus, the result is quite similar to that of the discrete case with densities substituted for mass functions. In the discrete case the mass function was bounded by unity; no such bound exists for this density function. However

$$\int_{-\infty}^{\infty} f(x \mid y)\, dx = \int_{-\infty}^{\infty} \frac{f(x, y)}{f(y)}\, dx = \frac{f(y)}{f(y)} = 1.$$

These results can easily be extended to the general case without any complication other than notation.

Exercise 4-7
Derive the fact that

$$f(x_i, x_j, \ldots, x_k \mid x_\mu, x_\nu, \ldots, x_\rho) = \frac{f(x_i, x_j, \ldots, x_k, x_\mu, x_\nu, \ldots, x_\rho)}{f(x_\mu, x_\nu, \ldots, x_\rho)}$$

at the points of continuity of $x_\mu, x_\nu, \ldots, x_\rho$.

The evaluation of conditional expectations now becomes what we expect it to be, namely

$$E[g(X) \mid y] = \lim_{\varepsilon \to 0} E[g(X) \mid y-\varepsilon \leqq Y \leqq y+\varepsilon]$$

$$= \lim_{\varepsilon \to 0} \frac{\int_{-\infty}^{\infty} g(x) \int_{y-\varepsilon}^{y+\varepsilon} f(x, t)\, dt\, dx}{\int_{y-\varepsilon}^{y+\varepsilon} f(t)\, dt}$$

$$= \int_{-\infty}^{\infty} g(x) \frac{f(x, y)}{f(y)}\, dx = \int_{-\infty}^{\infty} g(x) f(x \mid y)\, dx.$$

This too can be extended to the general case.

Exercise 4-8

Show that

$$E[g(X_i, X_j, \dots, X_k) \mid x_\mu, x_\nu, \dots x_\rho] = \int_{x_i}\int_{x_j}\dots\int_{x_k} g(x_i, x_j, \dots, x_k) f(x_i, x_j, \dots, x_k \mid x_\mu, x_\nu, \dots, x_\rho) dx_i dx_j \dots dx_k.$$

Exercise 4-9

Verify for continuous X and Y that

a) $E\{E[g(X) \mid Y]\} = E\{g(X)\}$

and

b) $\text{Var } X = \text{Var}[E(X \mid Y)] + E[\text{Var}(X \mid Y)]$.

Exercise 4-10

Show that if $X_i, X_j, \dots, X_k$ and $X_\mu, X_\nu, \dots, X_\rho$ are independent, the $f(x_i, x_j, \dots, x_k \mid x_\mu, x_\nu, \dots, x_\rho) = f(x_i, x_j, \dots, x_k)$.

THE BIVARIATE NORMAL DISTRIBUTION

The bivariate normal density function is continuous in both variables and will therefore serve well as an example of the results of this section. Since

$$f(x, y) = \frac{1}{2\pi\sigma_x\sigma_y\sqrt{(1-\rho^2)}} \exp\left\{-\frac{1}{2(1-\rho^2)}\left[\left(\frac{x-\mu_x}{\sigma_x}\right)^2 - 2\rho\left(\frac{x-\mu_x}{\sigma_x}\right)\left(\frac{y-\mu_y}{\sigma_y}\right) + \left(\frac{y-\mu_y}{\sigma_y}\right)^2\right]\right\} \qquad \begin{matrix} -\infty < x < \infty \\ -\infty < y < \infty \end{matrix}$$

and

$$f(y) = \frac{1}{\sqrt{(2\pi)}\sigma_y} \exp\left\{-\frac{(y-\mu_y)^2}{2\sigma_y^2}\right\} \qquad -\infty < y < \infty$$

we obtain that

$$f(x \mid y) = \frac{1}{\sqrt{(2\pi)}\sigma_x\sqrt{(1-\rho^2)}} \exp\left\{-\frac{1}{2(1-\rho^2)}\left[\left(\frac{x-\mu_x}{\sigma_x}\right)^2 - 2\rho\left(\frac{x-\mu_x}{\sigma_x}\right)\left(\frac{y-\mu_y}{\sigma_y}\right) + \left(\frac{y-\mu_y}{\sigma_y}\right)^2 - (1-\rho^2)\left(\frac{y-\mu_y}{\sigma_y}\right)^2\right]\right\}$$

$$= \frac{1}{\sqrt{(2\pi)}\sigma_x\sqrt{(1-\rho^2)}} \exp\left\{-\frac{1}{2(1-\rho^2)}\left[\left(\frac{x-\mu_x}{\sigma_x}\right)^2 - 2\rho\left(\frac{(x-\mu_x}{\sigma_y}\right)\left(\frac{y-\mu_y}{\sigma_y}\right) + \rho^2\left(\frac{y-\mu_y}{\sigma_y}\right)^2\right]\right\}$$

$$= \frac{1}{\sqrt{(2\pi)}\sigma_x\sqrt{(1-\rho^2)}} \exp\left\{-\frac{1}{2(1-\rho^2)}\left[\left(\frac{x-\mu_x}{\sigma_x}\right) - \rho\left(\frac{y-\mu_y}{\sigma_y}\right)\right]^2\right\}$$

$$= \frac{1}{\sqrt{(2\pi)}\sigma_x\sqrt{(1-\rho^2)}} \exp\left\{-\frac{1}{2\sigma_x^2(1-\rho^2)}\left[\left\{x-\left[\mu_x+\frac{\rho\sigma_x}{\sigma_y}(y-\mu_y)\right]\right\}^2\right]\right\} \qquad -\infty < x < \infty.$$

From the form of this density we see that X is still normally distributed in the conditional distribution, but now with mean

$$E(X \mid y) = \mu_x+\frac{\rho\sigma_x}{\sigma_y}(y-\mu_y)$$

and variance

$$\text{Var}(X \mid y) = \sigma_x^2(1-\rho^2).$$

From the symmetry of the densities we can easily see that

$$f(y \mid x) = \frac{1}{\sqrt{(2\pi)}\sigma_y\sqrt{(1-\rho^2)}} \exp\left\{-\frac{1}{2\sigma_y^2(1-\rho^2)}\left[\left\{y-\left[\mu_y+\frac{\rho\sigma_y}{\sigma_x}(x-\mu_x)\right]\right\}^2\right]\right\} \qquad -\infty < y < \infty$$

with

$$E(Y \mid x) = \mu_y+\frac{\rho\sigma_y}{\sigma_x}(x-\mu_x)$$

and

$$\text{Var}(Y \mid x) = \sigma_y^2(1-\rho^2).$$

It is interesting to note that $E(Y \mid x)$ depends on the value x, whereas $\text{Var}(Y \mid x)$ does not. Note also that

$$E[E(X \mid Y)] = \mu_x+\frac{\rho\sigma_x}{\sigma_y}(EY-\mu_y) = \mu_x+\frac{\rho\sigma_x}{\sigma_y}(\mu_y-\mu_y) = \mu_x$$

and similarly that

$$E[E(Y \mid X)] = \mu_y+\frac{\rho\sigma_y}{\sigma_x}(\mu_x-\mu_x) = \mu_y$$

as they should. We can also obtain

$$\text{Var}[E(X \mid Y)] = \rho^2\frac{\sigma_x^2}{\sigma_y^2}\text{Var } Y = \rho^2\sigma_x^2$$

and

$$\text{Var}[E(Y \mid X)] = \rho^2\frac{\sigma_y^2}{\sigma_x^2}\text{Var } X = \rho^2\sigma_y^2.$$

Noting that the conditional variances are constants we obtain

$$\text{Var } X = E[\text{Var}(X \mid Y)] + \text{Var}[E(X \mid Y)] = \sigma_x^2(1-\rho^2) + \rho^2\sigma_x^2 = \sigma_x^2$$

and

$$\text{Var } Y = E[\text{Var}(Y \mid X)] + \text{Var}[E(Y \mid X)] = \sigma_y^2(1-\rho^2) + \rho^2\sigma_y^2 = \sigma_y^2$$

as they should.

4-7 Problems

1. For the trinomial distribution with $n = 100$, $p = 0.3$ and $q = 0.5$ find
   a. EX
   b. Var X
   c. EY
   d. Var Y
   e. Cov(X, Y).
2. For $P(x, y) = \binom{7-x}{y}\binom{7}{x}(.2)^x(.3)^y(.5)^{7-x-y}\; 0 \leqq x \times y \leqq 7$, find
   a. EX
   b. Var X
   c. EY
   d. Var Y
   e. Cov(X, Y).
3. For a trinomial distribution with $EX = 3$, $EY = 2$, and Cov$(X, Y) = -0.6$, find the probability mass function.
4. For a trinomial distribution with $EX = 2$, $EY = 4$, and Var $X = 1.8$, find the probability mass function.
5. Assume

   $$P(x, y) = \binom{n-x}{y}\binom{n}{x}p^x q^y(1-p-q)^{n-x-y} \quad 0 \leqq x+y \leqq n$$

   with p, q, and n known. Define $Z = n - X - Y$. Find
   a. EZ
   b. Var Z
   c. Cov (X, Z)
   d. Cov (Y, Z).
6. For the distribution in Problem 5 define $W = X + Y$ and find
   a. EW
   b. Var W
   c. Cov (X, W)
   d. Cov (Y, W)
   e. Cov(W, Z).
7. For the mass function of Problem 2 find the values of parts a–e of Problem 6.

8. For the distribution of Problems 5 and 6 define $V = W+Z$ and find
   a. EV
   b. Var V.
9. For the bivariate normal distribution with $\mu_x = 2$, $\mu_y = 3$, $\sigma_x = 2$, $\sigma_y = 1$, and $\rho = 0.5$ find
   a. EX
   b. Var X
   c. EY
   d. Var Y
   e. Cov(X, Y).
10. If $f(x, y) = 2e^{-[4\pi^2x^2-4\pi xy+2y^2]}$ $\quad -\infty < x, y < \infty$, find
   a. EX
   b. Var X
   c. EY
   d. Var Y
   e. Cov(X, Y).
11. If $f(x, y) = \dfrac{5}{\pi\sqrt{6}} \exp\left(-\dfrac{100x^2}{48} - \dfrac{100y^2}{48} - \dfrac{10xy}{12} + \dfrac{30x}{12} + \dfrac{90y}{12} - \dfrac{55}{8}\right)$ $-\infty < x, y < \infty$,
   find
   a. EX
   b. Var X
   c. EY
   d. Var Y
   e. Cov(X, Y).
12. For the distribution of Problem 1 find
   a. $E(X \mid y)$
   b. Var($X \mid y$)
   c. $E(Y \mid x)$
   d. Var($Y \mid x$)
   e. Var[$E(X \mid Y)$]
   f. E[Var($X \mid Y$)]
   g. Var[$E(Y \mid X)$]
   h. E[Var($Y \mid X$)].
13. Using the results of Problem 12 verify that
   a. Var $X = E$[Var($X \mid Y$)]+Var[$E(X \mid Y)$]
   b. Var $Y = E$[Var($Y \mid X$)]+Var[$E(Y \mid X)$].
14. For the distribution of Problem 2 find
   a. $E(X \mid y)$
   b. Var($X \mid y$)
   c. $E(Y \mid x)$
   d. Var($Y \mid x$)
   e. Var[$E(X \mid Y)$]
   f. E[Var($X \mid Y$)]
   g. Var[$E(Y \mid X)$]
   h. E[Var($Y \mid X$)].
15. Using the results of Problem 14 verify that
   a. Var $X = E$[Var($X \mid Y$)]+Var[$E(X \mid Y)$]
   b. Var $Y = E$[Var($X \mid Y$)]+Var[$E(X \mid Y)$]

16. For the distribution of Problem 5 find
    a. $E(X \mid z)$
    b. $E(Z \mid x)$
    c. $\text{Var}(X \mid z)$
    d. $\text{Var}(Z \mid x)$
    e. $\text{Var}[E(X \mid Z)]$
    f. $\text{Var}[E(Z \mid X)]$
    g. $E[\text{Var}(X \mid Z)]$
    h. $E[\text{Var}(Z \mid X)]$.
17. Using the results of Problem 16 verify that
    a. $\text{Var } X = E[\text{Var}(X \mid Z)] + \text{Var}[E(X \mid Z)]$
    b. $\text{Var } Y = E[\text{Var}(Z \mid X)] + \text{Var}[E(Z \mid X)]$.
18. For the distribution of Problem 9 find
    a. $E(X \mid y)$
    b. $\text{Var}(X \mid y)$
    c. $E(Y \mid x)$
    d. $\text{Var}(Y \mid x)$
    e. $\text{Var}[E(X \mid Y)]$
    f. $E[\text{Var}(X \mid Y)]$
    g. $\text{Var}[E) Y \mid X)]$
    h. $E[\text{Var}(Y \mid X)]$.
19. Using the results of Problem 18 verify that
    a. $\text{Var } X = E[\text{Var}(X \mid Y)] + \text{Var}[E(X \mid Y)]$
    b. $\text{Var } Y = E[\text{Var}(Y \mid X)] + \text{Var}[E(Y \mid X)]$.
20. For the distribution of Problem 10 find
    a. $E(X \mid y)$
    b. $\text{Var}(X \mid y)$
    c. $E(Y \mid x)$
    d. $\text{Var}(Y \mid x)$
    e. $\text{Var}[E(X \mid Y)]$
    f. $E[\text{Var}(X \mid Y)]$
    g. $\text{Var}[E(Y \mid X)]$
    h. $E[\text{Var}(Y \mid X)]$.
21. Using the results of Problem 20 verify that
    a. $\text{Var } X = E[\text{Var}(X \mid Y)] + \text{Var}[E(X \mid Y)]$
    b. $\text{Var } Y = E[\text{Var}(Y \mid X)] + \text{Var}[E(Y \mid X)]$.
22. For the distribution of Problem 11 find
    a. $E(X \mid y)$
    b. $\text{Var}(X \mid y)$
    c. $E(Y \mid x)$
    d. $\text{Var}(Y \mid x)$
    e. $\text{Var}[E(X \mid Y)]$
    f. $E[\text{Var}(X \mid Y)]$
    g. $\text{Var}[E(Y \mid X)]$
    h. $E(\text{Var}(Y \mid X)]$.
23. Using the results of Problem 22 verify that
    a. $\text{Var } X = E[\text{Var}(X \mid Y)] + \text{Var}[E(X \mid Y)]$
    b. $\text{Var } Y = E[\text{Var}(Y \mid X)] + \text{Var}[E(Y \mid X)]$.

CHAPTER 5

Moment Generating Functions

In this chapter we will find some properties of a function called *the moment generating function* which, when it exists, is equivalent to the mass or density function. By equivalent here we mean that one is determined by the other, and vice versa. The functions themselves are quite different in nature. Another function similar to the moment generating function, *the characteristic function*, can also be found, but requires the use of mathematics not assumed as a prerequisite for this text. This function will be treated in the exercises. Exercises requiring this advanced mathematics will be marked with an asterisk and should be omitted by those who are not familiar with integration in the complex plane.

We now define the moment generating function for any univariate random variable having any distribution.

Definition The moment generating function of the random variable X is defined by

$$m(t) = E\, e^{tX}$$

when it exists.

From this, we see that in the discrete case

$$m(t) = \sum_{x} e^{tx} P(x)$$

whereas in the continuous case

$$m(t) = \int_{-\infty}^{\infty} e^{tx} f(x)\, dx$$

assuming they exist.

*EXERCISE 5-1

The characteristic function of any random variable X is defined by

$$\phi(t) = Ee^{itx}, \quad \text{where} \quad i = \sqrt{(-1)}.$$

Show that this always exists and has absolute value no larger than unity.

Remark *There is only one moment generating function for each particular distribution. Therefore, this function can be inverted, at least theoretically,* *to obtain the mass or density function. This procedure is usually cumbersome in practice, and will not be attempted here. The method of inversion that is most useful in statistics is simply to list the moment generating functions of known mass and density functions. Then, we can invert by using this table in reverse.*

For the multivariate situation we have a similar definition.

Definition If $X_1, X_2, X_3, \ldots, X_n$ are any jointly distributed random variables, then their joint moment generating function is given by

$$m(t_1, t_2, t_3, \ldots, t_n) = E\left[\exp\left(\sum_{i=1}^{n} t_i X_i\right)\right]$$

when it exists.

*Exercise 5-2

The joint characteristic function is defined by

$$\phi(t_1, t_2, t_3, \ldots, t_n) = E[\exp(\sum_{\nu=1}^{n} it_\nu X_\nu)].$$

Show that it always exists.

5-1 Properties of the Moment Generating Function

When the moment generating function exists it can be evaluated at any t-point. In particular

$$m(0) = E\, e^0 = E1 = 1.$$

This property will be used as a check to see if we have calculated the expectation properly.

In general, for the types of functions used as density or mass functions, it is proper to interchange the order of differentiation and integration or summation. Therefore, if the moment generating function exists, then by taking a derivative we find

$$m'(t) = \frac{d}{dt} m(t) = \frac{d}{dt} E\, e^{tX} = E \frac{d}{dt} e^{tX} = EX\, e^{tX}.$$

(That is

$$\frac{d}{dt} \sum_x e^{tx} p(x) = \sum_x \frac{d}{dt} e^{tx} p(x) = \sum_x x\, e^{tx} p(x),$$

or

$$\frac{d}{dt} \int e^{tx} f(x)\, dx = \int \frac{d}{dt} e^{tx} f(x)\, dx = \int x\, e^{tx} f(x)\, dx).$$

Now if we evaluate this at $t = 0$, we obtain

$$m'(0) = EX,$$

the mean. If we take another derivative

$$m''(t) = \frac{d^2}{dt^2} E\, e^{tX} = E \frac{d^2}{dt^2} e^{tX} = EX^2 e^{tX}.$$

On evaluating this at $t = 0$, we obtain

$$m''(0) = EX^2,$$

the second moment. In general for the rth derivative

$$m^{(r)}(t) = \frac{d^r}{dt^r} E\, e^{tX} = E \frac{d^r}{dt^r} e^{tX} = EX^r e^{tX}.$$

Evaluating this at $t = 0$ we obtain

$$m^{(r)}(0) = EX^r$$

the rth moment. Thus, we find that by using the moment generating function we can actually generate the moments of the distribution by taking derivatives and then evaluating at $t = 0$. Taking derivatives is often more easily done than performing the corresponding integrations. This then simplifies finding moments. We will see examples in the next two sections, however, which illustrate that this procedure is not always a simplification.

It will be useful to consider here the situation in which the random variable in question is the sum of independent random variables. Say

$$Y = X_1 + X_2 + X_3 + \dots + X_n$$

where $X_1, X_2, X_3, \dots, X_n$ are mutually independent. Then

$$\begin{aligned} m_Y(t) &= E\, e^{tY} = E\, e^{t(X_1+X_2+X_3+\dots+X_n)} = E\, e^{tX_1+tX_2+tX_3+\dots+tX_n} \\ &= E\, e^{tX_1} e^{tX_2} e^{tX_3} \dots e^{tX_n} = E\, e^{tX_1} E\, e^{tX_2} E\, e^{tX_3} \dots E\, e^{tX_n}. \end{aligned}$$

Since the joint probability factors into a product of marginal probabilities, we can say, as we have done here, that the expectation of a product is the product of the expectations.

Thus,

$$m_Y(t) = m_{X_1}(t) m_{X_2}(t) m_{X_3}(t) \dots m_{X_n}(t)$$

(where subscripts are used to distinguish the moment generating functions of the random variables).

If now $X_1, X_2, X_3, \dots, X_n$ all have the same moment generating function $m(t)$, then

$$m_Y(t) = [m(t)]^n.$$

Another useful relation occurs when one random variable is a linear function of another, say $Y = aX + b$. If $m_X(t)$ is the moment generating function for X, then the moment generating function for $Y, m_Y(t)$ can be found directly. Notice that

$$m_Y(t) = E\,e^{tY} = E\,e^{atX+bt} = E\,e^{bt}e^{atx}$$
$$= e^{bt}\,E\,e^{atX}.$$

But if $u = at$, then

$$m_X(u) = E\,e^{uX}.$$

Thus,

$$m_Y(t) = e^{bt}m_X(at)$$

where at is substituted for t in the moment generating function for X.

Results similar to those for the univariate situation hold for the multivariate situation. In particular

$$m(0, 0, 0, \ldots, 0) = E\,e^0 = 1.$$

Taking a derivative with respect to t_i we obtain

$$\frac{d}{dt_i}m(t_1, t_2, t_3, \ldots, t_n) = \frac{d}{dt_i}E\left[\exp\left(\sum_{j=1}^{n} t_jX_j\right)\right] = E\left[\frac{d}{dt_i}\exp\left(\sum_{j=1}^{n} t_jX_j\right]\right)$$
$$= E\left[X_i \exp\left(\sum_{j=1}^{n} t_jX_j\right)\right]$$

giving

$$\frac{d}{dt_i}m(t_1, t_2, t_3, \ldots, t_n)\Bigg|_0 = EX_i$$

where the zero indicates that $t_1, t_2, t_3, \ldots,$ and t_n have been set equal to zero. Thus we can obtain the expectation of any of the variables by taking a derivative with respect to the corresponding t and then evaluating at the origin. If we now take r derivatives with respect to t_i, we obtain

$$\frac{d^r}{dt_i^r}m(t_1, t_2, t_3, \ldots, t_n) = \frac{d^r}{dt_i^r}E\left[\exp\left(\sum_{j=1}^{n} t_jX_j\right)\right] = E\left[\frac{d^r}{dt_i^r}\exp\left(\sum_{j=1}^{n} t_jX_j\right)\right]$$
$$= E\left[X_i^r \exp\left(\sum_{j=1}^{n} t_jX_j\right)\right].$$

Thus

$$\frac{d^r}{dt_i^r}m(t_1, t_2, t_3, \ldots, t_n)\Bigg|_0 = EX_i^r$$

which is the rth moment of the ith variable.

It is also possible to take derivatives with respect to more than one t. For example,

$$\frac{d}{dt_i}\frac{d}{dt_j} m(t_1, t_2, t_3, \ldots, t_n) = \frac{d}{dt_i}\frac{d}{dt_j} E\left[\exp\left(\sum_{k=1}^{n} t_k x_k\right)\right]$$

$$= E\left[\frac{d}{dt_i}\frac{d}{dt_j}\exp\left(\sum_{k=1}^{n} t_k X_k\right)\right] = E\left[X_i X_j \exp\left(\sum_{k=1}^{n} t_k X_k\right)\right].$$

Therefore,

$$\frac{d}{dt_i}\frac{d}{dt_j} m(t_1, t_2, t_3, \ldots, t_n)\bigg|_0 = EX_i X_j$$

which can be used in determining the covariance between X_i and X_j. In general, it can be shown that

$$\frac{d^r}{dt_i^r}\frac{d^s}{dt_j^s}\cdots\frac{d^u}{dt_k^u} m(t_1, t_2, t_3, \ldots, t_n)\bigg|_0 = EX_i^r X_j^s \ldots X_k^u.$$

EXERCISE 5-3

Prove the above result.

*EXERCISE 5-4

Show that

$$\frac{d^r}{dt^r}\phi(t)\bigg|_0 = i^r EX^r$$

and

$$\frac{d^r}{dt_i^r}\frac{d^s}{dt_j^s}\cdots\frac{d^u}{dt_k^u}\phi(t_1, t_2, t_3, \ldots, t_n)\bigg|_0 = i^{r+s+\cdots+u} EX_i^r X_j^s \ldots X_k^u.$$

where $\phi(t)$ is the characteristic function.

5-2 Moment Generating Functions for Some Discrete Random Variables

THE BERNOULLI DISTRIBUTION

For the bernoulli distribution

$$P(x) = p^x(1-p)^{1-x} \qquad x = 0, 1.$$

The moment generating function is then

$$m(t) = E\,e^{tX} = \sum_{x=0}^{1} e^{tx}p^x(1-p)^{1-x} = (1-p)+p\,e^t = 1-p(1-e^t).$$

Clearly $m(0) = 1$. Now

$$m'(t) = \frac{d}{dt}\,[1-p(1-e^t)] = p\,e^t$$

giving

$$EX = m'(0) = p.$$

In general

$$m^{(r)}(t) = p\,e^t$$

and therefore

$$EX^r = m^{(r)}(0) = p.$$

THE BINOMIAL DISTRIBUTION

In the binomial case

$$P(x) = \binom{n}{x} p^x (1-p)^{n-x} \qquad x = 0, 1, 2, \ldots, n.$$

The moment generating function is then

$$m(t) = E\,e^{tX} = \sum_{x=0}^{n} e^{tx}\,P(x) = \sum_{x=0}^{n} e^{tx}\binom{n}{x} p^x(1-p)^{n-x}$$

$$= \sum_{x=0}^{n} \binom{n}{x} (p\,e^t)^x (1-p)^{n-x} = (p\,e^t+1-p)^n$$

by the Binomial Theorem. If we notice that the binomial random variable is the sum of n independent bernoulli random variables, then we could have obtained this result directly as the product of the n (identical) bernoulli moment generating functions.

Letting $t = 0$ in $m(t)$ for the binomial distribution we obtain

$$m(0) = (p+1-p)^n = 1.$$

Now

$$m'(t) = \frac{d}{dt}(p\,e^t+1-p)^n = n(p\,e^t+1-p)^{n-1}p\,e^t.$$

Thus

$$EX = m'(0) = np.$$

Taking a second derivative we obtain

$$m''(t) = \frac{d}{dt}\,m'(t) = \frac{d}{dt}\,[n(p\,e^t+1-p)^{n-1}p\,e^t]$$

$$= n(n-1)(p\,e^t+1-p)^{n-2}p^2e^{2t}+n(p\,e^t+1-p)^{n-1}p\,e^t$$

giving

$$EX^2 = m''(0) = n(n-1)p^2+np.$$

*EXERCISE 5-5

Find the characteristic function $\phi(t)$ for the binomial distribution, and find the first two moments using it.

THE POISSON DISTRIBUTION

In the case of a poisson distribution

$$P(x) = \frac{e^{-\lambda}\lambda^x}{x!} \qquad x = 0, 1, 2, \ldots.$$

The moment generating function is then

$$m(t) = E\,e^{tX} = \sum_{x=0}^{\infty} e^{tx}\frac{e^{-\lambda}\lambda^x}{x!} = e^{-\lambda}e^{\lambda e^t}\sum_{x=0}^{\infty}\frac{e^{-\lambda e^t}(\lambda\,e^t)^x}{x!} = e^{-\lambda}e^{\lambda e^t}$$

since

$$\sum_{x=0}^{\infty}\frac{e^{-z}z^x}{x!} = 1$$

for any z. We can easily see that $m(0) = e^{-\lambda}e^{\lambda} = 1$. Continuing as usual, we obtain

$$m'(t) = \frac{d}{dt}E\,e^{tX} = e^{-\lambda}\frac{d}{dt}e^{\lambda e^t} = e^{-\lambda}\lambda\,e^t e^{\lambda e^t}$$

giving

$$EX = m'(0) = \lambda.$$

Taking another derivative, we see that

$$m''(t) = e^{-\lambda}\lambda\,e^t e^{\lambda e^t} + e^{-\lambda}\lambda^2 e^{2t}e^{\lambda e^t}$$

giving

$$EX^2 = m''(0) = \lambda + \lambda^2.$$

*EXERCISE 5-6

Find the characteristic function $\phi(t)$, for the poisson distribution, and use it to find the mean and variance of the distribution.

THE NEGATIVE BINOMIAL DISTRIBUTION

In the case of a negative binomial distribution

$$P(x) = \binom{x+r-1}{r-1}p^r(1-p)^x \qquad x = 0, 1, 2, \ldots$$

from which we obtain

$$m(t) = E\,e^{tX} = \sum_{x=0}^{\infty} e^{tx}\binom{x+r-1}{r-1} p^r(1-p)^x$$

$$= \sum_{x=0}^{\infty}\binom{x+r-1}{r-1} p^r[(1-p)\,e^t]^x = \frac{p^r}{[1-(1-p)\,e^t]^r}$$

by the method of Section 2-2. Testing this, we find $m(0) = p^r/[1-(1-p)]^r = 1$. Continuing as before, we obtain

$$m'(t) = \frac{d}{dt}\frac{p^r}{[1-(1-p)\,e^t]^r} = p^r\{r[1-(1-p)\,e^t]^{-r-1}(1-p)\,e^t\}$$

giving

$$EX = m'(0) = r(1-p)/p.$$

Then

$$m''(t) = p^r\{r(r+1)[1-(1-p)\,e^t]^{-r-2}(1-p)^2e^{2t}$$

$$+r[1-(1-p)\,e^t]^{-r-1}(1-p)\,e^t\}$$

giving

$$EX^2 = m''(0) = \frac{r(r+1)}{p^2}(1-p)^2 + r\frac{(1-p)}{p}.$$

*EXERCISE 5-7

Find the characteristic function for the negative binomial distribution.

THE DISCRETE UNIFORM DISTRIBUTION

This will serve as an example in which even finding the first moment is cumbersome. For this distribution

$$P(x) = \frac{1}{n} \qquad x = 1, 2, 3, \ldots, n.$$

Then

$$m(t) = E\,e^{tX} = \sum_{x=1}^{n}\frac{e^{tx}}{n} = \frac{e^t}{n}\left(\frac{e^{nt}-1}{e^t-1}\right)$$

$$\left(\text{using the equality } \sum_{x=1}^{n} a^x = a\left(\frac{a^n-1}{a-1}\right)\right)$$

$$= \frac{1}{n}\left(\frac{e^{nt}-1}{1-e^{-t}}\right).$$

Now when one sets $t = 0$ one obtains

$$m(0) = \frac{1}{n}\left(\frac{1-1}{1-1}\right) = \frac{1}{n}\left(\frac{0}{0}\right)$$

which is an indeterminate form. Using l'Hospital's rule, however, we obtain

$$m(0) = \frac{1}{n}\left(\frac{n\, e^{nt}}{e^{-t}}\right)\bigg|_0 = 1$$

as it should. Now we take a derivative of the moment generating function:

$$m'(t) = \frac{d}{dt}\, m(t) = \frac{1}{n}\frac{d}{dt}\left(\frac{e^{nt}-1}{1-e^{-t}}\right)$$

$$= \frac{n\, e^{nt}(1-e^{-t}) - e^{-t}(e^{nt}-1)}{n(1-e^{-t})^2} .$$

When $t = 0$ this becomes

$$m'(0) = \left(\frac{0}{0}\right)$$

which is again indeterminate. Using l'Hospital's rule, we obtain

$$m'(0) = \frac{n^2 e^{nt}(1-e^{-t}) + e^{-t}(e^{nt}-1)}{2n(1-e^{-t})\, e^{-t}}\bigg|_0 = \left(\frac{0}{0}\right)$$

which is still indeterminate. Using l'Hospital's rule again, we obtain

$$m'(0) = \frac{n^3 e^{nt}(1-e^{-t}) + n^2 e^{nt} e^{-t} - e^{-t}(e^{nt}-1) + n\, e^{nt} e^{-t}}{2n\, e^{-t}[2\, e^{-t}-1]}\bigg|_0 = \frac{n+1}{2} .$$

This is correct, but compared with

$$EX = \sum_{x=0}^{n} \frac{x}{n} = \frac{n+1}{2}$$

it is extremely tedious.

5-3 Moment Generating Functions for Some Continuous Random Variables

We will now consider some continuous random variables. We start with another example illustrating that the moment generating function may not be the easiest way to generate moments.

THE UNIFORM DISTRIBUTION

For this distribution

$$f(x) = \frac{1}{\beta-\alpha} \qquad \alpha < x < \beta$$
$$= 0 \qquad \text{otherwise}$$

from which we obtain

$$m(t) = E\,e^{tX} = \frac{1}{\beta-\alpha}\int_{\alpha}^{\beta} e^{tx}dx = \frac{e^{t\beta}-e^{t\alpha}}{t(\beta-\alpha)}.$$

Here again

$$m(0) = \left(\frac{0}{0}\right)$$

an indeterminate form which using l'Hospital's rule becomes

$$m(0) = \left.\frac{\beta\, et^{t\beta}-\alpha\, e^{t\alpha}}{(\beta-\alpha)}\right|_0 = \frac{\beta-\alpha}{\beta-\alpha} = 1.$$

Taking a derivative of $m(t)$, we obtain

$$m'(t) = \frac{d}{dt}\frac{e^{t\beta}-e^{t\alpha}}{t(\beta-\alpha)} = \frac{t(\beta-\alpha)(\beta\, e^{t\beta}-\alpha\, e^{t\alpha})-(e^{t\beta}-e^{t\alpha})(\beta-\alpha)}{t^2(\beta-\alpha)^2}$$

giving

$$m'(0) = \left(\frac{0}{0}\right)$$

which is again indeterminate. As in the discrete case, it will require the use of l'Hospital's rule twice to clear the indeterminacy. This will be left as an exercise.

EXERCISE 5-8

Find EX using $m'(t)$ for the above example.

Once again we see that this method is quite cumbersome compared with finding the moments directly.

THE NORMAL DISTRIBUTION

For a normal random variable with mean μ and variance σ^2, we have

$$f(x) = \frac{1}{\sqrt{(2\pi)}\sigma}\, e^{-(x-\mu)^2/2\sigma^2} \qquad -\infty < x < \infty.$$

From this we find

$$m(t) = E\,e^{tX} = \frac{1}{\sqrt{(2\pi)}\sigma}\int_{-\infty}^{\infty} e^{tx}e^{-(x-\mu)^2/2\sigma^2}dx$$

$$= \frac{1}{\sqrt{(2\pi)}\sigma}\int_{-\infty}^{\infty} \exp[(2tx\sigma^2/2\sigma^2)-(x^2-2\mu x+\mu^2)/2\sigma^2]\,dx$$

$$= \frac{1}{\sqrt{(2\pi)}\sigma}$$

$$\int_{-\infty}^{\infty} \exp(-\{x^2-2(\mu+t\sigma^2)x+(\mu+t\sigma^2)^2-(\mu+t\sigma^2)+\mu^2\}/2\sigma^2)\,dx$$

$$= [\exp((\mu+t\sigma^2)^2-\mu^2/2\sigma^2)]\frac{1}{\sqrt{(2\pi)}\sigma}\int_{-\infty}^{\infty}\exp(-[x-(\mu+t\sigma^2)]^2/2\sigma^2)\,dx$$

$$= [\exp(t\mu+t^2\sigma^2/2)]\frac{1}{\sqrt{(2\pi)}\sigma}\int_{-\infty}^{\infty}\exp(-(x-\mu')^2/2\sigma^2)\,dx$$

$$= \exp(t\mu+t^2\sigma^2/2)$$

where $\mu' = \mu+t\sigma^2$.

In Chapter Six we find that if Y has a standard normal distribution, then $X = \sigma Y+\mu$ has a normal distribution with mean μ and variance σ^2. Knowing this, we could have derived $m(t)$ for the standard normal distribution, and used the result that if $Y = aX+b$, then $m_Y(t) = e^{bt}m(at)$.

EXERCISE 5-9

Obtain the moment generating function for the bivariate normal distribution using the relation in the preceding paragraph.

Clearly $m(0) = 1$. Taking a derivative, we obtain

$$m'(t) = (\mu+t\sigma^2)\,e^{t\mu+t^2\sigma^2/2}$$

giving

$$EX = m'(0) = \mu.$$

Taking another derivative we obtain

$$m''(t) = (\mu+t\sigma^2)^2\,e^{t\mu+t^2\sigma^2/2}+\sigma^2\,e^{t\mu+t^2\sigma^2/2}$$

giving

$$EX^2 = m''(0) = \mu^2+\sigma^2.$$

In contrast to the previous example, this illustrates the utility of the moment generating function.

*EXERCISE 5-10

Find the characteristic function for a normal random variable. Use the characteristic function to find the mean and variance.

THE GAMMA DISTRIBUTION

The general form for the gamma distribution is given by the density

$$f(x) = \frac{1}{\Gamma(\alpha)\beta^\alpha} x^{\alpha-1} e^{-x/\beta} \qquad 0 \leqq x \leqq \infty$$

$$\alpha, \beta > 0$$

$$= 0 \qquad \text{otherwise.}$$

Using this, we find

$$m(t) = E\, e^{tX} = \frac{1}{\Gamma(\alpha)\beta^\alpha} \int_0^\infty e^{tx} x^{\alpha-1} e^{-x/\beta} dx$$

$$= \frac{1}{\Gamma(\alpha)\beta^\alpha} \int_0^\infty x^{\alpha-1} e^{-x(1/\beta - t)} dx$$

$$= \frac{(1/\beta - t)^{-\alpha}}{\beta^\alpha} \frac{1}{\Gamma(\alpha)(1/\beta - t)^{-\alpha}} \int_0^\infty x^{\alpha-1} e^{-x/(1/\beta - t)^{-1}} dx$$

$$= \frac{(1/\beta - t)^{-\alpha}}{\beta^\alpha} \frac{1}{\Gamma(\alpha)\gamma^\alpha} \int_0^\infty x^{\alpha-1} e^{-x/\gamma} dx$$

[where $\gamma = (1/\beta - t)^{-1}$]

$$= \frac{(1/\beta - t)^{-\alpha}}{\beta^\alpha} = (1 - \beta t)^{-\alpha}.$$

Again $m(0) = 1$ as it should. Taking a derivative, we obtain

$$m'(t) = \alpha\beta(1-\beta t)^{-\alpha-1}$$

with

$$EX = m'(0) = \alpha\beta.$$

Taking another derivative, we obtain

$$m''(t) = \alpha(\alpha+1)\beta^2(1-\beta t)^{-\alpha-2}$$

with

$$EX^2 = m''(0) = \alpha(\alpha+1)\beta^2.$$

Clearly

$$m^{(r)}(t) = \alpha(\alpha+1)(\alpha+2)\ldots(\alpha+r-1)\beta^r(1-\beta t)^{-\alpha-r}$$

giving

$$EX^r = m^{(r)}(0) = \alpha(\alpha+1)(\alpha+2)\ldots(\alpha+r-1)\beta^r.$$

This is only a slight improvement over computing the rth moment directly. However, we shall see in Section 9-2 that the moment generating function is useful for other reasons.

*EXERCISE 5-11

Find the characteristic function for the gamma distribution, and use it to find the rth moment.

5-4 Multivariate Moment Generating Functions

THE TRINOMIAL DISTRIBUTION

The trinomial mass function for the random variables X and Y was given as

$$P(x, y) = \binom{n}{x}\binom{n-x}{y} p^x q^y (1-p-q)^{n-x-y} \qquad 0 \leqq x+y \leqq n$$

$$0 < p+q < 1.$$

We now find the joint moment generating function

$$m(t_1, t_2) = E\, e^{t_1X+t_2Y} = \sum_{x=0}^{n}\sum_{y=0}^{n-x} e^{t_1x+t_2y}\binom{n}{x}\binom{n-x}{y} p^x q^y (1-p-q)^{n-x-y}$$

$$= \sum_{x=0}^{n}\sum_{y=0}^{n-x}\binom{n}{x}\binom{n-x}{y}(p\,e^{t_1})^x(q\,e^{t_2})^y(1-p-q)^{n-x-y}$$

$$= \sum_{x=0}^{n}\binom{n}{x}\left(p\,e^{t_1}\right)^x \sum_{y=0}^{n-x}\binom{n-x}{y}\left(q\,e^{t_2}\right)^y(1-p-q)^{n-x-y}$$

$$= \sum_{x=0}^{n}\binom{n}{x}\left(p\,e^{t_1}\right)^x(1-p-q+q\,e^{t_2})^{n-x}$$

(by the Binomial Theorem)

$$= (1-p-q+p\,e^{t_1}+q\,e^{t_2})^n$$

again by the Binomial Theorem. Clearly $m(0, 0) = 1$. Taking a derivative with respect to t_1, we obtain

$$\frac{d}{dt_1} m(t_1, t_2) = np\,e^{t_1}(1-p-q+p\,e^{t_1}+q\,e^{t_2})^{n-1}$$

giving

$$EX = \frac{d}{dt_1} m(t_1, t_2)\bigg|_0 = np.$$

A second derivative with respect to t_1 gives

$$\frac{d^2}{dt_1{}^2} m(t_1, t_2) = \frac{d}{dt_1}\left[\frac{d}{dt_1} m(t_1, t_2)\right]$$

$$= \left[\frac{d}{dt_1} np\, e^{t_1}(1-p-q+p\, e^{t_1}+q\, e^{t_2})^{n-1}\right]$$

$$= np\, e^{t_1}(1-p-q+p\, e^{t_1}+q\, e^{t_2})^{n-1}$$

$$+\ n(n-1)p^2e^{2t_1}(1-p-q+p\, e^{t_1}+q\, e^{t_2})^{n-2}$$

giving

$$EX^2 = \frac{d^2}{dt_1{}^2} m(t_1, t_2)\bigg|_0 = n(n-1)p^2+np.$$

The derivatives with respect to t_2 are similar with q replacing p. It is interesting to find

$$\frac{d}{dt_2}\frac{d}{dt_1} m(t_1, t_2) = \frac{d}{dt_2} [np\, e^{t_1}(1-p-q+p\, e^{t_1}+q\, e^{t_2})^{n-1}]$$

$$= n(n-1)p\, e^{t_1}q\, e^{t_2}(1-p-q+p\, e^{t_1}+q\, e^{t_2})^{n-2}$$

giving

$$E(XY) = \frac{d}{dt_2}\frac{d}{dt_1} m(t_1, t_2)\bigg|_0 = n(n-1]pq.$$

This can now be used to find the covariance, indicating again the utility of the moment generating function.

*Exercise 5-12

Find the characteristic function for the trinomial distribution.

THE BIVARIATE NORMAL DISTRIBUTION

Another example of the utility of the moment generating function is afforded us by the bivariate normal distribution. Here

$$f(x, y) = \frac{1}{2\pi\sigma_x\sigma_y\surd(1-\rho^2)} \exp\left\{-\frac{1}{2(1-\rho^2)}\left[\left(\frac{x-\mu_x}{\sigma_x}\right)^2 - 2\rho\left(\frac{x-\mu_x}{\sigma_x}\right)\right.\right.$$

$$\left.\left.\left(\frac{y-\mu_y}{\sigma_y}\right)+\left(\frac{y-\mu_y}{\sigma_y}\right)^2\right]\right\} \qquad -\infty < x, y < \infty.$$

In order to find the joint moment generating function more easily we will use a relation which is often very useful, namely

$$\begin{aligned} E\exp(t_1X+t_2Y) &= \exp(t_1\mu_x+t_2\mu_x)\exp(-t_1\mu_x-t_2\mu_y)E\exp(t_1X+t_2Y) \\ &= \exp(t_1\mu_x+t_2\mu_y)E\exp(-t_1\mu_x-t_2\mu_y)\exp(t_1X+t_2Y) \\ &= \exp(t_1\mu_x+t_2\mu_y)E\exp(t_1(X-\mu_x)+t_2(Y-\mu_y)). \end{aligned}$$

Therefore

$$m(t_1, t_2) = \exp(t_1\mu_x+t_2\mu_y)\,E\exp[t_1(X-\mu_x)+t_2(Y-\mu_y)].$$

We will now find the expectation.

$$E\exp[t_1(X-\mu_x)+t_2(Y-\mu_y)]$$

$$= \frac{1}{2\pi\sigma_x\sigma_y\sqrt{(1-\rho^2)}}\iint \exp[t_1(x-\mu_x)+t_2(y-\mu_y)]$$

$$\exp\left\{-\frac{1}{2(1-\rho^2)}\left[\left(\frac{x-\mu_x}{\sigma_x}\right)^2-2\rho\left(\frac{x-\mu_x}{\sigma_x}\right)\left(\frac{y-\mu_y}{\sigma_y}\right)+\left(\frac{y-\mu_y}{\sigma_y}\right)^2\right]\right\}dx\,dy$$

$$= \frac{1}{2\pi\sigma_x\sigma_y\sqrt{(1-\rho^2)}}\iint \exp(t_1\mu+t_2v)$$

$$\exp\left\{\frac{1}{-2(1-\rho^2)}\left[\left(\frac{u}{\sigma_x}\right)^2-2\rho\left(\frac{u}{\sigma_x}\right)\left(\frac{v}{\sigma_y}\right)+\left(\frac{v}{\sigma_y}\right)^2\right]\right\}du\,dv$$

(where $u = x-\mu_x$ and $v = y-\mu_y$)

$$= \frac{1}{2\pi\sqrt{(1-\rho^2)}}\iint \exp(t_1\sigma_x w+t_2\sigma_y z)$$

$$\exp\left\{-\frac{1}{2(1-\rho^2)}[w^2-2\rho wz+z^2]\right\}dw\,dz$$

(where $v = u/\sigma_x$ and $z = v/\sigma_y$)

$$= \frac{1}{2\pi\sqrt{(1-\rho^2)}}\iint \exp\left\{-\frac{1}{2(1-\rho^2)}[w^2-2(1-\rho^2)t_1\sigma_x w-2\rho wz \right.$$

$$\left. -2(1-\rho^2)t_2\sigma_y z+z^2]\right\}dw\,dz.$$

We now *complete the square* by multiplying out

$$(w-A)^2-2\rho(w-A)(z-B)+(z-B)^2$$

recording only the terms in w and in z, and equating them to the corresponding terms in the brackets. By equating coefficients of w we obtain

$$-2A+2\rho B = -2(1-\rho^2)t_1\sigma_x$$

and by equating coefficients for z we obtain

$$-2B+2\rho A = -2(1-\rho^2)t_2\sigma_y$$

from which we find

$$A = t_1\sigma_x+\rho t_2\sigma_y$$

and

$$B = t_2\sigma_y+\rho t_1\sigma_x.$$

Completing the square we obtain for the bracketed term

$$[(w-t_1\sigma_x-t_2\sigma_y\rho)^2-2\rho(w-t_1\sigma_x-t_2\sigma_y\rho)(z-t_2\sigma_y-t_1\sigma_x\rho)$$
$$+(z-t_2\sigma_y-t_1\sigma_x\rho)^2-(1-\rho^2)t_1{}^2\sigma_x{}^2-2\rho(1-\rho^2)t_1t_2\sigma_x\sigma_y$$
$$-(1-\rho^2)t_2{}^2\sigma_y{}^2].$$

Therefore

$$E\exp[t_1(X-\mu_x)+t_2(Y-\mu_y)]$$
$$= \frac{1}{2\pi\sqrt{(1-\rho^2)}}\int_{-\infty}^{\infty}\!\!\int \exp(-[1/2(1-\rho^2)]\{[(w-A)^2-2\rho(w-A)(z-B)$$
$$+(z-B)^2]\})$$
$$= \exp(t_1{}^2\sigma_x{}^2+2\rho t_1t_2\sigma_x\sigma_y+t_2{}^2\sigma_y{}^2/2)\ dw\ dz$$
$$= \exp(t_1{}^2\sigma_x{}^2/2)+\rho t_1t_2\sigma_x\sigma_y+(t_2{}^2\sigma_y{}^2/2)$$
$$= \frac{1}{2\pi\sqrt{(1-\rho^2)}}\int_{-\infty}^{\infty}\!\!\int \exp(-(1/2(1-\rho^2))[(w-A)^2-2\rho(w-A)(z-B)]$$
$$+(z-B)^2])\ dw\ dz$$
$$= \exp(t_1{}^2\sigma_x{}^2/2)+\rho t_1t_2\sigma_x\sigma_y+(t_2{}^2\sigma_y{}^2/2).$$

Thus

$$m(t_1, t_2) = E\exp(t_1X+t_2Y) = \exp(t_1\mu_x+t_2\mu_y+t_1{}^2\sigma_x{}^2/2+\rho t_1t_2\sigma_x\sigma_y$$
$$+t_2{}^2\sigma_y{}^2/2).$$

EXERCISE 5-13

We will find that if U and V have a standard bivariate normal distribution with correlation ρ, then $X=(U-\mu_x)/\sigma_x$ and $Y=(V-\mu_y)/\sigma_y$ have the bivariate normal distribution given in this section. Derive the moment generating function for U, V and then obtain the moment generating function for X, Y by using the relation

$$E\,e^{t_1(aU+b)+t_2(cV+d)}=e^{bt_1+dt_2}E\,e^{at_1U+ct_2V}.$$

Derivatives of the bivariate normal moment generating function are not difficult to calculate. In particular

$$\frac{d}{dt_1}m(t_1,t_2)=(\mu_x+t_1\sigma_x^{\,2}+\rho t_2\sigma_x\sigma_y)$$
$$\exp(t_1\mu_x+t_2\mu_y+t_1^{\,2}\sigma_x^{\,2}/2+\rho t_1t_2\sigma_x\sigma_y+t_2^{\,2}\sigma_y^{\,2}/2)$$

which gives

$$EX=\frac{d}{dt_1}m(t_1,t_2)\bigg|_0=\mu_x.$$

Taking a second derivative, we obtain

$$\frac{d^2}{dt_2^{\,1}}m(t_1,t_2)=\frac{d}{dt_1}\frac{d}{dt_1}m(t_1,t_2)$$
$$=[\sigma_x^{\,2}+(\mu_x+t_1\sigma_x^{\,2}+\rho t_2\sigma_x\sigma_y)^2]$$
$$\exp(t_1\mu_x+t_2\mu_y+t_1^{\,2}\sigma_x^{\,2}/2+\rho t_1t_2\sigma_x\sigma_y+t_2^{\,2}\sigma_y^{\,2}/2)$$

giving

$$EX^2=\frac{d^2}{dt_1^{\,2}}m(t_1,t_2)\bigg|_0=\sigma_x^{\,2}+\mu_x^{\,2}.$$

Similar results hold when derivatives are taken with respect to t_2. If we now take one derivative with respect to each of t_1 and t_2, we obtain

$$\frac{d}{dt_2}\frac{d}{dt_1}m(t_1,t_2)=[\rho\sigma_x\sigma_y+(\mu_x+t_1\sigma_x^{\,2}+\rho t_2\sigma_x\sigma_y)$$
$$(\mu_y+t_2\sigma_y^{\,2}+\rho t_1\sigma_x\sigma_y)]$$
$$\exp(t_1\mu_x+t_2\mu_y+t_1^{\,2}\sigma_x^{\,2}/2+\rho t_1t_2\sigma_x\sigma_y+t_2^{\,2}\sigma_y^{\,2}/2)$$

which gives

$$E(XY)=\frac{d}{dt_2}\frac{d}{dt_1}m(t_1t_2)\bigg|_0=\rho\sigma_x\sigma_y+\mu_x\mu_y.$$

*EXERCISE 5-14

Find the characteristic function for the bivariate normal distribution.

5-5 Problems

1. Using the moment generating function, find the third moment of a binomial random variable.
2. Using the moment generating function, find the third moment of a poisson random variable.
3. Using the moment generating function, find the third moment of a negative binomial random variable.
4. A truncated poisson distribution is given by the mass function

$$P(x) = \frac{1}{1-e^{-\lambda}} \frac{e^{-\lambda}\lambda^x}{x^\lambda} \qquad x = 1, 2, 3, \ldots.$$
$$\lambda > 0.$$

   Find the moment generating function and the first two moments for this distribution.
5. Using the moment generating function, find the third moment of a normal random variable.
6. If $\mu = 0$, find the fourth moment of a normal random variable.
7. Find the moment generating function for the density

$$\begin{aligned} f(x) &= \lambda e^{-\lambda x} \qquad 0 < x < \infty \\ & \qquad\qquad\quad \lambda > 0 \\ &= 0 \qquad\qquad \text{otherwise.} \end{aligned}$$

8. Find the first two moments of the distribution of Problem 7 using the moment generating function.
9. Find the moment generating function for the density

$$\begin{aligned} f(x) &= x e^{-x} \qquad 0 < x < \infty \\ &= 0 \qquad\qquad \text{otherwise.} \end{aligned}$$

10. Find the first two moments of the distribution of Problem 9 using the moment generating function.
11. If X and Y have the trinomial mass function

$$P(x, y) = \binom{n}{x}\binom{n-x}{y} p^x q^y (1-p-q)^{n-x-y} \qquad 0 \leqq x+y \leqq n$$
$$0 < p+q < 1$$

   find the moment generating function of the random variable $Z = n-X-Y$.
12. Using the moment generating function find the first two moments of the variable Z in Problem 11.
13. Find the joint moment generating function of X and Z of Problem 11.
14. Find EZ using the joint moment generating function of Problem 13 and compare it with EZ found in Problem 12.
15. Using the moment generating function obtained in Problem 13 find the covariance between X and Z.
16. Using the moment generating function for the bivariate normal distribution, find $E(X^2Y)$.

CHAPTER 6

Distributions of Functions of Random Variables

The situation often arises wherein we are interested in probability statements about some function of X, say $g(X)$, rather than about X itself. It is, of course, possible to use the distribution of X directly to make probability statements about $g(X)$, but this is not always the most interesting or informative method. In this chapter we will explore several methods for finding the mass or density function for the random variable $Y = g(X)$. This mass or density function is informative, since one can frequently obtain answers merely by noting the form of this function.

6-1 Discrete Random Variables

WHEN $g(x)$ HAS AN INVERSE

If $g(x)$ has an inverse, then for each y_i there is at most one x_i such that $y_i = g(x_i)$. Therefore, whenever $X = x_i$, $Y = y_i$ and the frequency with which $Y = y_i$ must be the same as the frequency with which $X = x_i$. Thus, given the mass function $P(x_i)$, $i = 1, 2, 3, \ldots$ for X, we obtain the mass function for Y

$$P(y_i) \qquad i = 1, 2, 3, \ldots$$

by setting $P(y_i) = P(x_i)$ for each $i = 1, 2, 3, \ldots$, where $y_i = g(x_i)$.

EXAMPLE 6-1

Let X be such that $P(1) = 1/2$, $P(2) = 1/3$, and $P(3) = 1/6$. If $Y = X^2$ then the mass function for Y is given by $P(Y = 1) = P(X = 1) = 1/2$, $P(Y = 4) = P(X = 2) = 1/3$, and $P(Y = 9) = P(X = 3) = 1/6$.

WHEN $g(x)$ DOES NOT HAVE AN INVERSE

If $g(x)$ does not have an inverse then there may be more than one x value which will give the same value y according to the equation $y = g(x)$. Let

$\Omega(y) = \{x: y = g(x)\}$ be the set of all x values which give the value y. The event $\{Y = y\}$ is then equivalent to the event $\{X \in \Omega(y)\}$. From this it is clear that $P(Y = y) = P(X \in \Omega(y))$. Thus,

$$P(y) = \sum_{\Omega(y)} P(x).$$

EXAMPLE 6-2

Let X be such that $P(-1) = 1/3$, $P(0) = 1/2$ and $P(1) = 1/6$. If $Y = X^2$ then the mass function for Y is given by $P(Y = 0) = P(X = 0) = 1/2$ and $P(Y = 1) = P(X = 1) + P(X = -1) = 1/3 + 1/6 = 1/2$.

DISCRETIZING FUNCTIONS

The method used above can be extended to the case where X is a continuous random variable but $Y = g(X)$ is discrete. It should be clear that $\Omega(y)$ might very well be an interval of x values or a sum of intervals of x values. In this case, given that $f(x)$ is the probability density function for X, the mass function for Y is given by

$$P(Y = y) = P(X \in \Omega(y)) = \int_{\Omega(y)} f(x)\, dx.$$

EXAMPLE 6-3

Let X be uniform on the interval $-2.5 \leqq x \leqq +2.5$; that is, the density for X is given by

$$\begin{aligned} f(x) &= 1/5 \qquad -2.5 \leqq x \leqq +2.5 \\ &= 0 \qquad \text{otherwise.} \end{aligned}$$

If $y = [X]$ (i.e., the largest integer less than or equal to X), then the mass function for Y is given by

$$P(-3) = 1/5 \int_{-2.5}^{-2} dx = 1/10$$

$$P(-2) = 1/5 \int_{-2}^{-1} dx = 1/5$$

$$P(-1) = 1/5 \int_{-1}^{0} dx = 1/5$$

$$P(0) = 1/5 \int_{0}^{1} dx = 1/5$$

$$P(1) = 1/5 \int_{1}^{2} dx = 1/5$$

$$P(2) = 1/5 \int_{2}^{2.5} dx = 1/10.$$

6-2 Use of Distribution Functions

In this section and in succeeding sections of this chapter the random variables considered will be continuous.

The method of distribution functions is often easy to use, especially when the function $g(x)$ is not complicated. It can simply be explained by the following. If $Y = g(X)$ then $P(Y \leqq y) = P(g(X) \leqq y)$. If the set $g(X) \leqq y$ can now be shown equivalent to some set $g_1(y) \leqq X \leqq g_2(y)$, then

$$P(Y \leqq y) = P(g_1(y) \leqq X \leqq g_2(y)) = \int_{g_1(y)}^{g_2(y)} f(x)\,dx.$$

If this last integration can be transformed so that

$$P(Y \leqq y) = \int_{-\infty}^{y} f(t)\,dt$$

(where $f(t)$ need not be the same function as $f(x)$) then $f(t)$ is the probability density function for Y which we can call $f(y)$.

EXAMPLE 6-4

Let X be such that

$$f(x) = \frac{1}{\sqrt{(2\pi)}} e^{-x^2/2} \qquad -\infty < x < \infty$$

and let $Y = |X|$. Here

$$\begin{aligned} P(Y \leqq y) &= P(-y \leqq X \leqq y) \\ &= \frac{1}{\sqrt{(2\pi)}} \int_{-y}^{y} e^{-x^2/2}\,dx. \\ &= \sqrt{(2/\pi)} \int_{0}^{y} e^{-x^2/2}\,dx. \end{aligned}$$

Therefore

$$\begin{aligned} f(y) &= \sqrt{(2/\pi)}\, e^{-y^2/2} && 0 \leqq y < \infty \\ &= 0 && \text{otherwise.} \end{aligned}$$

The restricted range $0 \leqq y$ is an obvious consequence of the equation $Y = |X|$.

6-3 Integral Transformations

WHEN $g(x)$ HAS AN INVERSE

In this subsection we assume that $Y = g(X)$ has a unique inverse relation given by $X = g^{-1}(Y)$. What we want is the integrand $f(y)$ which satisfies

$$P(a \leqq Y \leqq b) = \int_{a}^{b} f(y)\,dy$$

for any values a and b.

Let us assume for the moment that $g(x)$ is an increasing function of x. Then

$$P(a \leqq Y \leqq b) = P(a \leqq g(x) \leqq b)$$

$$= P(g^{-1}(a) \leqq X \leqq g^{-1}(b)) = \int_{g^{-1}(a)}^{g^{-1}(b)} f(x)\,dx$$

where $f(x)$ is the density function for X.

Making the transformation $y = g(x)$, we obtain

$$P(a \leqq Y \leqq b) = \int_a^b f(g^{-1}(y)) \frac{dg^{-1}(y)}{dy}\,dy$$

where $f(g^{-1}(y))$ is still the density function for X. From this it is clear that

$$f(y) = f(g^{-1}(y)) \frac{dg^{-1}(y)}{dy}.$$

If $g(x)$ had been a decreasing function rather than an increasing function, the procedure would have been the following:

$$P(a \leqq Y \leqq b) = P(a \leqq g(x) \leqq b) = P(g^{-1}(b) \leqq X \leqq g^{-1}(a))$$

(note the reversed order in the inequality)

$$= \int_{g^{-1}(b)}^{g^{-1}(a)} f(x)\,dx \qquad = -\int_{g^{-1}(a)}^{g^{-1}(b)} f(x)\,dx.$$

Proceeding as before, we obtain

$$f(y) = -f(g^{-1}(y)) \frac{dg^{-1}(y)}{dy}.$$

The two can be unified by noting that when $g(x)$ is decreasing

$$\frac{dg^{-1}(y)}{dy} = -\left|\frac{dg^{-1}(y)}{dy}\right|.$$

The general result is therefore

$$f(y) = f(g^{-1}(y)) \left|\frac{dg^{-1}(y)}{dy}\right|$$

wherever the inverse exists. The range of y will be determined by $g(x)$ and the range of x. It is, of course, important to remember that $f(g^{-1}(y))$ is the density for X with $g^{-1}(y)$ replacing x.

EXAMPLE 6-5

Let X be a chi-squared random variable such that

$$f(x) = \frac{1}{\Gamma(n/2)2^{n/2}\sigma^n} x^{n/2-1} e^{-x/2\sigma^2} \qquad 0 < x < \infty$$
$$= 0 \qquad \text{otherwise.}$$

Let us find the density function for $Y = \sqrt{X}$.

The range of Y is clearly the positive real axis, while the inverse function is $X = Y^2$. The solution is therefore

$$f(y) = f(y^2)\left|\frac{dy^2}{dy}\right| = \frac{1}{\Gamma(n/2)2^{n/2}\sigma^n}(y^2)^{n/2-1} e^{-y^2/2\sigma}(2y)$$

giving

$$f(y) = \frac{2}{\Gamma(n/2)2^{n/2}\sigma^n} y^{n-1} e^{-y^2/2\sigma^2} \qquad 0 < y < \infty$$
$$= 0 \qquad \text{otherwise.}$$

Thus Y is a chi random variable.

EXERCISE 6-1

If X has any density function $f(x)$ $-\infty < x < \infty$ and $Y = aX+b$ where a and b are constants with $a \neq 0$, find the density for Y.

WHEN $g(x)$ DOES NOT HAVE AN INVERSE

If $Y = g(X)$ and $g(x)$ does not have an inverse then it cannot be a strictly increasing or a strictly decreasing function. In particular $g(x)$ looks something like the function depicted in Figure 6-1.

Remark *We will assume that there are no horizontal lines in this function. If this were not the case, then Y would have a positive probability of being at a particular value. Continuous random variables which have positive probability at particular values are called* mixed random variables *and will be discussed in Chapter Seven.*

When $g(x)$ is continuous (or at least piecewise continuous) and does not have horizontal segments it is possible to divide the range of x into segments on which $g(x)$ is either strictly increasing or strictly decreasing. In each of these segments $g(x)$ does have an inverse. We can therefore write the inverse as

$$\begin{aligned}
x &= g_1^{-1}(y) \qquad && x_0 < x \leqq x_1 \\
x &= g_2^{-1}(y) \qquad && x_1 < x \leqq x_2 \\
&\vdots && \vdots \\
x &= g_n^{-1}(y) \qquad && x_{n-1} < x \leqq x_n \\
&\vdots && \vdots \quad .
\end{aligned}$$

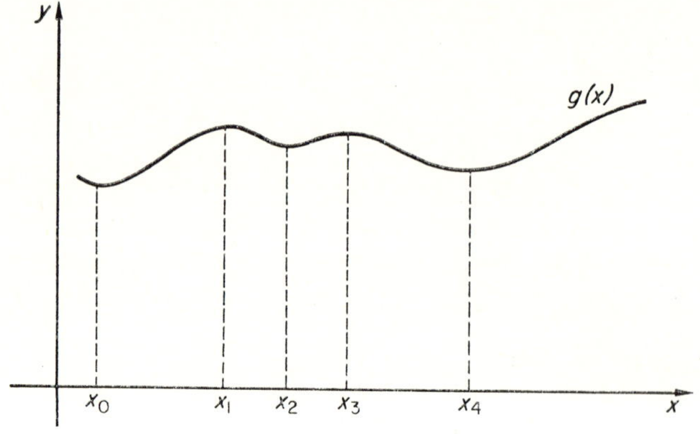

Figure 6-1

When Y lies in some interval, say $a \leqq Y \leqq b$, then X can be in any one of many intervals, each of which is a subinterval of one of the intervals on which the inverse is defined. Since these intervals must be disjoint

$$P(a \leqq Y \leqq b) = \sum_i P(X \in g_i^{-1}(a, b))$$

where $g_i^{-1}(a, b)$ is used here to represent that part of the interval (x_{i-1}, x_i) which lies between $g_i^{-1}(a)$ and $g_i^{-1}(b)$ [or between $g_i^{-1}(b)$ and $g_i^{-1}(a)$ when $g_i^{-1}(y)$ is strictly decreasing]. One must be careful here since $g_i^{-1}(a)$ or $g_i^{-1}(b)$ may lie outside the interval (x_{i-1}, x_i). To avoid this problem let us multiply each $g_i^{-1}(y)$ by a truncating function. We can define the truncating function for each $i = 1, 2, 3, \ldots$ by

$$\begin{aligned} \delta_i(y) &= 1 \quad \text{if} \quad g(x_{i-1}) < y \leqq g(x_i) \\ &\qquad\quad \text{or} \quad g(x_i) \leqq y < g(x_{i-1}) \\ &= 0 \qquad\qquad \text{otherwise.} \end{aligned}$$

Now we rewrite the inverse function as

$$\begin{aligned} x &= g_1^{-1}(y)\delta_1(y) & x_0 &< x \leqq x_1 \\ x &= g_2^{-1}(y)\delta_2(y) & x_1 &< x \leqq x_2 \\ &\ \vdots & &\ \vdots \\ x &= g_n^{-1}(y)\delta_n(y) & x_{n-1} &< x \leqq x_n \\ &\ \vdots & &\ \vdots \quad . \end{aligned}$$

The reason for this apparent complication will become clear presently.

Returning to the problem, we have

$$P(a \leqq Y \leqq b) = \sum_{i=1}^{n} P(X \in g_i^{-1}(a, b))$$

$$= \sum_i \int_{g_i^{-1}(a,b)} f(x)\, dx$$

$$= \sum_i \int_a^b \delta_i(y) f(g_i^{-1}(y)) \left| \frac{dg_i^{-1}(y)\delta_i(y)}{dy} \right| dy$$

(where the absolute value is introduced to accommodate both the strictly increasing and the strictly decreasing functions)

$$= \sum_i \int_a^b \delta_i(y) f(g_i^{-1}(y)) \left| \frac{dg_i^{-1}(y)}{dy} \right| dy.$$

Now if we interchange the order of summation and integration, we obtain

$$P(a \leqq Y \leqq b) = \int_a^b \sum_i f(g_i^{-1}(y)) \left| \frac{dg_i^{-1}(y)}{dy} \right| \delta_i(y)\, dy.$$

But, by definition

$$P(a \leqq Y \leqq b) = \int_a^b f(y)\, dy$$

and, therefore

$$f(y) = \sum_i f(g_i^{-1}(y)) \left| \frac{dg_i^{-1}(y)}{dy} \right| \delta_i(y).$$

Example 6-6

Let X be such that

$$f(x) = \frac{1}{\sqrt{(2\pi)}\sigma} e^{-x^2/2\sigma^2} \qquad -\infty < x < \infty$$

and let $Y = X^2$. Then

$$X = -\sqrt{Y} \qquad -\infty < x \leqq 0$$

$$X = +\sqrt{Y} \qquad 0 < x \leqq \infty$$

and, therefore

$$f(y) = \frac{1}{\sqrt{(2\pi)}\sigma} e^{-y/2\sigma^2} \left| -\frac{1}{2\sqrt{y}} \right| \qquad \text{for } -\infty < x \leqq 0$$

$$+ \frac{1}{\sqrt{(2\pi)}\sigma} e^{-y/2\sigma^2} \left| \frac{1}{2\sqrt{y}} \right| \qquad \text{for } 0 < x \leqq \infty$$

$$= \frac{1}{\sqrt{(2\pi)}\sigma} y^{-1/2} e^{-y/2\sigma^2} \qquad 0 \leqq y < \infty$$

$$= 0 \qquad \text{otherwise.}$$

6-4 The Use of Moment Generating Functions

In some cases the simplest method for finding the distribution of $Y = g(X)$ is to use the one to one correspondence between the moment generating function and the probability distribution. That is, we find the moment generating function $E\,e^{tY} = E\,e^{tg(X)}$ and recognize it as the moment generating function of some known distribution. The method is best explained by an example.

EXAMPLE 6-7

Let X have density

$$f(x) = \frac{1}{\sqrt{(2\pi)}} e^{-x^2/2} \qquad -\infty < x < \infty$$

and let $Y = X^2$. Then

$$E\,e^{tY} = E\,e^{tX^2} = \frac{1}{\sqrt{(2\pi)}} \int_{-\infty}^{\infty} e^{tx^2} e^{-x^2/2} dx = \frac{1}{\sqrt{(2\pi)}} \int_{-\infty}^{\infty} e^{-(1-2t)x^2/2} dx$$

$$= (1-2t)^{-1/2} \left[\frac{1}{\sqrt{(2\pi)(1-2t)^{-1/2}}} \int e^{-x^2/2(1-2t)^{-1}} dx \right]$$

$$= (1-2t)^{-1/2}.$$

Now $(1-\beta t)^{-\alpha}$ is the moment generating function for a gamma distribution. Therefore Y has a gamma distribution with $\beta = 2$, $\alpha = 1/2$, and density function

$$f(y) = \frac{1}{\sqrt{(2\pi)}} y^{-1/2} e^{-y/2} \qquad 0 < y < \infty$$

$$= 0 \qquad \text{otherwise}$$

which is a chi-squared distribution with one degree of freedom.

THE MULTIVARIATE SITUATION

If $X_1, X_2, X_3, \ldots, X_n$ are any n jointly distributed random variables, then we can ask for the distribution of any function of them. The function which is of interest to us in this section is the sum of these variables $Y = \Sigma_{i=1}^n X_i$. The moment generating function of this sum is

$$E\,e^{tY} = E\left[\exp\left(t \sum_{i=1}^n X_i\right)\right] = E\left[\prod_{i=1}^n e^{tX_i}\right].$$

If we now assume that the random variables are *independent*, then the expectation of the product becomes the product of the expectations, or

$$E\,e^{tY} = \prod_{i=1}^n E\,e^{tX_i}.$$

If we assume further that the random variables $X_1, X_2, X_3, \ldots, X_n$ each have the same marginal distribution (we say they are *identically distributed*) then they also have the same moment generating function; that is

$$E\, e^{tX_i} = E\, e^{tX_j} \qquad \text{all } i, j = 1, 2, 3, \ldots, n.$$

Therefore

$$E\, e^{tY} = (E\, e^{tX})^n$$

where $E\, e^{tX}$ represents the common moment generating function. Thus if $X_1, X_2, X_3, \ldots, X_n$ are independent and identically distributed, then all we need do to find the moment generating function of the sum is to take the common moment generating function and raise it to the nth power.

EXAMPLE 6-8

If $X_1, X_2, X_3, \ldots, X_n$ are independently and identically distributed with the density function

$$\begin{aligned} f(x) &= \frac{1}{\sqrt{2\pi}}\, x^{-1/2} e^{-x/2} \qquad && 0 < x < \infty \\ &= 0 && \text{otherwise} \end{aligned}$$

find the density function for

$$Y = \sum_{i=1}^{n} X_i.$$

The X variables have a gamma density with common moment generating function $(1-2t)^{-1/2}$, and therefore the moment generating function of their sum is $(1-2t)^{-n/2}$. This is the moment generating function of a gamma distribution with $\beta = 2$ and $\alpha = n/2$. Therefore

$$\begin{aligned} f(y) &= \frac{1}{\Gamma(n/2)2^{n/2}}\, y^{n/2-1} e^{-y/2} \qquad && 0 < y < \infty \\ &= 0 && \text{otherwise.} \end{aligned}$$

*EXERCISE 6-2

Characteristic functions can be used just as the moment generating functions were used. Derive the procedure for finding the characteristic function of a sum of independent, identically distributed random variables.

6-5 Multivariate Integral Transformations

Let us assume that we have n continuous random variables $X_1, X_2, X_3, \ldots, X_n$ which have a joint density function given by $f(x_1, x_2, x_3, \ldots, x_n)$.

We now ask for the joint probability density function of the random variables $Y_1, Y_2, Y_3, \ldots, Y_r$, where

$$Y_1 = g_1(X_1, X_2, X_3, \ldots, X_n)$$
$$Y_2 = g_2(X_1, X_2, X_3, \ldots, X_n)$$
$$\vdots$$
$$Y_r = g_r(X_1, X_2, X_3, \ldots, X_n).$$

WHEN THERE IS A UNIQUE INVERSE

When $r = n$, it may be possible to invert this system of equations and solve for $X_1, X_2, X_3, \ldots, X_n$, that is, we may find

$$X_1 = g_1^{-1}(Y_1, Y_2, Y_3, \ldots, Y_n)$$
$$X_2 = g_2^{-1}(Y_1, Y_2, Y_3, \ldots, Y_n)$$
$$\vdots$$
$$X_n = g_n^{-1}(Y_1, Y_2, Y_3, \ldots, Y_n).$$

Remark *The set of functions g_i^{-1}, $i = 1, 2, 3, \ldots, n$ is the inverse of the set g_i, $i = 1, 2, 3, \ldots, n$, but it should be noted that this does not imply that g_i^{-1} is in any sense an inverse of g_i for any i.*

Remark *We could treat the situation in which there was no unique inverse by finding disjoint volume sections in which there was an inverse. The results can then be obtained in a manner similar to that used in the univariate case. This, however, leads to complicated notation and will not be treated here.*

A joint probability statement on the random variables $Y_1, Y_2, Y_3, \ldots, Y_n$, can now be written

$$P(a_1 \leqq Y_1 \leqq b_1, a_2 \leqq Y_2 \leqq b_2, \ldots, a_n \leqq Y_n \leqq b_n)$$
$$= P(X_1 \in g_1^{-1}(a, b), X_2 \in g_2^{-1}(a, b), \ldots, X_n \in g_n^{-1}(a, b))$$

(where $X_i \in g_i^{-1}(a, b)$ is the set of x_i values for which

$$g_i^{-1}(a_1, a_2, \ldots, a_n) \leqq X_i \leqq g_i^{-1}(b_1, b_2, \ldots b_n)$$

or

$$g_i^{-1}(b_1, b_2, \ldots, b_n) \leqq X_i \leqq g_i^{-1}(a_1, a_2, \ldots, a_n)$$

whichever is in the correct order)

$$= \int_{g_1^{-1}(a,b)} \int_{g_2^{-1}(a,b)} \cdots \int_{g_n^{-1}(a,b)} f(x_1, x_2, \ldots, x_n)\, dx_1 dx_2 \ldots dx_n.$$

If we now make the transformation to

$$\begin{aligned} y_1 &= g_1(x_1, x_2, x_3, \dots, x_n) \\ y_2 &= g_2(x_1, x_2, x_3, \dots, x_n) \\ &\vdots \\ y_n &= g_n(x_1, x_2, x_3, \dots, x_n) \end{aligned}$$

we obtain

$$P(a_1 \leqq Y_1 \leqq b_1, a_2 \leqq Y_2 \leqq b_2, \dots, a_n \leqq Y_n \leqq b_n)$$

$$= \int_{a_1}^{b_1}\int_{a_2}^{b_2} \cdots \int_{a_n}^{b_n} f(g_1^{-1}(y_1, y_2, \dots, y_n), g_2^{-1}(y_1, y_2, \dots, y_n), \dots,$$

$$g_n^{-1}(y_1, y_2, \dots, y_n)) \mid J \mid dy_1 dy_2 \dots dy_n,$$

where J (the Jacobian of the transformation) is given by the determinant of derivatives

$$J = \begin{vmatrix} \dfrac{\partial g_1^{-1}(y_1, y_2, \dots, y_n)}{\partial y_1} & \dfrac{\partial g_2^{-1}(y_1, y_2, \dots, y_n)}{\partial y_1} & \cdots & \dfrac{\partial g_n^{-1}(y_1, y_2, \dots, y_n)}{\partial y_1} \\ \dfrac{\partial g_1^{-1}(y_1, y_2, \dots, y_n)}{\partial y_2} & \dfrac{\partial g_2^{-1}(y_1, y_2, \dots, y_n)}{\partial y_2} & \cdots & \dfrac{\partial g_n^{-1}(y_1, y_2, \dots, y_n)}{\partial y_2} \\ \vdots & \vdots & & \vdots \\ \dfrac{\partial g_1^{-1}(y_1, y_2, \dots, y_n)}{\partial y_n} & \dfrac{\partial g_2^{-1}(y_1, y_2, \dots, y_n)}{\partial y_n} & \cdots & \dfrac{\partial g_n^{-1}(y_1, y_2, \dots, y_n)}{\partial y_n} \end{vmatrix}$$

Remark *A derivation of multivariate integral transformations is beyond the scope of this text. It can only be assumed here that the student is either familiar with multivariate integral transformations, or that he will accept these procedures on faith until he becomes familiar with them. It should be pointed out, however, that we have already used this procedure in Section* 2-6 *where we transformed from u, v to r, θ with $u = r\cos\theta$ and $v = r\sin\theta$. There $du\,dv$ was replaced by $r\,dr\,d\theta$ with $\mid J \mid = r$.*

The absolute value of the Jacobian is used for the same reason $\mid dx/dy \mid$ was used in the univariate transformation, i.e., to allow the limits of integration to be written as they are, even when an inverse function is a decreasing function.

From the above equation it is clear that the joint density function for $Y_1, Y_2, Y_3, \dots, Y_n$ is given by

$$f(y_1, y_2, y_3, \dots y_n) = f(g_1^{-1}(y_1, y_2, \dots, y_n), g_2^{-1}(y_1, y_2, \dots, y_n), \dots, g_n^{-1}(y_1, y_2, \dots, y_n)) \mid J \mid$$

where the density function on the right-hand side is the joint density function for $X_1, X_2, X_3, \ldots, X_n$.

A very useful relation is the fact that the Jacobian can be expressed as the reciprocal of a determinant of the derivatives of the functions g_i, $i = 1, 2, 3, \ldots, n$; that is

$$J^{-1} = \begin{vmatrix} \dfrac{\partial g_1(x_1, x_2, \ldots, x_n)}{\partial x_1} & \dfrac{\partial g_2(x_1, x_2, \ldots, x_n)}{\partial x_1} & \cdots & \dfrac{\partial g_n(x_1, x_2, \ldots, x_n)}{\partial x_1} \\ \dfrac{\partial g_1(x_1, x_2, \ldots, x_n)}{\partial x_2} & \dfrac{\partial g_2(x_1, x_2, \ldots, x_n)}{\partial x_2} & \cdots & \dfrac{\partial g_n(x_1, x_2, \ldots, x_n)}{\partial x_2} \\ \vdots & \vdots & & \vdots \\ \dfrac{\partial g_n(x_1, x_2, \ldots, x_n)}{\partial x_n} & \dfrac{\partial g_2(x_1, x_2, \ldots, x_n)}{\partial x_n} & \cdots & \dfrac{\partial g_n(x_1, x_2, \ldots, x_n)}{\partial x_n} \end{vmatrix}$$

EXAMPLE 6-9

Let X_1, X_2 have joint density function

$$f(x_1, x_2) = \frac{1}{2\pi} e^{-(x_1{}^2+x_2{}^2)/2} \qquad -\infty < x_1, x_2 < \infty$$

and let

$$Y_1 = X_1 + X_2 = g_1(X_1, X_2)$$
$$Y_2 = X_1 - X_2 = g_2(X_1, X_2).$$

Since

$$\frac{\partial g_1(x_1, x_2)}{\partial x_1} = 1, \qquad \frac{\partial g_2(x_1, x_2)}{\partial x_1} = 1$$
$$\frac{\partial g_1(x_1, x_2)}{\partial x_2} = 1, \quad \text{and} \quad \frac{\partial g_2(x_1, x_2)}{\partial x_2} = -1$$

we have

$$J^{-1} = \begin{vmatrix} 1 & 1 \\ 1 & -1 \end{vmatrix} = -2.$$

Now we find that

$$x_1{}^2 + x_2{}^2 = \tfrac{1}{2}[(x_1+x_2)^2 + (x_1-x_2)^2]$$
$$= \tfrac{1}{2}(y_1{}^2 + y_2{}^2)$$

and therefore

$$f(y_1, y_2) = \frac{1}{4\pi} e^{-(y_1{}^2+y_2{}^2)/4} \qquad -\infty < y_1, y_2 < \infty.$$

EXAMPLE 6-10

Let X_1, X_2 have the joint density function

$$f(x_1, x_2) = \frac{1}{2\pi\sigma_1\sigma_2\sqrt{(1-\rho^2)}} \exp\left\{-\frac{1}{2(1-\rho^2)}\left[\left(\frac{x_1-\mu_1}{\sigma_1}\right)^2 - 2\rho\left(\frac{x_1-\mu_1}{\sigma_1}\right)\left(\frac{x_2-\mu_2}{\sigma_2}\right)\right.\right.$$
$$\left.\left. + \left(\frac{x_2-\mu_2}{\sigma_2}\right)^2\right]\right\} \qquad -\infty < x_1, x_2 < \infty$$

and let

$$Y_1 = \frac{X_1 - \mu_1}{\sigma_1} = g_1(X_1, X_2)$$

$$Y_2 = \frac{X_2 - \mu_2}{\sigma_2} = g_2(X_1, X_2).$$

Since

$$\frac{\partial g_1(x_1, x_2)}{\partial x_1} = \frac{1}{\sigma_1}, \qquad \frac{\partial g_2(x_1, x_2)}{\partial x_2} = 0$$

$$\frac{\partial g_1(x_1, x_2)}{\partial x_2} = 0, \quad \text{and} \quad \frac{\partial g_2(x_1, x_2)}{\partial x_2} = \frac{1}{\sigma_2}$$

we have

$$J^{-1} = \begin{vmatrix} \frac{1}{\sigma_1} & 0 \\ 0 & \frac{1}{\sigma_2} \end{vmatrix}$$

and therefore

$$J = \sigma_1 \sigma_2$$

and

$$f(y_1, y_2) = \frac{1}{2\pi\sqrt{(1-\rho^2)}} \exp\left\{-\frac{1}{2(1-\rho^2)}(y_1{}^2 - 2\rho y_1 y_2 + y_2{}^2)\right\} \qquad -\infty < y_1, y_2 < \infty.$$

TRANSFORMATIONS TO FEWER VARIABLES

It is often the case that $r < n$. That is, we are interested in only a few functions of the original n random variables. For example, we may ask for the distribution of the sum (one variable) of the n random variables.

Theoretically this poses no new problem. We could imagine using any $n-r$ other functions to complete the system of equations to n equations (as long as we had a unique inverse), finding the density for the n new variables and integrating out the unwanted $n-r$ variables. This marginal probability of the r random variables would be the same, regardless of what $n-r$ functions were chosen to fill out the transformation. This procedure involves finding an n-by-n determinant in order to obtain the Jacobian. In practice, we can reduce this problem to one of finding an r-by-r determinant, as will now be shown.

Assume that we are given r equations

$$Y_1 = g_1(X_1, X_2, X_3, \ldots, X_n)$$

$$Y_2 = g_2(X_1, X_2, X_3, \ldots, X_n)$$

$$Y_r = g_r(X_1, X_2, X_3, \ldots, X_n).$$

Augment these with the $n-r$ equations

$$\begin{aligned} Y_{r+1} &= X_{r+1} \\ Y_{r+2} &= X_{r+2} \\ &\vdots \\ Y_n &= X_n. \end{aligned}$$

(Actually we can choose any set of $n-r$ X's to use in these last $n-r$ equations without changing the results.)

We will now assume that these n equations have the unique inverse set of equations

$$\begin{aligned} X_1 &= g_1^{-1}(Y_1, Y_2, Y_3, \ldots, Y_n) \\ X_2 &= g_2^{-1}(Y_1, Y_2, Y_3, \ldots, Y_n) \\ &\vdots \\ X_r &= g_r^{-1}(Y_1, Y_2, Y_3, \ldots, Y_n) \\ &\vdots \\ X_{r+1} &= Y_{r+1} \\ X_{r+2} &= Y_{r+2} \\ &\vdots \\ X_n &= Y_n. \end{aligned}$$

In order to use the result of the previous subsection we must find the Jacobian. Note that

$$J = \left|\begin{array}{ccc|ccc} \dfrac{\partial g_1^{-1}(y_1, y_2, \ldots, y_n)}{\partial y_1} & \cdots & \dfrac{\partial g_r^{-1}(y_1, y_2, \ldots, y_n)}{\partial y_1} & 0 & \cdots & 0 \\ \vdots & & \vdots & & & \\ \dfrac{\partial g_1^{-1}(y_1, y_2, \ldots, y_n)}{\partial y_r} & \cdots & \dfrac{\partial g_r^{-1}(y_1, y_2, \ldots, y_n)}{\partial y_r} & 0 & \cdots & 0 \\ \hline 0 & \cdots & 0 & \dfrac{\partial y_{r+1}}{\partial y_{r+1}} & \cdots & 0 \\ \vdots & & \vdots & 0 & \ddots & \vdots \\ \vdots & & \vdots & \vdots & & \\ 0 & \cdots & 0 & 0 & \cdots & \dfrac{\partial y_n}{\partial y_n} \end{array}\right|$$

But since the submatrix at the lower right is a matrix with 0 off the diagonal and unity on the diagonal

$$J = \begin{vmatrix} \dfrac{\partial g_1^{-1}(y_1, y_2, \ldots, y_n)}{\partial y_1} & \cdots & \dfrac{\partial g_r^{-1}(y_1, y_2, \ldots, y_n)}{\partial y_1} \\ \vdots & & \vdots \\ \dfrac{\partial g_1^{-1}(y_1, y_2, \ldots, y_n)}{\partial y_r} & \cdots & \dfrac{\partial g_r^{-1}(y_1, y_2, \ldots, y_n)}{\partial y_r} \end{vmatrix}$$

or

$$J^{-1} = \begin{vmatrix} \dfrac{\partial g_1(x_1, x_2, \ldots, x_n)}{\partial x_1} & \cdots & \dfrac{\partial g_r(x_1, x_2, \ldots, x_n)}{\partial x_1} \\ \vdots & & \vdots \\ \dfrac{\partial g_1(x_1, x_2, \ldots, x_n)}{\partial x_r} & \cdots & \dfrac{\partial g_r(x_1, x_2, \ldots, x_n)}{\partial x_r} \end{vmatrix}$$

which, as predicted, requires calculating only an r-by-r determinant. We, therefore, have

$$f(y_1, y_2, y_3, \ldots, y_n) = f[g_1^{-1}(y_1, y_2, \ldots, y_n), g_2^{-1}(y_1, y_2, \ldots, y_n), \ldots \\ g_r^{-1}(y_1, y_2, \ldots, y_n), y_{r+1}, y_{r+2}, \ldots, y_n] \, |J| \,.$$

To obtain the marginal joint density for $Y_1, Y_2, Y_3, \ldots, Y_r$, we integrate out the unwanted variables obtaining

$$f(y_1, y_2, y_3, \ldots, y_r) = \int\int \cdots \int f[g_1^{-1}(y_1, y_2, \ldots, y_n), \\ g_2^{-1}(y_1, y_2, \ldots, y_n), \ldots, g_r^{-1}(y_1, y_2, \ldots, y_n), y_{r+1}, y_{r+2}, \ldots, y_n] \\ |J| \, dy_{r+1} dy_{r+2} \ldots dy_n.$$

It now becomes apparent that we should use those X variables for the $n-r$ last equations which are the easiest to integrate out in the equation above.

EXAMPLE 6-11

Let X_1, X_2 have the joint density function given by

$$f(x_1, x_2) = \frac{1}{2\pi\sqrt{(1-\rho^2)}} \exp\left\{-\frac{1}{2(1-\rho^2)}(x_1^2 - 2\rho x_1 x_2 + x_2^2)\right\} \qquad -\infty < x_1, x_2 < \infty$$

and let $Y = X_1 - \rho X_2$. Find the density for Y. If we use the auxiliary equation $Y_2 = X_2$, we can invert to obtain

$$X_1 = Y + \rho Y_2$$
$$X_2 = Y_2$$

with

$$J = \frac{\partial(y+\rho y_2)}{\partial y} = 1$$

giving

$$f(y) = \frac{1}{2\pi\sqrt{(1-\rho^2)}}\int \exp\left\{-\frac{1}{2(1-\rho^2)}((y+\rho y_2)^2 - 2\rho(y+\rho y_2)x_2 + y_2{}^2)\right\} dy_2$$

$$= \frac{1}{2\pi\sqrt{(1-\rho^2)}}\int \exp\left\{-\frac{1}{2}\left(\frac{y^2}{1-\rho^2}+y_2{}^2\right)\right\} dy_2 = \frac{1}{\sqrt{(2\pi)}\sqrt{(1-\rho^2)}}\exp\left\{-\frac{y^2}{2(1-\rho^2)}\right\}$$

$$-\infty < y < \infty.$$

6-6 Problems

1. If X has a binomial distribution with $n = 10$ and $p = 1/2$ find the mass function for $Y = X^2$.
2. If X has a poisson distribution with mean $\lambda = 1$ find the mass function for $Y = X^2$.
3. If X has a poisson distribution with mean λ what is the probability mass function for $Y = X[\text{mod } r]$, where $X[\text{mod } r]$ means the remainder after X is divided by 4 (e.g., $18[\text{mod } 4] = 2$)?
4. If X has a poisson distribution with mean λ what is the probability mass function for $Y = X[\text{mod } 2]$?
5. If X is normal with mean 0 and variance 1, find the mass function for $Y = [X]$, where $[x]$ is defined as the integer no larger than x but larger than $x-1$.
6. If X has a uniform distribution on the interval (0, 1) find the mass function for $Y = [10X]$.
7. If X is uniformly distributed on (1, 2), use the method of distributions to find the density function for $Y = \ln X$.
8. If X is uniformly distributed on (1, 2), use the method of distributions to find the density function for $Y = X^2$.
9. If X is uniformly distributed on (1, 2), use the method of integral transformations to find the density function for $Y = \ln X$.
10. If X is uniformly distributed on (1, 2), use the method of integral transformations to find the density function for $Y = X^2$.
11. If X is gamma distributed with parameters α and β, find the density function for $Y = X^2$.
12. If X is normally distributed with mean μ and variance σ^2, find the density function for $Y = (X-\mu)/\sigma$ using the result of Exercise 6-3.
13. If X is normal with mean μ and variance σ^2 find the density function for $Y = |X-\mu|$.
14. If X has a t distribution, find the density of $Y = |X|$.
15. If X has a t distribution, find the density function for $Y = aX+b$.
16. If X has the density function

$$f(x) = \frac{1}{2\Gamma(n/2)2^{n/2}\sigma^n}|x|^{n/2-1}e^{-|x|/2\sigma^2} \qquad -\infty < x < \infty$$

find the density function for $Y = \sqrt{|X|}$.

17. If X_i $i = 1, 2, \ldots, n$ are independent and identically distributed each with a normal distribution with mean μ and variance σ^2, find the distribution of

$$Y = \frac{1}{n} \sum_{i=1}^{n} X_i$$

by the method of moment generating functions.

18. If X_1, X_2, X_3, X_4, X_5 are independently distributed with the common density function

$$\begin{aligned} g(x) &= xe^{-x} \qquad 0 < x < \infty \\ &= 0 \qquad \text{otherwise} \end{aligned}$$

find the density function for $Y = \sum_{i=1}^{5} X_i$.

19. If X and Y are independent and have the same standard normal distribution, find the joint density function for $U = X + Y$ and $V = X - Y$.

20. If X and Y are independent and have the same standard normal distribution, find the density function for $Z = X^2 + Y^2$.

21. If X and Y have joint density function

$$\begin{aligned} f(x, y) = 10\, xy^2 \qquad & 0 < x < y \\ & 0 < y < 1 \end{aligned}$$

find the density function for $Z = XY$.

CHAPTER 7

Some More-Advanced Concepts

7-1 A Unified Notation for Probability Distributions

When we knew that X was a discrete random variable, we used one notation, and when we knew that X was a continuous random variable, we used a completely different notation. Properties were proved separately in the two notations and were generally shown to be the same except for notation. It would therefore seem advisable to unify the notation, thereby allowing us to treat both cases simultaneously. Moreover, this unified notation will also allow us to handle mixed distributions.

MIXED DISTRIBUTIONS

A discrete random variable has probabilities defined only at particular values, whereas a continuous random variable has probabilities defined only for values in an interval and not at particular values. It is possible for a random variable to have probabilities defined at particular values and for values in intervals not containing those particular values. As a simple example of this let us say

$$P(X = 0) = 1/2$$

$$P(\alpha \leqq X \leqq \beta) = \frac{\beta-\alpha}{2} \qquad 0 < \alpha \leqq \beta \leqq 1.$$

For this situation we do not have a probability mass function nor do we have a density function. The unified notation will, however, allow us to handle this situation.

In general, the distribution function for a mixed distribution will be zero at $x = -\infty$, unity at $x = +\infty$ and be nondecreasing. A typical graph is shown in Figure 7-1.

Any function which is differentiable except at a countable number of points and satisfies the properties above can represent a distribution function of a *mixed distribution*. [This definition includes distributions for discrete and continuous random variables as special cases.] To be consistent, we

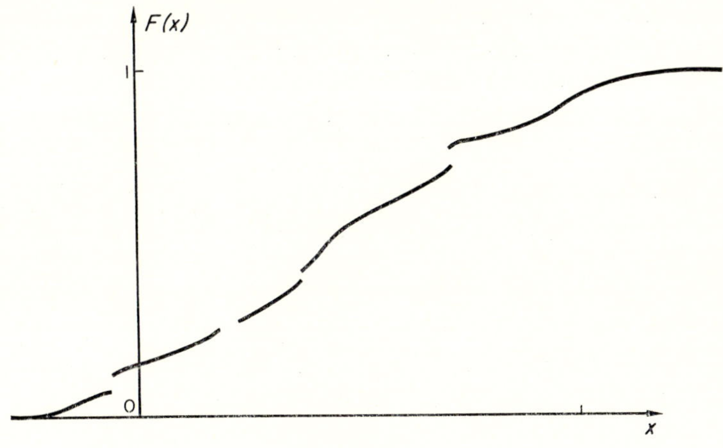

Figure 7-1

will define the value of the distribution function at a point of discontinuity to be its upper value.

The cumulative distribution function $F(z)$, as was the case for discrete and for continuous random variables, must be equivalent to $P(X \leqq z)$. This means that for any interval in which $F(z)$ is differentiable

$$F(b)-F(a) = \int_a^b f(x)\,dx$$

where $f(x)$ is any function which satisfies the equation. Since $F(x)$ is differentiable in (a, b), this implies that

$$f(x) = \frac{d}{dx} F(x)$$

is such a function. Since $F(x)$ is differentiable except at a countable number of points, we will define

$$f(x) = \frac{d}{dx} F(x)$$

all x except the points where the derivative does not exist. At these points (as in the discrete case) we define

$$P(x) = \Delta F(x)$$

where $\Delta F(x)$ represents the jump (difference between the upper and lower values) in $F(x)$ at x. This will be zero if $F(x)$ is continuous at X but not differentiable, and will be positive if $F(x)$ is discontinuous at x. We will use ω to represent the set of all x values at which $F(x)$ is differentiable and $\bar{\omega}$ to represent the remaining x values. Then it is clear that

$$\int_{\omega} f(x)\,dx + \sum_{\bar{\omega}} \Delta F(x) = 1.$$

We also find that

$$F(z) = \int_{\substack{x \in \omega \\ \text{such that} \\ x \leqq z}} f(x)\,dx + \sum_{\substack{x \in \bar{\omega} \\ \text{such that} \\ x \leqq z}} P(x).$$

Here $f(x)$ and $P(x)$ act like density and mass functions, respectively, but they do not separately integrate or sum to unity. They may be considered the density or mass part of the probability distribution.

THE UNIFIED NOTATION

Using the notation already developed in this chapter (i.e., the definitions of a mixed distribution function $F(x)$ and the corresponding $f(x)$ and $P(x)$ functions) we make the following definitions.

Definition In the notation developed above, let

$$\begin{aligned} dF(x) &= f(x)\,dx && \text{all } x \in \omega \\ &= P(x) && \text{all } x \in \bar{\omega} \end{aligned}$$

and let

$$\begin{aligned} F'(x) &= f(x) && \text{all } x \in \omega \\ &= P(x) && \text{all } x \in \bar{\omega}. \end{aligned}$$

Also for any piecewise continuous function $g(x)$ and any set A of x values, let

$$\int_A g(x)\,dF(x) = \int_{x \in \omega \cap A} g(x) f(x)\,dx + \sum_{x \in \bar{\omega} \cap A} g(x) P(x).$$

Definition For any piecewise continuous function $g(x)$, we define the expectation of $g(x)$ by

$$Eg(X) = \int g(x)\,dF(x)$$

(where omission of limits of integration indicates integration over all values of x).

From these definitions we see that

$$Eg(X) = \int_{\omega} g(x) f(x)\,dx + \sum_{\bar{\omega}} g(x) P(x).$$

EXERCISE 7-1

Show that the two preceding definitions are completely compatible with the definitions of expectations in previous chapters by showing the correspondence

a) when X is a continuous random variable and

b) when X is a discrete random variable.

MIXED JOINT DISTRIBUTIONS

Remark *The notation and results of this subsection will not be used in the subsequent text, and can be omitted without loss in continuity.*

If two random variables, X and Y, each have a mixed distribution, then $F(x, y) = P(X \leqq x, Y \leqq y)$ is a mixed joint distribution. There will now be four bivariate sets of interest: ω_1 is the set of all x, y at which $F(x, y)$ has a derivative with respect to both x and y; ω_2 is the set of all x, y at which $F(x, y)$ has a derivative with respect to x but not y; ω_3 is the set of all x, y at which $F(x, y)$ has a derivative with respect to y but not x; and ω_4 is the set of all x, y at which $F(x, y)$ has a derivative with respect to neither x nor y. Then we define

$$f(x, y) = \frac{\partial}{\partial x}\frac{\partial}{\partial y} F(x, y) \qquad x, y \in \omega_1,$$

$$P_2(x, y) = \Delta_y \frac{\partial}{\partial x} F(x, y) \qquad x, y \in \omega_2.$$

(This is the jump in the y direction of the function $(\partial/\partial x)F(x, y)$ at the point x, y.)

$$P_3(x, y) = \Delta_x \frac{\partial}{\partial y} F(x, y) \qquad x, y \in \omega_3$$

(This is the jump in the x direction of the function $(\partial/\partial y)F(x, y)$ at the point x, y) and

$$P(x, y) = \Delta_x \Delta_y F(x, y) \qquad x, y \in \omega_4.$$

(This is the jump in the function when x goes slightly below x to x and y goes from slightly below y to y, i.e., the jump at x, y.)

With these we can define

$$\begin{aligned} dF(x, y) &= f(x, y)\,dx\,dy && x, y \in \omega_1 \\ &= P_2(x, y)\,dx && x, y \in \omega_2 \\ &= P_3(x, y)\,dy && x, y \in \omega_3 \\ &= P(x, y) && x, y \in \omega_4 \end{aligned}$$

and for any piecewise continuous function in both variables $g(x, y)$ and any set A

$$\iint_A g(x, y)\, dF(x, y) = \iint_{x,y\in\omega_1\cap A} g(x, y) f(x, y)\, dx\, dy + \sum_{x,y\in\omega_2\cap A} \cap \int g(x, y) P_2(x, y)\, dx$$

$$+ \sum_{x,y\in\omega_3\cap A} \cap \int g(x, y) P_3(x, y)\, dy + \sum\sum_{x,y\in\omega_4\cap A} g(x, y) P(x, y).$$

The mixed joint expectation is then given by

$$Eg(X, Y) = \iint g(x, y)\, dF(x, y) = \iint_{x,y\in\omega_1} g(x, y) f(x, y)\, dx\, dy$$

$$+ \sum_{x,y\in\omega_2} \cap \int g(x, y) P_2(x, y)\, dx + \sum_{x,y\in\omega_3} \cap \int g(x, y) P_3(x, y)\, dy$$

$$+ \sum\sum_{x,y\in\omega_4} g(x, y) P(x, y).$$

The marginal mixed distribution for X can be found as

$$F(z) = P(X \leqq z) = \int_{-\infty}^{\infty}\int_{-\infty}^{z} dF(x, y) = \iint_{\substack{x,y\in\omega_1 \\ \text{such that} \\ x\leqq z}} f(x, y)\, dx\, dy$$

$$+ \sum_{\substack{x,y\in\omega_2 \\ \text{such that} \\ x\leqq z}}\int P_2(x, y)\, dx + \sum_{\substack{x,y\in\omega_3 \\ \text{such that} \\ x\leqq z}}\int P_3(x, y)\, dy$$

$$+ \sum\sum_{\substack{x,y\in\omega_4 \\ \text{such that} \\ x\leq z}} P(x, y)$$

with a symmetric result for the marginal distribution function for Y.

It should now be clear that in the marginal distribution of X there are two functions, $f(x)$ and $P(x)$. With a little effort, we find that

$$f(x) = \int_{x,y\in\omega_1} f(x, y)\, dy + \sum_{x,y\in\omega_2} P_2(x, y)$$

while

$$P(x) = \int_{x,y\in\omega_3} P_3(x, y)\, dy + \sum_{x,y\in\omega_4} P(x, y).$$

Symmetric relations hold for $f(y)$ and $P(y)$.

EXERCISE 7-2

Show that the relations above are compatible with the corresponding relations when both variables were either discrete or continuous.

The extension to n dimensions is straightforward but tedious and will not be attempted here.

CONDITIONAL DISTRIBUTIONS

Having the joint distribution and the marginal distributions, we can now easily define *conditional distributions*. The conditional distribution of Y, given X, is given by the functions

$$f(y \mid x) = \frac{f(x, y)}{f(x)} \qquad \text{for} \qquad x, y \in \omega_1$$

$$= \frac{P_3(x, y)}{P(x)} \qquad \text{for} \qquad x, y \in \omega_3$$

and

$$P(y \mid x) = \frac{P_2(x, y)}{f(x)} \qquad \text{for} \qquad x, y \in \omega_2$$

$$= \frac{P(x, y)}{P(x)} \qquad \text{for} \qquad x, y \in \omega_4.$$

Using the marginal for Y, we can define the conditional distribution of X, given Y, in a similar manner.

EXERCISE 7-3

Show that the relations above are compatible with previous relations obtained for conditional distributions by considering the cases where X and Y are both either discrete or continuous.

Conditional expectations are now defined by

$$E[g(Y) \mid x] = \int g(y)\, dF(y \mid x) = \int_{y \in \omega_1 \cup \omega_3} g(y) f(y \mid x)\, dx + \sum_{y \in \omega_2 \cup \omega_4} g(y) P(y \mid x)$$

with symmetrical results for a conditional expectation based on the X given Y distribution.

7-2 Inequalities and the Law of Large Numbers

MARKOV'S INEQUALITY

Let $g(x)$ be any continuous, positive, function symmetric about zero which is nondecreasing for $0 \leqq x < \infty$. We now find the expectation of $g(X)$, assuming it exists, with X any random variable.

$$Eg(X) = \int g(x)\, dF(x) = \int_{|x| \leqq \epsilon} (x)\, dF(x) + \int_{|x| > \epsilon} g(x)\, dF(x)$$

$$\geqq 0 + \int_{|x| > \epsilon} (\varepsilon)\, dF(x) = g(\varepsilon) \int_{|x| > \epsilon} dF(x) = g(\varepsilon) P(|X| > \varepsilon).$$

We thus obtain the result known as the *Markov inequality*, namely

$$P(|X| > \varepsilon) \leqq \frac{Eg(X)}{g(\varepsilon)}$$

where $\varepsilon > 0$ and $g(x)$ is any function satisfying the conditions given above.

When X is a random variable with zero mean we can let $g(x) = x^2$ and obtain for any $\varepsilon > 0$

$$P(|X| > \varepsilon) \leqq \frac{EX^2}{\varepsilon^2}$$

$$\leqq \frac{\text{Var } X}{\varepsilon^2}.$$

If Y is a random variable with mean μ, then $X = Y - \mu$ has mean zero and Var X = Var Y, giving for any $\varepsilon > 0$

$$P(|Y - \mu| > \varepsilon) \leqq \frac{\text{Var } Y}{\varepsilon^2}.$$

This inequality, which says that the probability that any random variable is further than ε from its mean is less than its variance divided by ε^2, is known as *Tchebychev's inequality*.

THE LAW OF LARGE NUMBERS

Let $X_i\; i = 1, 2, 3, \ldots$ be independent random variables with means $\mu_i\; i = 1, 2, 3, \ldots$ and variances σ_i^2, $i = 1, 2, 3, \ldots$. Define

$$Y_i = \frac{X_i - \mu_i}{\sigma_i} \quad \text{and} \quad Z_n = (1/n) \sum_{i=1}^{n} Y_i.$$

It is now easily seen that the mean of Z_n is zero.

Thus, by Tchebychev's inequality for any $\varepsilon > 0$

$$P(|Z| > \varepsilon) \leqq \frac{\text{Var } Z}{\varepsilon^2}.$$

However

$$\operatorname{Var} Z = \frac{1}{n^2} \operatorname{Var} \sum_{i=1}^{n} Y_i = \frac{1}{n^2} \sum_{i=1}^{n} \operatorname{Var} Y_i \quad \text{(by independence)}$$

$$= \frac{1}{n^2} \sum_{i=1}^{n} \frac{\operatorname{Var} X_i}{\sigma_i^2} = \frac{1}{n^2} \sum_{i=1}^{n} \frac{\sigma_i^2}{\sigma_i^2} = \frac{1}{n}.$$

We can therefore write

$$P(|Z| > \varepsilon) = P\left(\left|\frac{1}{n} \sum_{i=1}^{n} \left(\frac{X_i - \mu_i}{\sigma_i}\right)\right| > \varepsilon\right) \leqq \frac{1}{n\varepsilon^2}$$

and thus, for any $\varepsilon > 0$

$$\lim_{n \to \infty} P\left(\left|\frac{1}{n} \sum_{i=1}^{n} \left(\frac{X_i - \mu_i}{\sigma_i}\right)\right| > \varepsilon\right) = 0.$$

This says that if we standardize a sequence of random variables then the probability of their average converges to zero.

An important special case of this results when the X_i $i = 1, 2, 3, \ldots$ are independently distributed with the same mean and variance. In this case

$$P\left(\left|\frac{1}{n} \sum_{i=1}^{n} \frac{X_i - \mu}{\sigma}\right| > \varepsilon\right) = P\left(\left|\frac{1}{n} \sum_{i=1}^{n} X_i - \mu\right| > \sigma\varepsilon\right)$$

giving us

$$P\left(\left|\frac{1}{n} \sum_{i=1}^{n} X_i - \mu\right| > \sigma\varepsilon\right) \leqq \frac{1}{n\varepsilon^2}$$

or letting $\sigma\varepsilon = \delta$, we obtain

$$P\left(\left|\frac{1}{n} \sum_{i=1}^{n} X_i - \mu\right| > \delta\right) \leqq \frac{\sigma^2}{n\delta^2}.$$

Therefore, whenever $\sigma^2 < \infty$ for any $\delta > 0$

$$\lim_{n \to \infty} P\left(\left|\frac{1}{n} \sum_{i=1}^{n} X_i - \mu\right| > \delta\right) = 0.$$

This says that for a sequence of independent random variables with common mean and variance the probability that the average is not arbitrarily close to the mean converges to zero. This is called the *law of large numbers*.

7-3 The Central Limit Theorem

The normal distribution is the most often used distribution in statistics. The reason that many random variables have distributions resembling the normal can be found in what is known as the *central limit theorem.* Before stating the theorem we will derive a very useful result. Let $X_1, X_2, X_3, \ldots, X_n, \ldots$ be a sequence of independent random variables with means $\mu_1, \mu_2, \mu_3, \ldots, \mu_n, \ldots$ and variances $\sigma_1^2, \sigma_2^2, \sigma_2^3, \ldots, \sigma_n^2, \ldots$. Define

$$Z_n = \frac{1}{\sqrt{n}} \sum_{i=1}^{n} \left(\frac{X_i - \mu_i}{\sigma_i} \right).$$

Then

$$EZ_n = \frac{1}{\sqrt{n}} \sum_{i=1}^{n} \left(\frac{EX_i - \mu_i}{\sigma_i} \right) = 0$$

while

$$\operatorname{Var} Z_n = \frac{1}{n} \operatorname{Var} \sum_{i=1}^{n} \left(\frac{X_i - \mu_i}{\sigma_i} \right) = \frac{1}{n} \sum_{i=1}^{n} \operatorname{Var} \left(\frac{X_i - \mu_i}{\sigma_i} \right)$$

$$= \frac{1}{n} \sum_{i=1}^{n} \frac{\sigma_i^2}{\sigma_i^2} = 1.$$

Therefore, regardless of the distributions of the $X_i\, i = 1, 2, 3, \ldots, Z_n$ (for any n) has zero mean and unit variance. The central limit theorem now states that, under certain additional conditions, the distribution of Z_n approaches the normal distribution.

The Central Limit Theorem If Z_n (for any n) is $\sqrt{n}$ times the average of n standardized independent random variables whose moment generating functions exist and whose third moments are finite, then the distribution of Z_n converges to the standard normal distribution as n approaches infinity.

A Quick Proof We give a quick proof by using the method of moment generating functions. This proof is quick and not mathematically rigorous in that we will drop higher order terms as soon as they occur rather than carry them to the end and demonstrate that they can be neglected. These details will be left as an exercise. The moment generating function for

$$Z_n = \frac{1}{\sqrt{n}} \sum_{i=1}^{n} \left(\frac{X_i - \mu_i}{\sigma_i} \right)$$

is given by

$$E\,e^{tZ_n} = E\left\{\exp\left[\frac{t}{\sqrt{n}}\sum_{i=1}^{n}\left(\frac{X_i-\mu_i}{\sigma_i}\right)\right]\right\} = E\left\{\prod_{i=1}^{n}\exp\left[\frac{t}{\sqrt{n}}\left(\frac{X_i-u_i}{\sigma_i}\right)\right]\right\}$$

$$= \prod_{i=1}^{n} E\left\{\exp\left[\frac{t}{\sqrt{n}}\left(\frac{X_i-\mu_i}{\sigma_i}\right)\right]\right\}$$

which, on expanding the exponential in a power series and dropping terms that converge to zero faster than $1/n$

$$= \prod_{i=1}^{n} E\left[1+\frac{t(X_i-\mu_i)}{\sqrt{(n)}\sigma_i}+\frac{t^2(X_i-\mu_i)^2}{2n\sigma_i^{\,2}}\right]$$

$$= \prod_{i=1}^{n}\left[1+0+\frac{t^2\sigma_i^{\,2}}{2n\sigma_i^{\,2}}\right] = \prod_{i=1}^{n}(1+t^2/2n) = (1+t^2/2n)^n.$$

Now

$$\lim_{n\to\infty}(1+t^2/2n)^n = e^{t^2/2}$$

and therefore

$$\lim_{n\to\infty} E\,e^{tZ_n} = e^{t^2/2}$$

which is the moment generating function of the standard normal distribution. Therefore, the distribution of Z_n (as $n \to \infty$) approaches the normal distribution.

EXERCISE 7-4

Reprove the theorem taking the higher order terms into account.

An interesting special case occurs when the X_i, $i = 1, 2, 3, \ldots$ are identically distributed (at least when they have the same mean μ and variance σ^2). Then

$$Z_n = \sqrt{n}\left(\frac{1}{n}\sum_{i=1}^{n} X_i-\mu\right)\Big/\sigma.$$

If

$$\bar{X}_n = \frac{1}{n}\sum_{i=1}^{n} X_i$$

then we can solve for $\bar{X}_n$ as a function of Z_n giving $\bar{X}_n = \sigma Z_n/\sqrt{(n)}+\mu$. If *n is sufficiently large*, then Z_n will have a distribution which converges to the standard normal distribution. Since $\bar{X}_n$ is a linear function of Z_n, its distribution will converge to the normal distribution with mean and variance given by

$$E\bar{X}_n = \sigma EZ_n/\surd(n)+\mu = \mu$$

and

$$\text{Var}\,\bar{X}_n = \frac{\sigma^2}{n}\,\text{Var}\,Z_n = \sigma^2/n$$

as n approaches infinity.

It is not surprising then that we often find the distribution of an average of independent and identically distributed random variables with mean μ and variance σ^2 being approximated by a normal distribution with mean μ and variance σ^2/n. It should be pointed out, however, that the central limit theorem is an *asymptotic* result and the requirement that n approach infinity is often an important one. We should not be tempted into believing that the central limit theorem says that for moderate n any average of independent and identically distributed random variables has a normal distribution. The approximation does, however, improve as n becomes larger.

7-4 Types of Convergence

Let X be a random variable and let $X_1, X_2, X_3, \ldots, X_n, \ldots$ be a sequence of random variables. One may ask whether the sequence of random variables converges in some sense to X. Since the sequence is not a sequence of numbers, the usual concept of convergence cannot be used. In this section we will define three types of convergence used for a sequence of random variables.

CONVERGENCE IN DISTRIBUTIONS

We say X_n *converges in distribution to* X or write $X_n \xrightarrow{D} X$ if and only if the distribution functions $F_n(x)$ of X_n, $n = 1, 2, 3, \ldots$ converge to the distribution function $F(x)$ of X *at almost all* x. (The phrase *at almost all* x means that $F_n(x)$ and $F(x)$ differ significantly on at most a countable number of points on the real line.) This type of convergence can best be illustrated by example.

EXAMPLE 7-1

Let X be such that $P(X = 0) = 1/2$ and $P(X = 1) = 1/2$. Let X_n, for any integer n, be defined by

$$X_n = 1+1/n-X.$$

Since there is a one to one correspondence between X_n and X, we find that the mass function for X_n is given by

$$P(X_n = 1+1/n) = 1/2 \quad \text{and} \quad P(X_n = 1/n) = 1/2.$$

The distribution functions of X_n and of X are plotted in Figure 7-2. As $n \to \infty$ we see that $F_n(x) \to F(x)$ at all x except at the points $x = 0$ and $x = 1$. Therefore, we have $X_n \overset{D}{\to} X$ or the sequence $X_1, X_2, X_3, \ldots, X_n, \ldots$ approaches X in distribution.

EXAMPLE 7-2

Let $X_1, X_2, X_3, \ldots$ be independent random variables with means $\mu_1, \mu_2, \mu_3, \ldots$, variances $\sigma_1^2, \sigma_2^2, \sigma_3^2, \ldots$, and be such that their third moments are finite and their moment generating functions exist. Let

$$Z_n = \frac{1}{\sqrt{n}} \sum_{i=1}^{n} \frac{X_i - \mu_i}{\sigma_i}.$$

Consider the sequence of random variables $Z_1, Z_2, Z_3, \ldots$. The central limit theorem says that $Z_n \overset{D}{\to} Z$ where Z has a standard normal distribution.

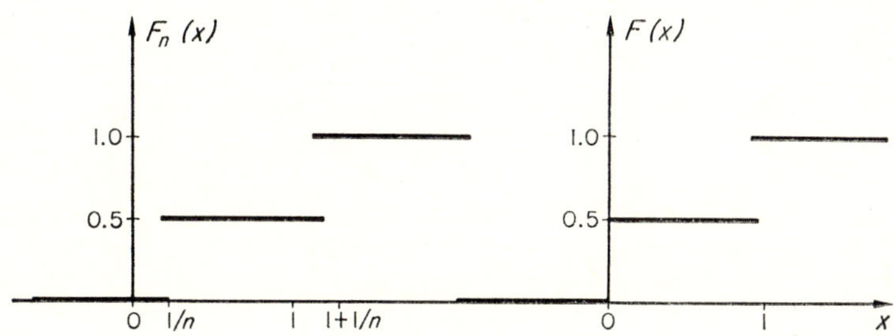

Figure 7-2

CONVERGENCE IN PROBABILITY

We say that X_n *converges in probability* to X or write $X_n \overset{P}{\to} X$ if and only if for any $\varepsilon > 0$

$$\lim_{n \to \infty} P(|X_n - X| > \varepsilon) = 0.$$

This says that as n approaches infinity the value of X_n will be arbitrarily close to the value of X, except, perhaps, at some x values whose total probability approaches zero. This too can best be illustrated by example.

EXAMPLE 7-3

Consider the half open unit interval (0, 1]. We will construct a sequence of subintervals in the following manner. First segment the unit interval into the ten half open intervals (0, .1], (.1, .2], (.2, .3], ..., (.9, 1]. Let the first ten subintervals in the sequence be these ten i.e.,

$$I_1 = (0, .1]$$
$$I_2 = (.1, .2]$$
$$\vdots$$
$$I_{10} = (.9, 1].$$

Now segment the unit interval into the one hundred half open intervals (0, .01], (.01, .02], ..., (.99, 1]. The intervals I_{11} through I_{111} in the sequence are now these half open intervals

$$I_{11} = (0, .01]$$
$$\vdots$$
$$I_{111} = (.99, 1].$$

Now segment the unit interval into one thousand half open intervals in a similar manner and these will represent the next thousand intervals in the sequence. Imagine doing this ad infinitum. Having done this we now begin the example.

Let X be a random variable uniformly distributed on the unit interval (0, 1]. Let $X_n = X + \delta(I_n)$ where $\delta(I_n) = 1$ if $x \in I_n$ and $\delta(I_n) = 0$ if $x \notin I_n$. Clearly, $|X_n - X| = 0$ except for $x \in I_n$. Therefore, for any $\in > 0$

$$P(|X_n - X| > \in) = P(x \in I_n)$$

and since X has a uniform distribution on a unit interval

$$= \text{length of } I_n.$$

Now as $n \to \infty$ the length of $I_n \to 0$ and, therefore, for any $\in > 0$

$$\lim_{n\to\infty} P(|X_n - X| > \in) = 0$$

which says that the sequence $X_1, X_2, X_3, \ldots, X_n, \ldots$ converges to X in probability, or $X_n \overset{P}{\to} X$.

Example 7-4

Let X be any random variable and let $X_1, X_2, X_3, \ldots$ be such that $X_n = X + (1/n)$. Then

$$P(|X_n - X| > \in) = P\left(\left|\frac{1}{n}\right| > \in\right).$$

Clearly for any $\in > 0$

$$\lim_{n\to\infty} P(|X_n - X| > \in) = 0$$

and

$$X_n \overset{P}{\to} X.$$

Example 7-5

Let X be a degenerate random variable with mass function given by

$$P(X = \mu) = 1.$$

Let $X_1, X_2, X_3, \ldots$ be independent identically distributed random variables with common mean μ and common variance σ^2. Let

$$Z_n = \frac{1}{n} \sum_{i=1}^{n} X_i.$$

By the law of large numbers for any $\in > 0$

$$\lim_{n\to\infty} P(|Z_n - \mu| > \in) = 0.$$

But

$$P(|Z_n - X| > \epsilon) = P(|Z_n - \mu| > \epsilon)$$

and, therefore,

$$\lim_{n\to\infty} P(|Z_n - X| > \epsilon) = 0$$

or

$$Z_n \xrightarrow{P} X.$$

For this degenerate case, we say

$$Z_n \xrightarrow{P} \mu.$$

It will pay to reconsider Example 7-1 and check to see whether the sequence given there converges in probability.

EXAMPLE 7-6

Consider the random variable X and the sequence of random variables $X_1, X_2, X_3, \ldots, X_n, \ldots$ described in Example 7-1. For these random variables $|X_n - X|$ either equals $1+1/n$ or $1-1/n$ each with probability 1/2. It, therefore, is no smaller than $1-1/\mathrm{n}$ with probability one. Thus, for any $\epsilon < 1-1/n$ or any $n > 1/(1-\epsilon)$ we have $P(|X_n - X| > \epsilon) = 1$. This implies that for any $\epsilon > 0$ $\lim P(|X_n - X| > \epsilon) = 1 \neq 0$, or that X_n does not converge in probability to X.

CONVERGENCE ALMOST SURELY

The type of convergence of a sequence of random variables $X_1, X_2, X_3, \ldots, X_n, \ldots$ to a random variable X, known as almost sure convergence, is even stronger than convergence in probability. We say that X_n *converges almost surely* (a.s.) to X or write $X_n \to X$ a.s., if for any

$$\varepsilon > 0 \lim_{n\to\infty} P(\bigcup_{k \geqq n} \{|X_k - X| > \varepsilon\}) = 0.$$

The set

$$\bigcup_{k \geqq n} \{|X_k - X| > \varepsilon\}$$

is the set of all values x where X_k and X differ by more than ε for any $k > n$. This too can best be seen by example.

EXAMPLE 7-7

Let

$$\begin{aligned} \delta_n(x) &= 1 \quad \text{if} \quad 0 < x < 1/n \\ &= 0 \qquad \text{otherwise.} \end{aligned}$$

Let X be a uniform random variable on the unit interval, and let $X_1, X_2, X_3, \ldots$ be such that

$$X_n = X + \delta_n(X).$$

Here

$$\bigcup_{k \geq n} \{|X_n - X| > \epsilon\} = \{0 < X < 1/n\}$$

and for any $\epsilon > 0$

$$P[\bigcup_{k \geq n} \{|X_n - X| > \epsilon\}] = P(0 < X < 1/n)$$
$$= 1/n.$$

Therefore,

$$\lim_{n\to\infty} P[\bigcup_{k \geq n} \{|X_n - X| > \epsilon\}] = 0$$

and

$$X_n \to X \quad \text{a.s.}$$

EXERCISE 7-5

Demonstrate that for the situation in Example 7-3 X_n does not converge to X a.s.

It can easily be seen that convergence almost surely is stronger than convergence in probability, since for any

$$\varepsilon > 0 \, \{| X_n - X | > \varepsilon\} \subset \bigcup_{k \geqq n} \{| X_k - X | > \varepsilon\},$$

from which we obtain that

$$P(| X_n - X | > \varepsilon) \leqq P(\bigcup_{k \geqq n} \{| X_k - X | > \varepsilon\}).$$

This, of course, implies that if

$$\lim_{n\to\infty} P(\bigcup_{k \geqq n} \{| X_k - X | > \varepsilon\}) = 0, \quad \text{then} \quad \lim_{n\to\infty} P(| X_n - X | > \varepsilon) = 0,$$

or

$$X_n \to X \quad \text{a.s.} \quad \Rightarrow X_n \overset{P}{\to} X.$$

7-5 Problems

1. Let X be a mixed random variable such that

$$f(x) = \frac{e^{-x/\lambda}}{2\lambda} \qquad 0 < x < \infty \text{ omitting the integers}$$
$$= 0 \qquad \text{otherwise}$$

and

$$P(x) = \frac{e^{-\lambda}\lambda^x}{2(x!)} \qquad x = 0, 1, 2, \ldots.$$

Find EX and Var X.

2. Let X be a mixed random variable such that

$$f(x) = \frac{r}{\Gamma(\alpha)\beta^\alpha} x^{\alpha-1} e^{-x/\beta} \qquad \begin{array}{l} 0 < x < \infty \\ \text{omitting the integers} \\ 1, 2, 3, \ldots, n \\ 0 < \alpha, \beta \\ 0 < r < 1 \end{array}$$
$$= 0 \qquad \text{otherwise}$$

and

$$P(x) = \binom{n}{x} p^x(1-p)^{n-x}(1-r) \qquad \begin{array}{l} x = 0, 1, 2, \ldots, n \\ n > 0 \\ 0 < p < 1. \end{array}$$

Find EX and Var X.

3. Let X be a mixed random variable such that

$$f(x) = \frac{p}{\sqrt{(2\pi)}\sigma} e^{-(x-\mu)^2/2\sigma^2} \qquad \begin{array}{l} -\infty < x < \infty \\ \text{omitting the non-negative integers} \\ -\infty < \mu < \infty \\ 0 < \sigma \\ 0 < p < 1 \end{array}$$

$$= 0 \qquad \text{otherwise}$$

and

$$P(x) = (1-p)\frac{e^{-\lambda}\lambda^x}{x!} \qquad x = 0, 1, 2, \ldots.$$

Find EX and Var X.

4. Let X be a mixed random variable such that

$$f(x) = s \qquad \begin{array}{l} 0 < x < 1 \\ 0 < s < 1 \end{array}$$

$$= 0 \qquad \text{otherwise}$$

and

$$P(x) = \binom{x+r-1}{r-1} p^r(1-p)^x(1-s) \qquad \begin{array}{l} X = 0, 1, 2, \ldots. \\ r \text{ an integer} \\ 0 < p < 1. \end{array}$$

Find EX and Var X.

5. Find the moment generating function for the random variable of Problem 1.

6. Find the moment generating function for the random variable of Problem 2.

7. Find the moment generating function for the random variable of Problem 3.

8. Find the moment generating function for the random variable of Problem 4.

9. Let X and Y both be mixed random variables such that

$$f(x, y) = \frac{r}{2\pi\sigma_x\sigma_y\sqrt{(1-\rho^2)}} \exp\left\{-\frac{1}{2(1-\rho^2)}\left[\left(\frac{x-\mu_x}{\sigma_x}\right)^2 - 2\rho\left(\frac{x-\mu_x}{\sigma_x}\right)\left(\frac{y-\mu_y}{\sigma_y}\right) + \left(\frac{y-\mu_y}{\sigma_y}\right)^2\right]\right\}$$

$$\begin{array}{l} -\infty < x, y < \infty \\ \text{omitting the integers} \\ 0, 1, 2, \ldots, n \\ -\infty < \mu_x, \mu_y < \infty \\ 0 < \sigma_x, \sigma_y \qquad -1 < \rho < 1 \\ 0 < r < 1 \end{array}$$

$$= 0 \qquad \text{otherwise}$$

and

$$P(x, y) = \binom{n}{x}\binom{n-x}{y} p^x q^y (1-p-q)^{n-x-y}(1-r) \qquad \begin{array}{l} x, y \text{ integers} \\ 0 \leqq x+y \leqq n \\ n \text{ an integer} \\ 0 < p, q \\ 0 < p+q < 1. \end{array}$$

Find EX and Var X.

10. For the random variables X and Y of Problem 9 find EY and Var Y.
11. For the random variables of Problem 9 find $E(X \mid y)$ and Var $(X \mid y)$.
12. For the random variables of Problem 9 find $E(Y \mid x)$ and Var $(Y \mid x)$.
13. For the random variables of Problem 9 find Cov (X, Y).
14. Let X and Y be such that

$$f(x, y) = P_3(x, y) = P(x, y) = 0$$

and

$$P_2(x, y) = \frac{\lambda^y}{\sqrt{(2\pi)}\sigma y!} e^{-\lambda-(x-\mu)^2/2\sigma^2} \qquad \begin{array}{l} -\infty < x < \infty \\ y = 0, 1, 2, \ldots \\ -\infty < \mu < \infty \\ 0 < \sigma, \lambda. \end{array}$$

Find EX, Var X, EY and Var Y.

15. For the random variables of Problem 14 find $E(X \mid y)$, Var $(X \mid y)$, $E(Y \mid x)$ and Var $(Y \mid x)$. Could you have predicted these results?
16. For the random variables of Problem 14 find Cov (X, Y). Could you have predicted this?
17. Assume that X has a normal distribution with mean zero and variance σ^2. Give an upper bound for $P(|X| > 3\sigma)$ first by using Tchebychev's inequality and then by using the standard normal tables and compare the results. Can you explain the difference?
18. Using the central limit theorem find an approximation for the distribution of X/n where X is a binomial random variable with probability p for success. *Hint*: X can be considered to be the sum of bernoulli random variables.
19. Let X be a random variable with mass function

$$P(x) = \frac{e^{-\lambda}\lambda^x}{x!} \qquad \begin{array}{l} x = 0, 1, 2, \ldots \\ \lambda > 0. \end{array}$$

Let $X_1, X_2, X_3, \ldots, X_n, \ldots$ be a sequence of random variables such that $X_n = X+1/n$. Show that $X_n \overset{D}{\rightarrow} X$.

20. For the random variables of Problem 19 show that $X_n \overset{P}{\rightarrow} X$.
21. For the random variables of Problem 19 show that $X_n \rightarrow X$ a.s.
22. Let X be a random variable with any mass function

$$P(x) \qquad x = 0, 1, 2, 3, \ldots.$$

If $X_1, X_2, X_3, \ldots, X_n \ldots$ are such that $X_n = X+1/n$ show that $X_n \overset{D}{\rightarrow} X$.

23. For the random variables of Problem 22 show that $X_n \overset{P}{\rightarrow} X$.
24. For the random variables of Problem 22 show that $X_n \rightarrow X$ a.s.

CHAPTER 8

An Introduction to Statistical Inference

Statistical inference takes many forms and attempts to answer many different kinds of questions. The only feature common to all statistical inference is the following. We observe the values of some random variables, and on the basis of these observed values, we are to make some statement or take some action. Without observed values, there can be no statistical inference, only probability theory. Confusion concerning the borders between the two fields, statistical inference and probability theory, is often unavoidable, but can be minimized by using the criterion that statistical inference pertains to inferences made as a result of observations.

The inferences that can be made are varied in type. We can make inferences about probabilities, about parameters in the distribution, about averages or expectations, or about the "best" action to take, such as the decision to allow a particular contractor to build our home.

8-1 A Simple Example of Statistical Inferences Concerning Probabilities

If we knew the probability distribution completely, we could ask for, and find, $P(a < X < b)$ for any values a and b. If we have only incomplete information concerning the distribution, then it is not generally possible to find this probability. When the form of the distribution is known, except for the value of one or more parameters, we say that we are in a *parametric* situation. For example, we may know that the random variable has a normal distribution, but we do not know its mean or variance.

When the form of the distribution is completely unknown, then we say that we are in a nonparametric situation. At first, this latter situation seems hopeless, but we should keep in mind that we will have observed values to aid us in our inferences. For example, if we wanted the probability that $X < -10$ and observed 1,000 independent values of X all of which were over 100, we might feel (and justly so) that $P(X < -10)$ was rather small.

We will now consider a parametric example.

A PARAMETRIC EXAMPLE

Let us assume that we know that the random variable X has a probability density function

$$f(x) = \frac{1}{\beta - \alpha} \qquad \alpha < x < \beta \text{ (with } \alpha, \beta \text{ fixed but unknown)}$$
$$= 0 \qquad \text{otherwise.}$$

We want to say something about $P(a < X < b)$ for some pair of values a, b. But all we can say without any further information is that $0 \leqq P(a < X < b) \leqq 1$, which is not very informative.

Let us assume that several values of X have been observed. By this we mean that we have observed the value x_1 of the random variable X_1, which has the distribution given for X. Then we observe the value x_2 of the random variable X_2, which is independent of X_1 but distributed identically. This continues until we have the observed values $x_1, x_2, x_3, \ldots, x_n$ from the *independent identically distributed* random variables $X_1, X_2, X_3, \ldots, X_n$. To be more explicit, let us assume that these values were 5.2, 6.3, 4.7, 2.1, 4.6, 3.2. What can now be said concerning $P(a < X < b)$?

It should be clear that if $M = \text{Max}(x_1, x_2, \ldots, x_n)$ and $m = \text{Min}(x_1, x_2, \ldots, x_n)$, then $\alpha \leqq m$ and $\beta \geqq M$ or in our example, $\alpha \leqq m = 2.1$ and $\beta \geqq M = 6.3$. From this one can infer that

$$f(x) = \frac{1}{\beta - \alpha} \leqq \frac{1}{M - m} \qquad \text{for } m < x < M.$$

In our example

$$f(x) \leqq \frac{1}{4.2} = 0.238 \qquad \text{for any } 2.1 < x < 6.3.$$

If we want to find $P(a < X < b)$ where $m < a < b < M$, then it is clear that

$$P(a < X < b) \leqq \frac{b - a}{M - m}.$$

In the numerical example, if $a = 3$ and $b = 5$, we see that $P(3 < x < 5) \leqq 0.476$. We cannot give any lower limit other than zero even with our data.

What if $b > M$? There are only two possibilities: either $M \leqq \beta \leqq b$ or $\beta > b \geqq M$.

Case I Assume $M \leqq \beta \leqq b$.

It is not difficult to see that in this case

$$f(x) = \frac{1}{\beta - \alpha} \leqq \frac{1}{\beta - m} \qquad m < x < \beta.$$

Therefore, with $m < a$ but $b \geqq \beta$, we obtain

$$P(a < X < b) \leqq \frac{\beta-a}{\beta-m}.$$

This is an increasing function of β, and therefore

$$P(a < X < b) \leqq \frac{b-a}{b-m}.$$

In our numerical example if $a = 3$ and $b = 7$ we obtain

$$P(3 < X < 7) \leqq 4/4.9 = 0.816.$$

Case II Assume $\beta \geqq b \geqq M$.

In this case

$$f(x) = \frac{1}{\beta-\alpha} \leqq \frac{1}{b-m} \qquad m < x < b.$$

Therefore

$$P(a < X < b) \leqq \frac{b-a}{b-m}$$

which is the same as the result of Case I.

At this point we may observe that in any situation considered so far

$$P(a < X < b) \leqq \frac{b-a}{\text{Max}(b, M)-\text{Min}(a, m)}.$$

Exercise 8-1

Show that when $a < m$ we obtain

$$P(a < X < b) \leqq \frac{b-a}{\text{Max}(b, M)-\text{Min}(a, m)}.$$

Exercise 8-2

Show that even when $a < m < M < b$ we obtain

$$P(a < X < b) \leqq \frac{b-a}{\text{Max}(b, M)-\text{Min}(a, m)}.$$

Using our results, and those of the exercises above, we can conclude that if we know that X is such that

$$f(x) = \frac{1}{\beta-\alpha} \qquad \alpha < x < \beta$$

$$= 0 \qquad \text{otherwise}$$

and if $x_1, x_2, x_3, \ldots, x_n$ represents the values of independent random variables all with this distribution, then

$$0 \leqq P(a < X < b) \leqq \frac{b-a}{\text{Max}(b, M) - \text{Min}(a, m)}$$

where $M = \text{Max}(x_1, x_2, x_3, \ldots, x_n)$ and $m = \text{Min}(x_1, x_2, x_3, \ldots, x_n)$.

It should be pointed out that although these results require the values $x_1, x_2, x_3, \ldots, x_n$, these statements can be made before these values are obtained. In general we will derive inferences which depend on data, but we will derive them before we see any of the data.

It is interesting to note that in this example we did not say what we thought the values of α and β were. In this situation this was not required. In the usual parametric situation we would be interested in the values of the parameters involved.

8-2 The General Parametric Situation

The general parametric situation can be formulated as follows. There is a set of n random variables $X_1, X_2, X_3, \ldots, X_n$ which has a joint distribution function $F_\theta(x_1, x_2, x_3, \ldots, x_n)$. This distribution function is not completely known. However, it is known that F_θ is a member of the class of distribution functions given by $\{F_\theta(x_1, x_2, x_3, \ldots, x_n): \theta \in \Theta\}$ where Θ is a known set of possible values for θ and where F_θ for each given θ is a completely specified (*known*) distribution function.

Remark *We will treat θ as if it were a single real valued parameter. The problem can be extended to the case where θ is a sequence $\theta_1, \theta_2, \theta_3, \ldots, \theta_r$ of r parameters.*

EXAMPLE 8-1

Let $X_1, X_2, X_3, \ldots, X_n$ be independent, normally distributed random variables such that

$$f_\theta(x_1, x_2, x_3, \ldots, x_n) = \frac{1}{(2\pi)^{n/2}} \exp\left(-\sum_{i=1}^{n} (x_i - \theta)^2/2\right) \quad -\infty < \quad \text{all} \quad x_i < \infty;$$

with $-\infty < \theta < \infty$.

The distributions in the family are, therefore, the joint distributions of n independent and identically distributed normal random variables with unknown mean θ and known variance one; Θ, the possible values of θ, is the real line.

We might ask ourselves several different kinds of questions about the true value of the parameter θ. The questions we ask would depend on what the true value of θ actually represented. For example:

a θ might be the amount of a drug needed to cure a patient without causing aftereffects. In this case we would like to know the value of θ as accurately as possible.

b θ might be the proportion of defectives in a box of computer parts. In this case, we may have given a guarantee of, say, less than 1% defectives. If the customer finds more than 1%, he can sue us for the excess time spent in replacing the part and for the computer time lost. We would, therefore, be very much interested in knowing whether the true value of θ was less than 0.01.

c θ might be the distance to an artillery target. In this case we may want to have an interval of distances in which we can be confident θ lies so that we can cover this stretch with a barrage of artillery shells.

These three situations are examples of three types of questions which are often amenable to statistical procedures.

ESTIMATION

In the problem known generally as *estimation*, a guess at the true value of θ is desired. Of course, any guess is either correct or incorrect. In most situations we cannot hope to be correct. We, therefore, require that our guess (*or estimate*), be *as correct as possible*. What *as correct as possible* means will depend on the situation. In situation **a** we want to guess at (or estimate) θ. If an underdose will cause the patient to take a few minutes longer to recover, whereas an overdose will cause serious aftereffects, we would prefer that our estimate be an underestimate rather than an overestimate. If, on the other hand, an overdose caused little aftereffect, whereas an underdose would not cure the patient, then we would prefer to overestimate θ rather than underestimate it. If a small underdose was not more serious than a small overdose, then we would simply like to be as close as possible.

Our guess at the true value of θ will depend on the observed values of the random variables $X_1, X_2, X_3, \ldots, X_n$. This collection of observed values $x_1, x_2, x_3, \ldots, x_n$ is called the *sample* which may, for example, be the numbers of white blood cells found in drops of blood. As indicated previously, we can often formulate our guess in terms of a function of $X_1, X_2, X_3, \ldots, X_n$. This function is called the *estimator* for θ and the corresponding value obtained when the observed values $x_1, x_2, x_3, \ldots, x_n$ are substituted for $X_1, X_2, X_3, \ldots, X_n$ is called the *estimate* of θ.

Definition If $X_1, X_2, X_3, \ldots, X_n$ have a joint distribution function $F_\theta(x_1, x_2, x_3, \ldots, x_n)$ which is a member of the family of distributions $\{F_\theta(x_1, x_2, x_3, \ldots, x_n): \theta \in \Theta\}$ then any function $\hat{\theta}(X_1, X_2, X_3, \ldots, X_n)$

which will take on a numerical value when $X_1, X_2, X_3, \ldots, X_n$ are replaced by their observed values $x_1, x_2, x_3, \ldots, x_n$ is an *estimator* for θ.

Definition The value of an estimator for θ when observed values $x_1, x_2, x_3, \ldots, x_n$ replace $X_1, X_2, X_3, \ldots, X_n$ is an *estimate* of θ.

From the rather general definition it should be clear that there are very many estimators for θ. Which estimator to use will depend on what mathematical criterion we use to define *as correct as possible.* As already indicated, this must depend on the situation at hand. For situation **a** one might use some fraction of the total number of white blood cells.

Estimation will be reconsidered in more detail in Chapters Ten and Eleven.

TESTS OF HYPOTHESES

In situation **b** we are interested in deciding whether or not θ is in a particular subset, say $(0 \leqq \theta < .01)$, of $\Theta = (0 \leqq \theta \leqq 1)$. We do not need to estimate θ here, but only decide whether to accept or reject the hypothesis that it is in this subset. To decide, we use what is called a *test function,* which will now be defined.

Definition If $X_1, X_2, X_3, \ldots, X_n$ have a joint distribution function $F_\theta(x_1, x_2, x_3, \ldots, x_n)$ which is a member of the family of distributions $\{F_\theta(x_1, x_2, x_3, \ldots, x_n): \theta \in \Theta\}$, then any function $\delta(X_1, X_2, X_3, \ldots, X_n)$ which takes on the word values "*reject*" or "*accept*" when $X_1, X_2, X_3, \ldots, X_n$ are replaced by their observed values $x_1, x_2, x_3, \ldots, x_n$ is a *test function* for the hypothesis that θ is in a subset of Θ.

We will also have occasion to use what is called a randomized test function, which we now define.

Definition If $X_1, X_2, X_3, \ldots, X_n$ have joint distribution function $F_\theta(x_1, x_2, x_3, \ldots, x_n)$ which is a member of the family of distributions $\{F_\theta(x_1, x_2, x_3, \ldots, x_n): \theta \in \Theta\}$, and Z is a bernoulli random variable then any function $\delta(X_1, X_2, X_3, \ldots, X_n, Z)$ which takes on the word values "*reject*" or "*accept*" when $X_1, X_2, X_3, \ldots, X_n, Z$ are replaced by their observed values is a *randomized test function* for the hypothesis that θ is in a subset of Θ.

We occasionally have use for randomized test functions which depend on more than one bernoulli random variable. However, it is rare that more than two are required.

Definition The value "*accept*" or "*reject*" of a test function which occurs when observed values $x_1, x_2, x_3, \ldots, x_n$ replace $X_1, X_2, X_3, \ldots, X_n$ is a *test result* for the hypothesis that $\theta \in \Theta_0 \subset \Theta$.

The hypothesis that $\theta \in \Theta_0$ is usually called the null hypothesis with its alternative $\theta \in \Theta_1 = \Theta - \Theta_0$ called the alternative hypothesis. This situation is usually written

$$H_0\colon \theta \in \Theta_0$$

vs.

$$H_1\colon \theta \in \Theta_1.$$

When the subset of a hypothesis contains but one element, it is said to be a *simple* hypothesis. When it contains more than one element, it is said to be a *composite* hypothesis. We will be considering hypothesis testing situations of various kinds in Chapter Twelve.

Example 8-2

Assume that $X_1, X_2, X_3, \ldots, X_n$ are independent and identically distributed random variables with a normal distribution with unknown mean θ and known variance σ^2. It is known only that $-\infty < \theta < \infty$ and it is hypothesized that θ is negative, i.e. $(-\infty < \theta < 0)$.

One possible test function is

$$\begin{aligned}\delta(X_1, X_2, X_3, \ldots, X_n) &= \textit{accept} \quad \text{if} \quad \sum_{i=1}^{n} X_i < 0\\ &= \textit{reject} \quad \text{if} \quad \sum_{i=1}^{n} X_i \geqq 0.\end{aligned}$$

When $x_1, x_2, x_3, \ldots, x_n$ are observed we can check whether or not $\sum_{i=1}^{n} x_i < 0$. If it is, accept the hypothesis, if not, reject it.

Example 8-3

For situation **b** we might sample n computer parts, setting $X = 1$ if the part is defective, and $X = 0$ if not. One possible test function would be given by

$$\begin{aligned}\delta &= \textit{reject} \quad \text{if} \quad \sum_{i=1}^{n} X_i > n/100\\ &= \textit{accept} \quad \text{if} \quad \sum_{i=1}^{n} X_i < n/100\\ &= \textit{reject} \text{ with probability } 1/2 \quad \text{if} \quad \sum_{i=1}^{n} X_i = n/100.\end{aligned}$$

That is, reject the lot if there are more than 1% defectives found, accept it if there are less than 1% defectives found, and flip a coin if exactly 1% defectives are found.

INTERVAL ESTIMATION

In situation **c** we were interested in obtaining a subset of Θ, say Θ^*, which we feel actually contains the true value of θ. In example **c** as in most cases, Θ^* is an interval of values. In general, a subset of Θ which we assert contains the true value of θ is called a *confidence set* for θ.

Definition If $X_1, X_2, X_3, \ldots, X_n$ have a joint distribution function $F_\theta(x_1, x_2, x_3, \ldots, x_n)$ which is a member of the family of distributions $\{F_\theta(x_1, x_2, x_3, \ldots, x_n): \theta \in \Theta\}$, then any function $\Omega(X_1, X_2, X_3, \ldots, X_n)$ that becomes a subset of Θ when $X_1, X_2, X_3, \ldots, X_n$ are replaced by $x_1, x_2, x_3, \ldots, x_n$ is a *confidence set* for θ.

Definition The subset of Θ obtained when $X_1, X_2, X_3, \ldots, X_n$ in $\Omega(X_1, X_2, X_3, \ldots, X_n)$ is replaced by their observed values $x_1, x_2, x_3, \ldots, x_n$ is called a *numerical confidence set* for θ.

The most often used confidence sets are in the form of intervals. In this case there are two functions $\Omega_L(X_1, X_2, X_3, \ldots, X_n)$ and $\Omega_U(X_1, X_2, X_3, \ldots, X_n)$ such that the asserted subset of Θ is $\Omega_L(X_1, X_2, X_3, \ldots, X_n) \leqq \theta \leqq \Omega_U(X_1, X_2, X_3, \ldots, X_n)$ with a corresponding pair of numerical functions. These are called the lower and upper *confidence bounds* for θ respectively while the double inequality is called a *confidence interval* for θ. The inequalities may be strict, in which case the inequality is called an *open confidence interval* for θ.

As in the previous situations, a mathematical criterion must be decided upon for choosing a confidence set which is *as correct as possible.* This will be discussed in Chapter thirteen.

EXAMPLE 8-4

In situation **c** consider the random variables $X_1, X_2, X_3, \ldots, X_n$ to be the enemy positions as reported by n different observers. Then we might use

$$\Omega_U(X_1, X_2, X_3, \ldots, X_n) = \text{Max}(X_1, X_2, X_3, \ldots, X_n)$$

and

$$\Omega_L(X_1, X_2, X_3, \ldots, X_n) = \text{Min}(X_1, X_2, X_3, \ldots, X_n).$$

8-3 Problems

1. If 2.1, 3.6, 1.4, and 3.3 are independent observations taken from a common uniform distribution given by the density function

$$f(x) = \frac{1}{\beta - \alpha} \qquad \begin{array}{l} \alpha < x < \beta \\ \alpha, \beta \text{ unknown} \end{array}$$

give an upper bound for

a. $P(2 < X < 3)$

b. $P(1 < X < 3)$

c. $P(2 < X < 4)$

d. $P(1 < X < 4)$

e. $P(X < 4)$

f. $P(1 < X)$.

2. Using the observations in Problem 1 give a lower bound for the value of β.

3. Using the observations in Problem 1 give an upper bound for the value of α.

4. Assume $\{F_\theta(x)\colon \theta \in \Theta\}$ is such that $F_\theta(x)$ corresponds to

$$f_\theta(x) = \frac{1}{\sqrt{(2\pi)}} e^{-(x-\theta)^2/2} \qquad -\infty < x < \infty$$

with

$$\Theta = (-\infty, \infty).$$

Which of the following are estimators for θ?

a. $\hat{\theta}(X) = X$

b. $\hat{\theta}(X) = 2$

c. $\hat{\theta}(X) = X+\theta$

d. $\hat{\theta}(X) = EX$.

5. Assume $\{F_\theta(x_1, x_2, x_3)\colon \theta \in_\theta\}$ is such that $F_\theta(x_1, x_2, x_3)$ corresponds to the joint density

$$f_\theta(x_1, x_2, x_3) = \theta^3 e^{-\theta \sum_{i=1}^{3} x_i} \qquad 0 < x_1, x_2, x_3, < \infty$$

with $\Theta = (0, \infty)$. Which of the following are estimators for θ?

a. $\hat{\theta}(X_1, X_2, X_3) = \sum_{i=1}^{3} X_i$

b. $\hat{\theta}(X_1, X_2, X_3) = 2$

c. $\hat{\theta}(X_1, X_2, X_3) = \sum_{i=1}^{3} X_i + \theta$

d. $\hat{\theta}(X_1, X_2, X_3) = \sum_{i=1}^{3} EX_i$.

6. If $\{F_\theta(x)\colon \theta \in \Theta\}$ is as indicated in Problem 4, which of the following are test functions of the hypothesis that $\theta \leqq 0$?

a. *Reject* if $X \leqq 0$, *accept* otherwise.

b. *Reject* if $X \leqq \theta$, *accept* otherwise.

c. *Reject*.

d. *Reject* if $EX \leqq 0$, *accept* otherwise.

7. If $\{F_\theta(x_1, x_2, x_3)\colon \theta \in \Theta\}$ is as indicated in Problem 5 which of the following are test functions of the hypothesis that $\theta \leqq 0$?

a. *Reject* if $\sum_{i=1}^{3} X_i \leqq 10$; *accept* otherwise.

b. *Reject* if $X_1 \leqq X_2 + X_3$; *accept* otherwise.

c. *Reject* if $\sum_{i=1}^{3} X_i \leqq \theta$; *accept* otherwise.

d. *Reject* if $\sum_{i=1}^{3} EX_i \leqq 10$; *accept* otherwise.

8. If $\{F_\theta(x)\colon \theta \in \Theta\}$ is as indicated in Problem 4, which of the following are confidence intervals for θ?
   a. $X-2 \leqq \theta \leqq X+2$.
   b. $-\infty \leqq \theta \leqq X$.
   c. $X \leqq \theta \leqq +\infty$.
   d. $-2 \leqq \theta \leqq +2$.
   e. $\dfrac{X-\theta}{2} \leqq \theta \leqq \dfrac{X+\theta}{2}$.

9. If $\{F_\theta(x_1, x_2, x_3)\colon \theta \in \Theta\}$ is as indicated in Problem 5 which of the following are confidence intervals for θ?
   a. $-\sum_{i=1}^{3} X_i \leqq \theta \leqq +\sum_{i=1}^{3} X_i$.
   b. $X_1 \leqq \theta \leqq X_2$.
   c. $-2 \leqq \theta \leqq X_3$.
   d. $-2 \leqq \theta \leqq +2$.
   e. $\dfrac{X_1-X_2}{X_3} \leqq \theta \leqq \dfrac{X_1+X_2}{X_3}$.
   f. $\dfrac{\theta-X_1}{X_2} \leqq \theta \leqq \dfrac{\theta+X_1}{X_2}$.

CHAPTER 9

Sample Moments

Before we continue with statistical inference, it will be worthwhile to consider the properties of what are called *statistics*. A *statistic* is any function of observable random variables whose value becomes known when observations replace the corresponding random variables. Estimators, test functions, and confidence bounds are all statistics. A great number of statistics encountered in statistical inference involve what is known as the *sample moments*. For this reason we will devote this chapter to the properties of sample moments.

9-1 Sample Moments in General

POPULATION MOMENTS

In Section 2-3 the *population moments* μ_r', $r = 1, 2, 3, \ldots$ were defined by

$$\mu_r' = EX^r.$$

If there is a sequence of random variables $X_1, X_2, X_3, \ldots, X_n$ we will call the rth population moment of the ith random variable $\mu_{i,r}'$. Then

$$\mu_{i,r}' = EX_{ir}.$$

In the same section the *population central moments* were defined by

$$\mu_r = E(X-\mu)^r \qquad r = 1, 2, 3, \ldots.$$

we can now define the rth central moment for the ith random variable by

$$\mu_{i,r} = E(X_i - \mu_{i,1}')^r.$$

When the random variables are identically distributed we will drop the subscript i and use μ_r' and μ_r.

SAMPLE MOMENTS

Let us assume that there is a sequence of n random variables $X_1, X_2, X_3, \ldots, X_n$. The *first sample moment*, usually called the *average*, is defined by

$$\overline{X}_n = \frac{1}{n}\sum_{i=1}^{n} X_i.$$

Corresponding to this statistic is its numerical value $\bar{x}_n$ which is defined by

$$\bar{x}_n = \frac{1}{n} \sum_{i=1}^{n} x_i$$

where x_i represents the observed value of X_i. The *rth sample moment* for any r is defined by

$$\overline{X_n^r} = \frac{1}{n} \sum_{i=1}^{n} X_i^r.$$

This too has a numerical counterpart given by

$$\overline{x_n^r} = \frac{1}{n} \sum_{i=1}^{n} x_i^r.$$

PROPERTIES OF SAMPLE MOMENTS

Since

$$E \sum_{i=1}^{n} g(X_i) = \sum_{i=1}^{n} Eg(X_i)$$

it is clear that

$$E\overline{X_n^r} = \frac{1}{n} \sum_{i=1}^{n} EX_i^r = \frac{1}{n} \sum_{i=1}^{n} \mu_{i,r}'.$$

In the special case where the X's are identically distributed, we obtain

$$E\overline{X_n^r} = \frac{1}{n} \sum_{i=1}^{n} \mu_r' = \mu_r'.$$

This means that for identically distributed variables the expectation of a sample moment is the corresponding population moment.

It is interesting to note that

$$\operatorname{Var} \overline{X_n^r} = \operatorname{Var}\left[\frac{1}{n} \sum_{i=1}^{n} X_i^r\right] = \frac{1}{n^2} \operatorname{Var}\left[\sum_{i=1}^{n} X_i^r\right].$$

If now the random variables are assumed independent, then

$$\operatorname{Var} \overline{X_n^r} = \frac{1}{n^2} \sum_{i=1}^{n} \operatorname{Var} X_i^r$$

and if they are also identically distributed, then

$$\operatorname{Var} \overline{X_n^r} = \frac{1}{n} \operatorname{Var} X^r$$

where X is used to denote any of the random variables.

In particular, with $r = 1$, we have for the average of n independent and identically distributed random variables that

$$\operatorname{Var} \overline{X}_n = \sigma^2/n$$

where σ^2 is the common population variance.

SAMPLE CENTRAL MOMENTS

Let us define the *sample central moments* by

$$C_n^r = \frac{1}{n} \sum_{i=1}^{n} (X_i - \mu'_{i,1})^r \qquad r = 1, 2, 3, \ldots$$

Note that unless the first population moments are known these are not statistics and cannot be used in estimators, test functions, or confidence bounds. When the first population moments are known, however, these are statistics, and we now study their properties.

It is not the case (as one might expect) that $C_n^1 = 0$. It is easily seen, however, that

$$EC_n^1 = \frac{1}{n} \sum_{i=1}^{n} (EX_i - \mu'_{i,1}) = 0.$$

PROPERTIES OF SAMPLE CENTRAL MOMENTS

As the expectation of the rth central moment, we obtain

$$EC_n^r = \frac{1}{n} \sum_{i=1}^{n} E(X_i - \mu'_{i,1})^r = \frac{1}{n} \sum_{i=1}^{n} \mu_{i,r}.$$

If the X_i are identically distributed then

$$EC_n^r = \mu_r.$$

Now

$$\operatorname{Var} C_n^r = \frac{1}{n^2} \operatorname{Var} \sum_{i=1}^{n} (X_i - \mu'_{i,1})^r.$$

If the X_i are independently distributed, then

$$\operatorname{Var} C_n^r = \frac{1}{n^2} \sum_{i=1}^{n} \operatorname{Var}(X_i - \mu'_{i,1})^r.$$

If they are also identically distributed, then

$$\operatorname{Var} C_n^r = \frac{1}{n} \operatorname{Var}(X - \mu_1')^r.$$

In the special case where $r = 1$, we obtain that for independent identically distributed random variables

$$\operatorname{Var} C_n^{\,1} = \sigma^2/n.$$

EXERCISE 9-1

If $X_1, X_2, X_3, \ldots, X_n$ are independent and identically distributed random variables, find $\operatorname{Var} C_n^{\,2}$ in terms of the first four population moments.

SAMPLE MOMENTS ABOUT THE AVERAGE

Let us define the *rth sample moment about the average* by

$$M_n^{\,r} = \frac{1}{n}\sum_{i=1}^{n} (X_i - \overline{X}_n)^r \qquad r = 1, 2, 3, \ldots.$$

This is clearly a statistic. We will denote by $m_n^{\,r}$ the numerical value of this statistic. That is

$$m_n^{\,r} = \frac{1}{n}\sum_{i=1}^{n} (x_i - \bar{x}_n)^r.$$

In the special case where $r = 1$ we have

$$M_n^{\,1} = \frac{1}{n}\sum_{i=1}^{n} (X_i - \overline{X}_n) = \frac{1}{n}\sum_{i=1}^{n} X_i - \overline{X}_n = \overline{X}_n - \overline{X}_n = 0.$$

PROPERTIES OF SAMPLE MOMENTS ABOUT THE AVERAGE

Although expectations of sample moments about the average are generally much too tedious to obtain, we will find some properties of the most useful of these, namely $M_n^{\,2}$. Note that

$$M_n^{\,2} = \frac{1}{n}\sum_{i=1}^{n} (X_i - \overline{X}_n)^2 = \frac{1}{n}\sum_{i=1}^{n} [X_i^{\,2} - 2X_i\overline{X}_n + (\overline{X}_n)^2] = \frac{1}{n}\sum_{i=1}^{n} X_i^{\,2} - (\overline{X}_n)^2$$

and therefore

$$EM_n^{\,2} = \frac{1}{n} E\left[\sum_{i=1}^{n} X_i^{\,2}\right] - E[(\overline{X}_n)^2] = \frac{1}{n}\sum_{i=1}^{n} EX_i^{\,2} - \operatorname{Var} \overline{X}_n - (E\overline{X}_n)^2$$

$$= \frac{1}{n}\sum_{i=1}^{n} \mu'_{i,2} - \left(\frac{1}{n}\sum_{i=1}^{n} \mu'_{i,1}\right)^2 - \operatorname{Var} \overline{X}_n.$$

If the X_i are independent and identically distributed, then

$$EM_n^{\,2} = \mu_2' - \mu_1'^2 - \sigma^2/n = \sigma^2 - \sigma^2/n = \frac{n-1}{n}\sigma^2$$

where μ_2' and μ_1' are the second and first population moments and where σ^2 is the population variance.

Finding the variance of the second sample moment about the average is not quite so simple.

$$\operatorname{Var} M_n^2 = E[(M_n^2)^2] - (EM_n^2)^2.$$

The second term on the right can easily be obtained from the results above. We will concentrate here on the first term.

$$(M_n^2)^2 = \left(\frac{1}{n}\sum_{i=1}^{n}(X_i - \bar{X}_n)^2\right)^2 = \frac{1}{n^2}\left(\sum_{i=1}^{n} X_i^2 - n(\bar{X}_n)^2\right)^2$$

$$= \frac{1}{n^2}\left[\left(\sum_{i=1}^{n} X_i^2\right)^2 - 2n(\bar{X}_n)^2 \sum_{i=1}^{n} X_i^2 + n^2(\bar{X}_n)^4\right].$$

Therefore

$$E[(M_n^2)^2] = \frac{1}{n^2}\left\{E\left[\left(\sum_{i=1}^{n} X_i^2\right)^2\right] - 2nE\left[(\bar{X}_n)^2 \sum_{i} X_i^2\right] + n^2 E[(\bar{X}_n)^4]\right\}.$$

In general, this depends on the expectations of products of four random variables. Let us consider the special case where $X_1, X_2, X_3, \ldots, X_n$ are independent and identically distributed. In this case we may just as well assume that $\mu = 0$ since

$$M_n^2 = \frac{1}{n}\sum_{i=1}^{n}(X_i - \bar{X}_n)^2 = \frac{1}{n}\sum_{i=1}^{n}(X_i - \mu - \bar{X}_n + \mu)^2 = \frac{1}{n}\sum_{i=1}^{n}(Z_i - Z_n)^2$$

where $Z_i = X_i - \mu$ has mean zero, and population moments equal to the population central moments of X_i. Therefore, we can assume $\mu = 0$ and obtain results in terms of the central moments. Then we can readjust to the noncentral moments if desired.

We return now to $E[(M_n^2)^2]$, assuming that $X_1, X_2, X_3, \ldots, X_n$ are independent and identically distributed with mean zero and variance σ^2. We will reduce each term on the right-hand side of the equation separately, neglecting the coefficients, by expanding out and taking expectations. For the first term:

$$E\left(\sum_{i=1}^{n} X_i^2\right)^2 = E\sum_{i=1}^{n} X_i^2 \sum_{j=1}^{n} X_j^2 = E\left[\sum_{i=1}^{n} X_i^4 + \sum_{i \neq j}\sum X_i^2 X_j^2\right]$$

$$= \sum_{i=1}^{n} EX_i^4 + \sum_{i \neq j}\sum EX_i^2 EX_j^2 = n\mu_4 + n(n-1)(\mu_2)^2$$

$$= n\mu_4 + n(n-1)\sigma^4.$$

For the second term:

$$E\left[(\bar{X}_n)^2 \sum_{i=1}^{n} X_i^2\right] = \frac{1}{n^2} E\left[\sum_{j=1}^{n} X_j \sum_{k=1}^{n} X_k \sum_{i=1}^{n} X_i^2\right]$$

$$= \frac{1}{n^2} E\left[\sum_{i=1}^{n} X_i^4 + \sum_{i \neq j}\sum X_i^2 X_j^2 + \sum_{j \neq k}\sum X_j X_k \sum_{\substack{i \neq j \\ i \neq k}} X_i^2\right]$$

$$= \frac{1}{n^2}\left[\sum_{i=1}^{n} EX_i^4 + \sum_{i \neq j}^{n}\sum^{n} EX_i^2 EX_j^2 \right.$$

$$\left. + \sum_{j \neq k}^{n}\sum^{n} EX_j EX_k \sum_{\substack{i \neq j \\ i \neq k}} EX_i^2\right]$$

$$= \frac{1}{n^2}\,[n\mu_4 + n(n-1)(\mu_2{}^2) + 0] = \frac{1}{n^2}\,[n\mu_4 + n(n-1)\sigma^4].$$

For the last term:

$$E(\bar{X}_n)^4 = \frac{1}{n^4} E\left[\sum_{i=1}^{n} X_i \sum_{j=1}^{n} X_j \sum_{k=1}^{n} X_k \sum_{l=1}^{n} X_l\right]$$

$$= \frac{1}{n^4} E\left[\sum_{i=1}^{n} X_i^4 + \sum_{i \neq k}\sum X_i^2 X_k^2 \qquad (\text{where } i = j \neq k = l)\right.$$

$$+ \sum_{i \neq j}\sum X_i^2 X_j^2 \;(\text{where } i = k \neq j = l) + \sum_{i \neq j}\sum X_i^2 X_j^2 \quad (\text{where } i = l \neq j = k)$$

$$\left. + \text{ terms containing } X_i\right]$$

$$= \frac{1}{n^4}\left[\sum_{i=1}^{n} EX_i^4 + 3\sum_{i \neq j}\sum EX_i^2 EX_j^2 + \text{terms containing } EX_i\right]$$

$$= \frac{1}{n^4}\,[n\mu_4 + 3n(n-1)(\mu_2)^2 + 0] = \frac{1}{n^4}\,[n\mu_4 + 3n(n-1)\sigma^4].$$

Combining terms

$$E[(M_n{}^2)^2] = \frac{1}{n^2}\,\{n\mu_4 + n(n-1)\sigma^4 - \frac{2n}{n^2}\,[n\mu_4 + n(n-1)\sigma^4]$$

$$+ \frac{n^2}{n^4}\,[n\mu_4 + 3n(n-1)\sigma^4]\} = \frac{(n-1)^2}{n^3}\,\mu_4 + \frac{(n-1)(n^2-2n+3)}{n^3}\,\sigma^4.$$

Since

$$EM_n{}^2 = \frac{n-1}{n}\,\sigma^2$$

we have

$$(EM_n^{\ 2})^2 = \frac{(n-1)^2}{n^2}\,\sigma^4$$

giving

$$\text{Var } M_n^{\ 2} = \frac{(n-1)^2}{n^3}\,\mu_4 + \frac{(n-1)(n^2-2n+3)}{n^3}\,\sigma^4 - \frac{(n-1)^2}{n^2}\,\sigma^4$$

$$= \frac{(n-1)^2}{n^3}\,\mu_4 - \frac{(n-1)(n-3)}{n^3}\,\sigma^4.$$

SAMPLE VARIANCE

As we have already seen, for independent and identically distributed random variables

$$EM_n^{\ 2} = \frac{n-1}{n}\,\sigma^2.$$

It is often useful to consider a statistic whose expected value is equal to the population variance. For this reason one defines the *sample variance* as

$$S_n^{\ 2} = \frac{n}{n-1}\,M_n^{\ 2} = \frac{1}{n-1}\sum_{i=1}^{n}(X_i - \bar{X}_n)^2.$$

For independent and identically distributed random variables we find, from the general properties of expectations,

$$ES_n^{\ 2} = \frac{n}{n-1}\,EM_n^{\ 2} = \frac{n}{n-1}\cdot\frac{n-1}{n}\,\sigma^2 = \sigma^2$$

and

$$\text{Var } S_n^{\ 2} = \frac{n^2}{(n-1)^2}\,\text{Var } M_n^{\ 2} = \frac{\mu_4}{n} - \frac{(n-3)}{n(n-1)}\,\sigma^4.$$

The square root of $S_n^{\ 2}$ is called the *sample standard deviation*

$$S_n = \sqrt{(S_n^{\ 2})} = \sqrt{\left(\frac{1}{n-1}\sum_{i=1}^{n}(X_i - \bar{X}_n)^2\right)}$$

We will not find the properties of this statistic but will point out that in general $ES_n \neq \sigma$.

This can easily be seen since $\text{Var } S_n = ES_n^{\ 2} - (ES_n)^2$. If $ES_n^{\ 2} = \sigma^2$ and $ES_n = \sigma$, then $\text{Var } S_n = 0$. This could not be the case unless each X_i takes a single value with probability one. However, since we have assumed identically distributed random variables, this would imply that $S_n = 0$ and therefore $ES_n = ES_n^{\ 2} = 0$. Thus, if and only if $\sigma = 0$ could $ES_n = \sigma$.

9-2 Distributions of the Average

In the last section we found the mean and variance of an average of n random variables in terms of the moments of the parent distributions. In this section we will use the method of moment generating functions to find the distribution of the average in some interesting cases.

THE METHOD OF MOMENT GENERATING FUNCTIONS

The moment generating function of an average of n random variables can be written

$$E\,e^{t\bar{X}_n} = E\exp\left(\frac{t}{n}\sum_{i=1}^{n} X_i\right) = E\left(\prod_{i=1}^{n} e^{(t/n)X_i}\right)$$

If the random variables are mutually independent then

$$E\,e^{t\bar{X}_n} = \prod_{i=1}^{n} E\,e^{(t/n)X_i}.$$

If they are also identically distributed then

$$E\,e^{t\bar{X}_n} = [E\,e^{(t/n)X}]^n$$

where X represents any one of the random variables. Thus, we obtain the following rules.

a To obtain the moment generating function of an average of n independent random variables: replace t by t/n in the moment generating function of each random variable and then multiply the resulting n functions.

b To obtain the moment generating function of an average of n independent and identically distributed random variables: replace t by t/n in their common moment generating function and raise the resulting function to the nth power.

We will now use these rules in several special cases.

THE NORMAL DISTRIBUTION

Let us find the distribution of an average of independent random variables $X_1, X_2, X_3, \ldots, X_n$ which have normal distributions with means $\mu_1, \mu_2, \mu_3, \ldots, \mu_n$ and variances $\sigma_1{}^2, \sigma_2{}^2, \sigma_2{}^3, \ldots, \sigma_n{}^2$.

The moment generating function for X_i is given by

$$E\,e^{tX_i} = \exp(\mu_i t + \sigma_i{}^2 t^2/2).$$

Therefore, by rule **a**

$$E\,e^{t\bar{X}_n} = \prod_{i=1}^{n} \exp(\mu_i t/n + \sigma_i^2 t^2/2n^2) = \exp\left(\frac{1}{n}\sum_{i=1}^{n} u_i t + \frac{1}{2n^2}\sum_{i=1}^{n} \sigma_i^2 t^2\right).$$

This is the moment generating function of a normal variable with mean

$$\frac{1}{n}\sum_{i=1}^{n} \mu_i$$

and variance

$$\frac{1}{n^2}\sum_{i=1}^{n} \sigma_i^2.$$

The average of independent normal random variables is thus a normal random variable.

In the special case of independent and identically distributed normal random variables each with mean μ and variance σ^2 the density function of $\bar{X}_n$ is

$$f(\bar{x}_n) = \frac{1}{\sqrt{(2\pi\sigma^2/n)}} \exp[-(\bar{x}_n - \mu)^2/(2\sigma^2/n)] \qquad -\infty < \bar{x}_n < \infty.$$

EXERCISE 9-2

Show that the use of rule **b** gives this same density function.

THE GAMMA DISTRIBUTION

Let us find the distribution of the average of the independent random variables $X_1, X_2, X_3, \ldots, X_n$ which have gamma distributions with parameters $\alpha_1, \alpha_2, \alpha_3, \ldots, \alpha_n$ and $\beta_1, \beta_2, \beta_3, \ldots, \beta_n$.

The moment generating function for the ith random variable is given by

$$E\,e^{tX_i} = (1-\beta_i t)^{-\alpha_i}$$

giving by rule **a**

$$E\,e^{t\bar{X}_n} = \prod_{i=1}^{n} (1-\beta_i t/n)^{-\alpha_i}.$$

If now $X_1, X_2, X_3, \ldots, X_n$ have a common β value, i.e., $\beta_1 = \beta_2 = \beta_3 = \ldots = \beta_n = \beta$, then

$$E\,e^{tX_n} = \exp_{[1-(\beta/n)t]}\left(-\sum_{i=1}^{n} \alpha_i\right).$$

This is the moment generating function of a gamma distribution with parameters

$$\sum_{i=1}^{n} \alpha_i \qquad \text{and} \qquad \beta/n.$$

Therefore, in the special case of common β values

$$f(\bar{x}_n) = \frac{1}{\Gamma\left(\sum_{i=1}^{n} \alpha_i\right) \exp_{(\beta/n)}\left(\sum_{i=1}^{n} \alpha_i\right)} \exp_{(\bar{x}_n)}\left(\sum_{i=1}^{n} \alpha_i - 1\right) \exp(-n\bar{x}_n/\beta) \qquad 0 < \bar{x}_n < \infty$$

$$= 0 \qquad \text{otherwise.}$$

The average of independent gamma random variables with the same β value is thus a gamma random variable. In the even more special case of independent and identically distributed gamma random variables each with parameters α and β, we have

$$f(\bar{x}_n) = \frac{1}{\Gamma(n\alpha)(\beta/n)^{n\alpha}} (\bar{x}_n)^{n\alpha - 1} e^{-n\bar{x}_n/\beta} \qquad 0 < \bar{x}_n < \infty$$

$$= 0 \qquad \text{otherwise.}$$

EXERCISE 9-3

Show that in the case of independent and identically distributed random variables rule **b** gives the same result.

THE POISSON DISTRIBUTION

Let us find the distribution of the average of the independent random variables $X_1, X_2, X_3, \ldots, X_n$ which have poisson distributions with parameters $\lambda_1, \lambda_2, \lambda_3, \ldots, \lambda_n$.

The moment generating function for X_i is given by

$$E\, e^{tX_i} = e^{-\lambda_i} e^{\lambda_i e^t}$$

and therefore by rule **a**

$$E\, e^{t\bar{X}_n} = \exp\left(-\sum_{i=1}^{n} \lambda_i\right) \exp\left(\sum_{i=1}^{n} \lambda_i e^{t/n}\right).$$

This is not a recognizable moment generating function. However, the moment generating function for $Y = n\bar{X}_n$ can be obtained by replacing t by nt in the moment generating function for $\bar{X}_n$, that is

$$E\, e^{tY} = E\, e^{tn\bar{X}_n} = E\, e^{(tn)\bar{X}_n} = \exp\left(-\sum_{i=1}^{n} \lambda_i\right) \exp\left(\sum_{i=1}^{n} \lambda_i e^t\right).$$

This is the moment generating function of a poisson distribution with parameter $\Sigma_{i=1}^{n} \lambda_i$. Therefore

$$P(y) = \frac{\exp\left(-\sum_{i=1}^{n} \lambda_i\right)\left(\sum_{i=1}^{n} \lambda_i\right)^y}{y!} \qquad y = 0, 1, 2, \ldots.$$

We can now make the one to one transformation $\bar{X}_n = Y/n$ and obtain

$$P(\bar{x}_n) = \frac{\exp\left(-\sum_{i=1}^{n} \lambda_i\right)\left(\sum_{i=1}^{n} \lambda_i\right)^{n\bar{x}_n}}{(n\bar{x}_n)!} \qquad \bar{x}_n = 0, 1/n, 2/n, 3/n, \ldots.$$

Therefore the sum of independent poisson random variables is a poisson random variable but the average has the mass function above, which is not quite poisson.

In the special case of independent and identically distributed poisson random variables with parameter λ, we obtain

$$P(y) = \frac{e^{-n\lambda}(n\lambda)^y}{y!} \qquad y = 0, 1, 2, \ldots$$

and

$$P(\bar{x}_n) = \frac{e^{-n\lambda}(n\lambda)^{n\bar{x}_n}}{(n\bar{x}_n)!} \qquad \bar{x}_n = 0, 1/n, 2/n, 3/n, \ldots.$$

EXERCISE 9-4

Verify these results by using rule **b**.

THE BINOMIAL DISTRIBUTION

Let us find the distribution of the average of the independent random variables $X_1, X_2, X_3, \ldots, X_n$ which have binomial distributions with parameters $m_1, m_2, m_3, \ldots, m_n$ and $p_1, p_2, p_3, \ldots, p_n$. ($m$ is used to represent the number of trials since n has another meaning here.)

The moment generating function for X_i is given by

$$E\, e^{tX_i} = [p_i e^t + (1-p_i)]^{m_i}$$

and, therefore, by rule **a**

$$E\, e^{t\bar{X}_n} = \prod_{i=1}^{n} [p_i e^{t/n} + (1-p_i)]^{m_i}.$$

When $p_1 = p_2 = p_3 = \ldots = p_n = p$ then

$$E\, e^{t\bar{X}_n} = \exp_{[pe^{t/n}+(1-p)]}\left(\sum_{i=1}^{n} m_i\right).$$

This is still not a recognizable moment generating function. As in the poisson case, if we let $Y = n\bar{X}_n$, then

$$E\, e^{tY} = E\, e^{(tn)\bar{X}_n} = \exp_{[pe^t+(1-p)]}\left(\sum_{i=1}^{n} m_i\right)$$

which is the moment generating function of a binomial distribution with parameters

$$\sum_{i=1}^{n} m_i \quad \text{and} \quad p.$$

Therefore

$$P(y) = \binom{\sum_{i=1}^{n} m_i}{y} p^y \exp_{(1-p)}\left(\sum_{i=1}^{n} m_i - y\right) \qquad y = 0, 1, 2, \ldots \sum_{i=1}^{n} m_i.$$

Making the one to one transformation $\bar{X}_n = Y/n$ we obtain

$$P(\bar{x}_n) = \binom{\sum_{i=1}^{n} m_i}{n\bar{x}_n} p^{n\bar{x}_n} \exp_{(1-p)}\left(\sum_{i=1}^{n} m_i - n\bar{x}_n\right)$$

$$\bar{x}_n = 0, 1/n, 2/n, \ldots \sum_{i=1}^{n} m_i/n.$$

Thus, we see that the sum of independent binomial random variables each with the same parameter p is itself a binomial variable. The average, on the other hand, has the mass function given above.

In the special case of independent and identically distributed random variables with parameters m and p, we obtain

$$P(\bar{x}_n) = \binom{nm}{n\bar{x}_n} p^{n\bar{x}_n}(1-p)^{nm-n\bar{x}_n} \qquad \bar{x}_n = 0, 1/n, 2/n, \ldots, m.$$

THE NEGATIVE BINOMIAL DISTRIBUTION

Let us find the distribution of the average of independent random variables $X_1, X_2, X_3, \ldots, X_n$ which have negative binomial distributions with parameters $r_1, r_2, r_3, \ldots, r_n$ and $p_1, p_2, p_3, \ldots, p_n$.

The moment generating function for X_i is given by

$$E\, e^{tX_i} = p_i^{r_i}[1-(1-p_i)\, e^t]^{-r_i}$$

and, therefore, by rule **a**

$$E\, e^{t\bar{X}_n} = \prod_{i=1}^{n} p_i^{r_i}[1-(1-p_i)\, e^{t/n}]^{-r_i}.$$

When $p_1 = p_2 = p_3 = \ldots = p_n = p$ we have

$$E\, e^{t\bar{X}_n} = \exp_p\left(\sum_{i=1}^{n} r_i\right) \exp_{[1-(1-p)e^{t/n}]}\left(-\sum_{i=1}^{n} r_i\right).$$

This is not a recognizable moment generating function, but again using $Y = n\bar{X}_n$, we obtain

$$E\, e^{tY} = E\, e^{(tn)\bar{X}_n} = \exp_p\left(\sum_{i=1}^{n} r_i\right) \exp_{[1-(1-p)e^t]}\left(-\sum_{i=1}^{n} r_i\right).$$

This is the moment generating function of a negative binomial distribution with parameters

$$\sum_{i=1}^{n} r_i \qquad \text{and} \qquad p.$$

Thus

$$P(y) = \binom{y+\sum_{i=1}^{n} r_i-1}{\sum_{i=1}^{n} r_i-1} \exp_p\left(\sum_{i=1}^{n} r_i\right)(1-p)^y \qquad y = 0, 1, 2, \ldots.$$

Using the one to one transformation $\bar{X}_n = Y/n$ we obtain

$$P(\bar{x}_n) = \binom{n\bar{x}_n+\sum_{i=1}^{n} r_i-1}{\sum_{i=1}^{n} r_i-1} \exp_p\left(\sum_{i=1}^{n} r_i\right)(1-p)^{n\bar{x}_n} \qquad \bar{x}_n = 0, 1/n, 2/n, \ldots$$

Thus, the sum of independent negative binomial variables with common p value has a negative binomial distribution. The average, however, has the distribution given above.

In the special case of independent and identically distributed random variables with parameters r and p, we obtain

$$p(\bar{x}_n) = \binom{n\bar{x}_n+nr-1}{nr-1} p^{nr}(1-p)^{n\bar{x}_n} \qquad \bar{x}_n = 0, 1/n, 2/n, \ldots.$$

9-3 The Distribution of Sample Moments and of Certain Ratios of Them in the Normal Case

The most widely used distribution in statistics is the *normal distribution*. We will devote this section to finding the distributions of some sample moments and of certain ratios of them when the underlying distributions of the random variables are normal.

THE FIRST SAMPLE CENTRAL MOMENT

Let us assume that the independent random variables $X_1, X_2, X_3, \ldots, X_n$ have normal distributions with means $\mu_1, \mu_2, \mu_3, \ldots, \mu_n$ and variances $\sigma_1^2, \sigma_2^2, \sigma_3^2, \ldots, \sigma_n^2$. We will find, for this case, the distribution of the first sample central moment

$$C_n^1 = \frac{1}{n}\sum_{i=1}^{n} (X_i-\mu_i).$$

If we let $Z_i = X_i - \mu_i$ for $i = 1, 2, 3, \ldots, n$, then we have that $Z_1, Z_2, Z_3, \ldots, Z_n$ are independent (since functions of independent random variables are independent). Now since

$$f(x_i) = \frac{1}{\sqrt{(2\pi)}\sigma_i} \exp(-(x_i - \mu_i)^2/2\sigma_i^2) \qquad -\infty < x_i < \infty$$

we obtain

$$f(z_i) = \frac{1}{\sqrt{(2\pi)}\sigma_i} \exp(-z_i^2/2\sigma^2) \qquad -\infty < z_i < \infty.$$

Thus

$$C_n^1 = \frac{1}{n} \sum_{i=1}^{n} (X_i - \mu_i) = \frac{1}{n} \sum_{i=1}^{n} Z_i$$

is the average of independent normal variables with common mean zero and variances $\sigma_1^2, \sigma_2^2, \sigma_3^2, \ldots, \sigma_n^2$. From the result in Section 9-2, we have

$$f(c_n^1) = \frac{n}{\sqrt{(2\pi)}\left(\sqrt{\sum_{i=l}^{n} \sigma_i^2}\right)} \exp\left(-n^2(c_n^1)^2/2 \sum_{i=1}^{n} \sigma^2\right) \qquad \infty < c_n' < \infty.$$

When $X_1, X_2, X_3, \ldots, X_n$ are independent and identically distributed with common mean μ and variance σ^2 then

$$f(c_n^1) = \sqrt{\left(\frac{n}{2\pi}\right)} \frac{1}{\sigma} \exp(-n(c_n^1)^2/2\sigma^2) \qquad -\infty < c_n^1 < \infty$$

that is, c_n^1 has a normal distribution with mean zero and variance σ^2/n.

THE SECOND SAMPLE CENTRAL MOMENT

Let us again assume that the independent random variables $X_1, X_2, X_3, \ldots, X_n$ have normal distributions with means $\mu_1, \mu_2, \mu_3, \ldots, \mu_n$ and variances $\sigma_1^2, \sigma_2^2, \sigma_3^2, \ldots, \sigma_n^2$. We now ask for the distribution of the second sample central moment

$$C_n^2 = \frac{1}{n} \sum_{i=l}^{n} (X_i - \mu_i)^2.$$

If we once more let $Z_i = X_i - \mu_i$ for $i = 1, 2, 3, \ldots, n$, we have

$$C_n^2 = \frac{1}{n} \sum_{i=l}^{n} (X_i - \mu_i)^2 = \frac{1}{n} \sum_{i=l}^{n} Z_i^2$$

which is the average of independent random variables whose distributions we have yet to find. Z_i^2, for each i, is the square of a normal random variable.

We will therefore find the distribution of $Y = Z^2$ where Z is a normal random variable with zero mean and variance σ^2. Now

$$f(z) = \frac{1}{\sqrt{(2\pi)}\sigma} e^{-z^2/2\sigma^2} \qquad -\infty < z < \infty.$$

If we let $W = Z^2$, we obtain as the inverse transformation

$$Z = \sqrt{W} \qquad 0 \leqq Z < \infty$$

$$Z = -\sqrt{W} \qquad -\infty < Z < 0.$$

Since $|J| = 1/2\sqrt{w}$ in both intervals, we obtain

$$\begin{aligned} f(w) &= \frac{1}{\sqrt{(2\pi)}\sigma} w^{-1/2} e^{-w/2\sigma^2} && 0 < w < \infty \\ &= 0 && \text{otherwise.} \end{aligned}$$

This is the gamma density with $\alpha = 1/2$ and $\beta = 2\sigma^2$. (A chi-squared distribution with $n = 1$.) Thus $C_n{}^2$ is the average of gamma random variables with common parameter $\alpha = 1/2$ and different second parameters $\beta_1, \beta_2, \beta_3, \ldots, \beta_n$. As we saw in Section 9-2, if the random variables have the same variance, then $\beta_1 = \beta_2 = \beta_3 = \ldots = \beta_n = 2\sigma^2$ and the average has a gamma distribution with parameters $n/2$ and $2\sigma^2/n$. Therefore

$$\begin{aligned} f(c_n{}^2) &= \frac{n^{n/2}}{\Gamma(n/2)(2\sigma^2)^{n/2}} (c_n{}^2)^{n/2-1} e^{-nc_n{}^2/2\sigma^2} && 0 < c_n{}^2 < \infty \\ &= 0 && \text{otherwise.} \end{aligned}$$

This is a chi-squared density with parameters n and σ^2/n. n, which is also the number of random variables in the average $C_n{}^2$, is called the *number of degrees of freedom* of this distribution.

It is interesting to note that

$$Y = nC_n{}^2 = \sum_{i=l}^{n} (X_i - \mu_i)^2$$

is the sum of squares of centralized normal variables each with variance σ^2. Since $(dy/dc_n{}^2) = n$, we obtain

$$\begin{aligned} f(y) &= \frac{1}{\Gamma(n/2)(2\sigma^2)^{n/2}} y^{n/2-1} e^{-y/2\sigma^2} && 0 < y < \infty \\ &= 0 && \text{otherwise.} \end{aligned}$$

Thus the sum of n squares of centralized independent normal random variables with common variance σ^2 is a chi-squared random variable with n degrees of freedom and parameter σ^2.

EXERCISE 9-5

If $X_1, X_2, X_3, \ldots, X_n$ are independently distributed as normal random variables with means $\mu_1, \mu_2, \mu_3, \ldots, \mu_n$ and variances $\sigma_1^2, \sigma_2^2, \sigma_3^2, \ldots, \sigma_n^2$, show that the density of

$$Z = \sum_{i=1}^{n} \left(\frac{X_i - \mu_i}{\sigma_i}\right)^2$$

is given by

$$\begin{aligned} f(z) &= \frac{1}{\Gamma(n/2)2^{n/2}} z^{n/2-1} e^{-z/2} \qquad && 0 < z < \infty \\ &= 0 && \text{otherwise.} \end{aligned}$$

THE DISTRIBUTION OF THE SAMPLE VARIANCE

For this subsection we will assume that $X_1, X_2, X_3, \ldots, X_n$ are independent and identically distributed as normal variables with common mean μ and variance σ^2. We now ask for the distribution of

$$S_n^2 = \frac{1}{\mathrm{n}-1} \sum_{i=l}^{n} (X_i - \overline{X}_n)^2.$$

We will show that S_n^2 is equivalent to an average of $n-1$ squares of independent and identically distributed normal variables with common mean zero and variance σ^2. This will show that S_n^2 is a chi-squared random variable with $n-1$ degrees of freedom and parameter $\sigma^2/(n-1)$. We will also see that $\overline{X}_n$ and S_n^2 are independently distributed, a fact which will be useful later.

There are several ways in which this problem can be solved. We will use here the method of transformations.

Considering the joint distribution of $X_1, X_2, X_3, \ldots, X_n$, we write

$$f(x_1, x_2, \ldots, x_n) = \frac{1}{(2\pi)^{n/2}\sigma^n} \exp\left(-\sum_{i=1}^{n} (x_i - \mu)^2/2\sigma^2\right) \qquad -\infty < \text{all } x_i < \infty.$$

Let $Y_i = X_i - \mu$, $i = 1, 2, 3, \ldots, n$. Then

$$f(y_1, y_2, \ldots, y_n) = \frac{1}{(2\pi)^{n/2}\sigma^n} \exp\left(-\sum_{i=1}^{n} y_i^2/2\sigma^2\right) \qquad -\infty < \text{all } y_i < \infty.$$

Now let

$$\begin{aligned} Z_1 &= (Y_1 - Y_2)/\sqrt{2} \\ Z_2 &= (Y_1 + Y_2 - 2Y_3)/\sqrt{(2\cdot 3)} \\ &\vdots \\ Z_m &= (Y_1 + Y_2 + \ldots + Y_m - mY_{m+1})/\sqrt{[m(m+1)]} \\ &\vdots \\ Z_{n-1} &= (Y_1 + Y_2 + \ldots + Y_{n-1} - (n-1)Y_n)/\sqrt{[(n-1)n]} \\ Z_n &= (Y_1 + Y_2 + \ldots + Y_n)/\sqrt{n}. \end{aligned}$$

We will now show that $J^{-1} = 1$ and therefore $J = 1$ for this transformation.

$$
= \begin{vmatrix}
\frac{1}{\sqrt{2}} & \frac{-1}{\sqrt{2}} & 0 & \ldots\ 0 & \ldots\ 0 & 0 \\
\frac{1}{\sqrt{(2\cdot 3)}} & \frac{1}{\sqrt{(2\cdot 3)}} & \frac{-2}{\sqrt{(2\cdot 3)}} & \ldots\ 0 & \ldots\ 0 & 0 \\
\vdots & \vdots & \vdots & \vdots & \vdots & \vdots \\
\frac{1}{\sqrt{[m(m+1)]}} & \frac{1}{\sqrt{[m(m+1)]}} & \frac{1}{\sqrt{[m(m+1)]}} & \ldots\ \frac{-m}{\sqrt{[m(m+1)]}} & \ldots\ 0 & 0 \\
\vdots & \vdots & \vdots & \vdots & \vdots & \vdots \\
\frac{1}{\sqrt{[(n-1)n]}} & \frac{1}{\sqrt{[(n-1)n]}} & \frac{1}{\sqrt{[(n-1)n]}} & \ldots\ \frac{1}{\sqrt{[(n-1)n]}} & \ldots\ \frac{-n+1}{\sqrt{[(n-1)n]}} & 0 \\
\frac{1}{\sqrt{n}} & \frac{1}{\sqrt{n}} & \frac{1}{\sqrt{n}} & \frac{1}{\sqrt{n}} & \ldots\ \frac{1}{\sqrt{n}} & \frac{1}{\sqrt{n}}
\end{vmatrix}
$$

We now find the determinant by first finding the cofactors of the columns beginning with the last column and working inward toward the first column. When we do this, we obtain

$$J^{-1} = \frac{1}{\sqrt{n}} \cdot \frac{n}{\sqrt{[(n-1)n]}} \cdots \frac{m+1}{\sqrt{[m(m+1)]}} \cdots \frac{3}{\sqrt{(2\cdot 3)}} \cdot \frac{2}{\sqrt{2}} = \frac{n!}{\sqrt{(n!n!)}} = 1.$$

We will now show that

$$\sum_{i=l}^{n} Z_i = \sum_{i=l}^{n} Y_i$$

which will enable us to obtain the joint density for $Z_1, Z_2, Z_3, \ldots, Z_n$ easily. In order to show the equivalence we square each Z_i, obtaining

$$
\begin{aligned}
Z_1^2 &= Y_1^2/2 + Y_2^2/2 - 2Y_1Y_2/2 \\
&\ \vdots \\
Z_m^2 &= Y_1^2/m(m+1) + Y_2^2/m(m+1) + \ldots\ Y_m^2/m(m+1) \\
&\qquad + m^2 Y_{m+1}^2/m(m+1) + 2Y_1Y_2/m(m+1) + \ldots \\
&\qquad + 2Y_{m-1}Y_m/m(m+1) - 2mY_1Y_{m+1}/m(m+1) - \ldots \\
&\qquad - 2mY_mY_{m+1}/m(m+1) \\
&\ \vdots \\
Z_n^2 &= Y_1^2/n + Y_2^2/n + \ldots\ Y_n^2/n + 2Y_1Y_2/n + \ldots + 2Y_{n-1}Y_n/n.
\end{aligned}
$$

Summing, we obtain

$$\sum_{i=l}^{n} Z_i^2 = K_1 Y_1^2 + K_2 Y_2^2 \ldots + K_n Y_n^2 + K_{1,2} Y_1 Y_2 + \ldots K_{n-1,n} Y_{n-1} Y_n$$

where

$$K_1 = \frac{1}{2} + \frac{1}{2\cdot 3} + \cdots + \frac{1}{m(m+1)} + \cdots + \frac{1}{(n-1)n} + \frac{1}{n}$$

$$\vdots$$

$$K_m = \frac{(m-1)^2}{(m-1)m} + \frac{1}{m(m+1)} + \frac{1}{(m+1)(m+2)} + \cdots + \frac{1}{(n-1)n} + \frac{1}{n}$$

$$\vdots$$

$$K_n = \frac{(n-1)^2}{(n-1)n} + \frac{1}{n}$$

and

$$K_{1,2} = -\frac{4}{2} + \frac{2}{2\cdot 3} + \cdots + \frac{2}{m(m+1)} + \cdots + \frac{2}{(n-1)n} + \frac{2}{n}$$

$$\vdots$$

$$K_{r,m} = -\frac{2(m-1)}{(m-1)m} + \frac{2}{m(m+1)} + \cdots + \frac{2}{(n-1)n} + \frac{2}{n} \qquad r < m$$

$$\vdots$$

$$K_{r,n} = -\frac{2(n-1)}{(n-1)n} + \frac{2}{n}\cdot$$

EXERCISE 9-6

Demonstrate that

a. $K_m = 1 \qquad m = 1, 2, 3, \ldots, n$

b. $K_{r,m} = 0 \qquad m = 1, 2, 3, \ldots, n$

$\qquad r = 1, 2, 3, \ldots, m-1.$

From the results of this exercise we conclude that

$$\sum_{i=1}^{n} Z_i^2 = \sum_{i=1}^{n} Y_i^2.$$

The joint density of the Z_i is therefore

$$f(z_1, z_2, \ldots, z_n) = \frac{1}{(2\pi)^{n/2}\sigma^n} \exp\left(-\sum_{i=l}^{n} z_i^2/2\sigma^2\right) \qquad -\infty < \text{all } z_i < \infty.$$

Clearly the Z_i are independent and each is normally distributed with mean zero and variance σ^2.

We now note that

$$Z_n = \sqrt{(n)}\bar{Y}_n = \sqrt{(n)}\bar{X}_n - \sqrt{(n)}\mu.$$

Therefore

$$(n-1)S_n^2 = \sum_{i=1}^{n} (X_i - \bar{X}_{n)})^2 = \sum_{i=1}^{n} (X_i - \mu - \bar{X}_n - \mu)^2$$

$$= \sum_{i=1}^{n} (Y_i - \bar{Y}_{n)})^2 = \sum_{i=1}^{n} Y_i^2 - n\bar{Y}_n^2$$

$$= \sum_{i=1}^{n} Z_i^2 - Z_n^2 = \sum_{i=1}^{n-1} Z_i^2$$

or

$$S_n^2 = \frac{1}{n-1} \sum_{i=1}^{n-1} Z_i^2.$$

Therefore S_n^2 is the average of the squares of $n-1$ independent and identically distributed normal random variables with common mean zero and variance σ^2. Thus, from the previous subsection

$$f(s_n^2) = \frac{(n-1)^{(n-1)/2}}{\Gamma[(n-1)/2](2\sigma^2)^{(n-1)/2}} (s_n^2)^{(n-3)/2} \exp[-(n-1)s_n^2/2\sigma^2] \qquad 0 < s_n^2 < \infty.$$

This is the density of a chi-squared distribution with $n-1$ degrees of freedom and parameter $\sigma^2/(n-1)$. It is interesting to note that the reason for a lowering of n to $n-1$ was the fact that a function of the average was subtracted from the sum of n variables. This, however, is a special case and should not be considered a general rule.

We also note that $\bar{X}_n$ is a function of Z_n only, while S_n^2 is a function of Z_i, $i = 1, 2, \ldots, n-1$ only, and therefore $\bar{X}_n$ and S_n^2 are independently distributed. We can thus write the joint distribution of $\bar{X}_n$ and S_n^2 as

$$f(\bar{x}_n, s_n^2) = \frac{\sqrt{(n)}(n-1)^{(n-1)/2}}{\Gamma(n-1/2)(2\sigma^2)^{n/2}\sqrt{\pi}} (s_n^2)^{(n-3)/2} \exp[-(n-1)s_n^2/2\sigma^2]$$

$$\exp[-n(\bar{x}_n - \mu)^2/2\sigma^2] \qquad 0 < s_n^2 < \infty, \quad -\infty < \bar{x}_n < \infty$$

$$= 0 \qquad \text{otherwise.}$$

Another method commonly used to obtain this result is to find the joint moment generating function of $\bar{X}_n$ and S_n^2. That is, find $E \exp(t_1\bar{X}_n + t_2 S_n^2)$ and note that it is the product of the generating functions of a normal distribution with mean μ and variance σ^2/n and a chi-squared distribution with $n-1$ degrees of freedom and parameter $\sigma^2/(n-1)$.

EXERCISE 9-7

If $X_1, X_2, X_3, \ldots, X_n$ are independent and identically distributed normal random variables with common mean μ and variance σ^2, find $E\, e^{t_1\bar{X}_n + t_2 S_n^2}$.

THE DISTRIBUTION OF S_n

When $X_1, X_2, X_3, \ldots, X_n$ are independent and identically distributed normal random variables with common mean μ and variance σ^2, the density function for S_n^2 is given in the previous subsection. What then is the distribution of $S_n = \sqrt{(S_n^2)}$?

This can easily be obtained by the simple transformation $S_n = \sqrt{(S_n^2)}$. Since

$$\frac{\partial S_n}{\partial S_n^2} = \frac{1}{2\sqrt{(S_n^2)}}$$

we obtain

$$f(s_n) = \frac{(n-1)^{(n-1)/2}}{\Gamma[(n-1)/2]2^{[(n-1)/2]-1}\sigma^{(n-1)}} s_n^{n-2} \exp[-(n-1)s_n/2\sigma^2] \qquad 0 < s_n < \infty$$

$$= 0 \qquad \text{otherwise.}$$

This is the density of a chi random variable with n replaced by $n-1$ and σ replaced by $\sigma/\sqrt{(n-1)}$.

EXERCISE 9-8

Show that, in general, if X has a chi-squared distribution with n degrees of freedom and parameter σ^2, then $Y = \sqrt{X}$ has a chi distribution with n degrees of freedom and parameter σ.

THE DISTRIBUTION OF THE t RATIO

When $X_1, X_2, X_3, \ldots, X_n$ are independent and identically distributed normal random variables with common mean μ and variance σ^2, the ratio

$$t = \frac{\sqrt{(n)}(\bar{X}_n - \mu)}{S_n}$$

is called the *t ratio*. It should be noted that when μ is not known, the value of t cannot be determined by the observations and therefore t is not a statistic. It is in this case, however, that t is most useful.

The joint density of $\bar{X}_n$ and S_n^2 was given previously. We now make the transformation

$$t = \frac{\sqrt{(n)}(\bar{X}_n - \mu)}{S_n}$$

$$Z = S_n^2$$

with the inverse transformation

$$\bar{X}_n = \frac{t\sqrt{Z}}{\sqrt{n}} + \mu$$

$$S_n^2 = Z.$$

The Jacobian is $J = (\partial \bar{X}_n / \partial t) = \sqrt{(Z/n)}$ and therefore

$$f(t, z) = \frac{\sqrt{(n)}(n-1)^{(n-1)/2} z^{(n-3)/2}}{(2\sigma^2)^{n/2}\sqrt{(\pi)}\Gamma[(n-1)/2]} \exp(-zt^2/2\sigma^2)\exp[-(n-1)z/2\sigma^2]\sqrt{z/n}$$

$$= \frac{(n-1)^{(n-1)/2}}{(2\sigma^2)^{n/2}\sqrt{[(\pi)}\Gamma(n-1)/2]} z^{(n-2)/2} \exp[-(t^2+n-1)z/2\sigma^2] \qquad \begin{array}{l} -\infty < t < \infty \\ 0 < z < \infty \end{array}$$

$$= 0 \qquad \text{otherwise.}$$

Integrating out the z (which is a gamma variable), we obtain

$$f(t) = \frac{\Gamma(n/2)}{\sqrt{[(n-1)]}\Gamma(n-1/2)}\left(1+\frac{t^2}{n-1}\right)^{-n/2} \qquad -\infty < t < \infty.$$

This is the density of the t distribution with parameter $(n-1)$. We say this is a t distribution with $n-1$ degrees of freedom. This can often be confusing, since there were originally n (not $n-1$) random variables.

THE DISTRIBUTION OF THE F RATIO

Another ratio which is often used is the ratio of the square of two independent sample moments about the average when each is based on a set of independent normal random variables with the same variance.

Let $X_1, X_2, X_3, \ldots, X_m$ be independent and identically distributed normal random variables with common mean μ and variance σ^2. Further let $Y_1, Y_2, Y_3, \ldots, Y_n$ be independent and identically distributed normal random variables with common mean ν and variance σ^2. Assume further that the two sets of random variables are mutually independent. Using the two statistics

$$S_m^{\ 2} = \frac{1}{m-1} \sum_{i=1}^{m} (X_i - \bar{X}_m)^2$$

and

$$S_n^{\ 2} = \frac{1}{n-1} \sum_{i=1}^{n} (Y_i - \bar{Y}_n)^2$$

we form the ratio

$$F = \frac{S_m^{\ 2}}{S_n^{\ 2}} \cdot$$

This is known as the F statistic, or the *variance ratio* statistic.

The joint distribution of $S_m^{\ 2}$ and $S_n^{\ 2}$ is easily seen to be

$$f(s_m^{\ 2}, s_n^{\ 2}) = \frac{(m-1)^{(m-1)/2}(n-1)^{(n-1)/2}}{\Gamma[(m-1)/2]\Gamma[(n-1)/2](2\sigma^2)^{(m+n-2)/2}} (s_m^{\ 2})^{(m-3)/2}(s_n^{\ 2})^{(n-3)/2}$$

$$\exp(-[(m-1)s_m^{\ 2}/2\sigma^2]-[(n-1)s_n^{\ 2}/2\sigma^2]) \qquad 0 < s_m^{\ 2}, s_n^{\ 2} < \infty$$

$$= 0 \qquad \text{otherwise.}$$

If we now let

$$F = S_m^{\ 2}/S_n^{\ 2}$$
$$Z = S_n^{\ 2}$$

we obtain the inverse relations

$$S_m^{\ 2} = FZ$$
$$S_n^{\ 2} = Z.$$

The Jacobian of this transformation is $J = (\partial S_m^{\ 2}/\partial F) = Z$ and therefore

$$f(F, z) = \frac{(m-1)^{(m-1)/2}(n-1)^{(n-1)/2}}{\Gamma(m-1/2)\Gamma(n-1/2)(2\sigma^2)^{(m+n-2)/2}} F^{(m-3)/2} z^{(m+n-4)/2}$$

$$\exp-\left(\frac{n-1+(m-1)F}{2\sigma^2}\right)^z \qquad 0 < F, z < \infty$$

$$= 0 \qquad \text{otherwise.}$$

We now integrate out the z (which is a gamma variable) and obtain

$$f(F) = \frac{\Gamma[(m+n-2)/2]}{\Gamma[(m-1)/2]\Gamma[(n-1)/2]} \left(\frac{m-1}{n-1}\right)^{(m-1)/2} \frac{F^{[(m-1)/2]-1}}{[1+(m-1)/(n-1)F]^{(m+n-2)/2}}$$

$$0 < F < \infty$$

$$= 0 \qquad \text{otherwise.}$$

This is the density of the F distribution with $m-1$ and $n-1$ degrees of freedom. Note that there were originally m and n variables.

9-4 Problems

1. If $X_1, X_2, X_3, \ldots, X_n$ are distributed with gamma distributions with parameters $\alpha_1, \alpha_2, \alpha_3, \ldots, \alpha_n$ and $\beta_1, \beta_2, \beta_3, \ldots, \beta_n$, find $E\bar{X}_n$.
2. Assume that the random variables in Problem 1 are mutually independent and find $E\bar{X}_n$ and Var $\bar{X}_n$.
3. If $X_1, X_2, X_3, \ldots, X_n$ are distributed with binomial distributions with parameters $m_1, m_2, m_3, \ldots, m_n$ and $p_1, p_2, p_3, \ldots, p_n$ find $E\bar{X}_n$.
4. Assume that the random variables in Problem 3 are mutually independent and find $E\bar{X}_n$ and Var $\bar{X}_n$.
5. For the random variables of Problem 1 find EC_n^2.
6. For the random variables of Problem 3 find EC_n^2.
7. For the random variables of Problem 2 find EM_n^2.
8. For the random variables of Problem 4 find EM_n^2.
9. If $X_1, X_2, X_3, \ldots, X_n$ are independent and identically distributed with a gamma distribution with parameters α and β, find EM_n^2. Find ES_n^2.
10. If $X_1, X_2, X_3, \ldots, X_n$ are independent and identically distributed with a normal distribution with mean μ and variance σ^2, find EM_n^2. Find ES_n^2.
11. For the random variables of Problem 9 find Var M_n^2. Find Var S_n^2.
12. For the random variables of Problem 10 find Var M_n^2. Find Var S_n^2.
13. If $X_1, X_2, X_3, \ldots, X_n$ are independent and identically distributed with an exponential distribution with parameter λ, find $E\bar{X}_n$ and Var $\bar{X}_n$.
14. For the random variables of Problem 13 find the density function for $\bar{X}_n$.
15. If $X_1, X_2, X_3, \ldots, X_n$ are independent and distributed with chi-squared distributions with $m_1, m_2, m_3, \ldots, m_n$ degrees of freedom, respectively, and common parameter σ^2, find $E\bar{X}_n$ and Var $\bar{X}_n$.

17. If $X_1, X_2, X_3, \ldots, X_n$ are independently distributed with poisson distributions with parameters $\lambda_1 = 1, \lambda_2 = 2, \lambda_3 = 3, \ldots, \lambda_n = n$, find $E\bar{X}_n$ and Var $\bar{X}_n$.
18. For the random variables in Problem 17 find the mass function for $\bar{X}_n$.
19. If $X_1, X_2, X_3, \ldots, X_n$ are independent and identically distributed with density

$$f(x) = \frac{1}{16} x^2 e^{-x/2} \qquad 0 < x < \infty$$

$$= 0 \qquad \text{otherwise}$$

find the density function for $\bar{X}_n$.

20. If $X_1, X_2, X_3, \ldots, X_n$ are independent and identically distributed with mass function

$$P(x) = \frac{e^{-3+x \ln 3}}{x!} \qquad x = 0, 1, 2, \ldots$$

find the probability that $\bar{X}_n = 3$.

21. Prove that for independent and identically distributed random variables

a. $EC_n^2 = ES_n^2$,

and

b. $\text{Var } C_n^2 \leqq \text{Var } S_n^2$.

22. If $X_1, X_2, X_3, \ldots, X_n$ are independent and identically distributed with a standard normal distribution find the densities of $\bar{X}_n$, C_n^2 and S_n^2.

23. If in Problem 22 the random variables had mean μ, what would be the densities of $\bar{X}_n$, C_n^2 and S_n^2 ?

24. For the random variables of Problem 22 what is the joint density of $\bar{X}_n$ and S_n^2 ?

25. If X_1, X_2, X_3 are independent and each with a standard normal distribution find the density function for

$$\sqrt{(3)}\bar{X}_3/S_3.$$

26. For the random variables of Problem 25 find the density function for $\bar{X}_3/S_3$.

27. If $X_1, X_2, \ldots, X_{10}$ are independent standard normal variables find the density function for

$$Y = \frac{\sum_{i=j}^{5} (X_i - \frac{1}{5} \sum_{j=1}^{5} X_j)^2}{\sum_{i=6}^{10} (X_i - \frac{1}{5} \sum_{j=6}^{10} X_j)^2}.$$

28. If the random variables in Problem 27 each had variance σ^2, what would be the density function for the given statistic?

CHAPTER 10

Properties of Estimators

In estimation we are interested in proposing a value which we feel represents the value of an unknown parameter. Of course, we would like our proposed value, or estimate, to be equal to the value of the unknown parameter; but this is often too much to expect. We will, therefore, devote this chapter to discussing the properties that estimators can have in order that we may have a means for deciding which estimate to propose.

First, we introduce a property known as *invariance*. This property has nothing to do with the probabilities involved, but depends solely on the *form* of the estimator. It is often useful in evaluating estimators when units of measurement are important to the problem.

10-1 Invariance

In this section we assume that there is a sequence of n random variables $X_1, X_2, X_3, \ldots, X_n$, which *will* be observed. Also, $\hat{\theta}(X_1, X_2, X_3, \ldots, X_n)$ will be used to represent an estimator for the parameter θ of the joint distribution of $X_1, X_2, X_3, \ldots, X_n$.

LOCATION INVARIANCE

We say that the estimator $\hat{\theta}(X_1, X_2, X_3, \ldots, X_n)$ is *location invariant* if

$$\hat{\theta}(X_1+b, X_2+b, X_3+b, \ldots, X_n+b) = \hat{\theta}(X_1, X_2, X_3, \ldots, X_n).$$

This means that if we add the same quantity to all the variables, there is no change in the value of the estimate.

Location invariance is usually important when difference measurements are used to find distance traveled or elapsed time. For example, when taking a trip the time of departure is subtracted from the time of arrival to find out how long the trip lasted. The time of travel should not depend on whether standard time or daylight saving time was used. As another illustration, in taking a measurement of length, we often start at the first inch mark on the ruler rather than at the zero mark in order to avoid inaccuracy caused by fraying or rounding at the edge of the ruler. This should not affect the estimate of the length.

SCALE INVARIANCE

We say that an estimator $\hat{\theta}(X_1, X_2, X_3, \ldots, X_n)$ is *scale invariant* if

$$\hat{\theta}(aX_1, aX_2, aX_3, \ldots, aX_n) = \hat{\theta}(X_1, X_2, X_3, \ldots, X_n).$$

This means that if we multiply each variable by a constant, the value of the estimate will not be changed. This property becomes important when the value of the estimator does not depend on the units of measurement. For example, the ratio of the lengths of two rods, or the ratio of times of travel.

AN INVARIANT ESTIMATOR

An estimator is said to be *invariant* if it is both location and scale invariant. That is, if

$$\hat{\theta}(aX_1+b, aX_2+b, aX_3+b, \ldots, aX_n+b) = \hat{\theta}(X_1, X_2, X_3, \ldots, X_n).$$

An estimator for the ratio of lengths or times of travel should be invariant to both location and scale.

EXAMPLE 10-1

If X_1 and X_2 are the observed starting times of two autos in a race and X_3 and X_4 are the corresponding observed finishing times, then

$$\hat{\theta}(X_1, X_2, X_3, X_4) = \frac{X_3 - X_1}{X_4 - X_2}$$

is an invariant estimator for the relative times of travel of the two autos.

HIGHER ORDER LOCATION INVARIANCE

We say that an estimator $\hat{\theta}(X_1, X_2, X_3, \ldots, X_n)$ is *rth-order location invariant* if

$$\hat{\theta}(X_1+b, X_2+b, X_3+b, \ldots, X_n+b) = \hat{\theta}(X_1, X_2, X_3, \ldots, X_n)+rb.$$

That is, adding a constant to each variable merely adds r times that constant to the estimate.

If each X were measuring θ itself, then we would expect any shift in the values of X to be reflected in our estimate of θ. An average, for example, would be first order location invariant.

HIGHER ORDER SCALE INVARIANCE

We say that an estimator $\hat{\theta}(X_1, X_2, X_3, \ldots, X_n)$ is *rth-order scale invariant* if

$$\hat{\theta}(aX_1, aX_2, aX_3, \ldots, aX_n) = a^r\hat{\theta}(X_1, X_2, X_3, \ldots, X_n).$$

That is, multiplying each value by a constant multiplies the estimate by the same constant to the rth power.

This property is *important* when the units of measurement are an inherent part of what is being estimated. If, for example, we want the length of a rod in inches, the estimate should be the same if the rod were measured in inches or measured in feet and then multiplied by twelve. In this case, we would want a first order scale invariant estimator for the length. For area, we would want a second-order scale invariant estimator, while for volume we would want a third-order scale invariant estimator.

EXAMPLE 10-2

It is interesting to note that

$$\hat{\theta}(X_1, X_2, X_3, \ldots, X_n) = S_n^2 = \frac{1}{n-1}\sum_{i=1}^{n}(X_i - \bar{X}_n)^2$$

is location invariant and second-order scale invariant. This can be demonstrated by replacing X_i by aX_i+b, obtaining

$$\hat{\theta}(aX_1+b, aX_2+b, \ldots, aX_n+b = \frac{1}{n-1}\sum_{i=1}^{n}[aX_i+b-(a\bar{X}_n+b)]^2$$

$$= a^2\frac{1}{n-1}\sum_{i=1}^{n}(X_i-\bar{X}_n)^2 = a^2 S_n^2.$$

REFLECTION INVARIANCE

We say that an estimator $\hat{\theta}(X_1, X_2, X_3, \ldots, X_n)$ is *reflection invariant* if

$$\hat{\theta}(\pm X_1, \pm X_2, \pm X_3, \ldots \pm X_n) = \hat{\theta}(X_1, X_2, X_3, \ldots, X_n).$$

That is, using the value of X_i with a positive or negative sign does not change the value of the estimate.

This property would be required if the sign of the measurement had no relation to the parameter being estimated. For example, the heating effect of electric currents through n resistors depends on the amount of current in each resistor but not on the direction of flow.

PERMUTATION INVARIANCE

We say that the estimator $\hat{\theta}(X_1, X_2, X_3, \ldots, X_n)$ is *permutation invariant* if

$$\hat{\theta}(X_i, X_j, \ldots, X_k) = \hat{\theta}(X_1, X_2, \ldots, X_n)$$

where $i, j, \ldots, k$ is any permutation of the indices $1, 2, \ldots, n$. This means that the estimate depends only on the values of the random variables and not on their order.

When $X_1, X_2, X_3, \ldots, X_n$ are independent and identically distributed it might be thought that the order in which the values were observed has no significance at all. In this case it would be unreasonable for an estimator not to be permutation invariant.

It should be pointed out that all the invariance properties given here, and many others which can be imagined, are merely properties which an estimator could have. Whether or not an estimator should have any of these properties depends entirely on what kind of measurements are taken and just what is being estimated. The requirement of an invariance property should depend on the situation.

10-2 Mean Squared Error and Uniformly Minimum Variance Unbiased Estimators

The desire to be close to the true value of the parameter being estimated leads to the question of what is meant by "close". How do we evaluate the relative worth of two estimates which are both wrong?

An obvious criterion of closeness is the absolute deviation, defined by $|\hat{\theta}-\theta|$, where we use $\hat{\theta}$ as an abbreviated form of $\hat{\theta}(x_1, x_2, x_3, \ldots, x_n)$. This criterion turns out to be exceptionally difficult to work with in general, and is usually replaced with the squared error criterion of closeness, given by $(\hat{\theta}-\theta)^2$.

MINIMUM MEAN SQUARED ERROR

Let us say that we have an estimator $\hat{\theta}(X_1, X_2, X_3, \ldots, X_n)$ for θ. We define the mean squared error for $\hat{\theta}$ (for a given θ) by

$$\text{M.S.E.} \quad (\hat{\theta}, \theta) = E(\hat{\theta}(X_1, X_2, X_3, \ldots, X_n)-\theta)^2 = E(\hat{\theta}-\theta)^2$$

where the expectation is with respect to the joint distribution of $X_1, X_2, X_3, \ldots, X_n$ for given value θ.

From a mean squared error point of view, the best $\hat{\theta}$ is the one with the minimum mean squared error. That is, we should choose the $\hat{\theta}$ such that M.S.E. $(\hat{\theta}, \theta)$ is minimum.

This could be done trivially if θ were known. But since θ is unknown, we might ask for the *uniformly minimum mean squared error* estimator, if it exists.

UNIFORMLY MINIMUM MEAN SQUARED ERROR

Suppose that θ is known to be a member of the set of θ values represented by Θ. Perhaps it is possible to find one $\hat{\theta}$ which will have a minimum mean squared error for each $\theta \in \Theta$. If so, we will call this the uniformly minimum mean squared error estimator for θ, i.e., $\hat{\theta}^*$ is that estimator for which

$$E(\hat{\theta}^*-\theta)^2 = \text{minimum for each } \theta \in \Theta.$$

In general, no such $\hat{\theta}^*$ exists. Consider the simple case where $\Theta = \{1, 2\}$ with $\hat{\theta}_1 = 1$ and $\hat{\theta}_2 = 2$. The mean squared error for $\hat{\theta}_1$ would be zero at $\theta = 1$. On the other hand, $\hat{\theta}_2 = 2$ has a zero mean squared error at $\theta = 2$. In order for a uniformly minimum mean squared error estimator to exist, it would have to have zero mean squared error at both $\theta = 1$ and $\theta = 2$. This type of estimator usually does not exist.

It is, therefore, apparent that when the set of available estimators is large there may be no uniformly minimum mean squared error estimator. In an attempt to solve the problem, we often restrict the class of eligible estimators, and among this subclass look for the uniformly minimum mean squared error estimator.

Before describing the usual restrictions, let us reduce the mean squared error as follows:

$$\begin{aligned} E(\hat{\theta}-\theta)^2 &= E(\hat{\theta}^2-2\hat{\theta}\theta+\theta^2) \\ &= E\hat{\theta}^2-2\theta E\hat{\theta}+\theta^2 \\ &= E\hat{\theta}^2-(E\hat{\theta})^2+(E\hat{\theta})^2-2\theta E\hat{\theta}+\theta^2 \\ &= (E\hat{\theta}-\theta)^2+E(\hat{\theta}-E\hat{\theta})^2. \end{aligned}$$

The first term on the right is called the *squared bias* of $\hat{\theta}$, while the second term is clearly the variance of $\hat{\theta}$. The mean squared error is then the sum of the squared bias in $\hat{\theta}$ and the variance of $\hat{\theta}$. One problem which arises when looking for a uniformly minimum mean squared estimator stems from the fact that when the variance goes down, the squared bias may go up. If we consider only those estimators for which the squared bias term is zero, then we may find one estimator within that subset of estimators which has uniformly minimum mean squared error, i.e., minimum variance.

UNBIASED ESTIMATORS

We say that the estimator $\hat{\theta}(X_1, X_2, X_3, \ldots, X_n)$ is *unbiased* for θ if

$$E\hat{\theta}(X_1, X_2, X_3, \ldots, X_n) = \theta \qquad \text{all } \theta \in \Theta$$

where the expectation is taken with respect to the joint distribution of $X_1, X_2, X_3, \ldots, X_n$; i.e., the distribution $F_\theta(X_1, X_2, X_3, \ldots, X_n)$.

EXAMPLE 10-3

Let $X_1, X_2, X_3, \ldots, X_n$ be independently and identically distributed with density

$$f(x) = \frac{1}{\sqrt{(2\pi)}}e^{-(x-\theta)^2/2} \qquad \begin{array}{l} -\infty < x < \infty \\ \theta \in \Theta. \end{array}$$

Let Θ be the real line. Then $\theta, (X_1, X_2, X_3 \ldots, X_n) = \bar{X}_n$ is an unbiased estimator for θ, since $E\bar{X}_n = \theta$ for all $\theta \in \Theta$.

N

Let us be very careful in interpreting the meaning of this property. If we use an unbiased estimator many times for estimating the same value of θ, then the average value of the estimates would be near θ by the law of large numbers (assuming the variance of the estimator exists). What does this say about the present estimate of θ? Is it likely to be close to θ? The property of unbiasedness does not in any way say that the estimate is likely to be close. The value of unbiasedness is, therefore, at best dubious. In spite of this, the property of unbiasedness is often considered necessary by many statisticians. If it has nothing else to offer (and indeed this may be the case), unbiasedness at least reduces the class of estimators to a class that is more easily handled mathematically.

UNIFORMLY MINIMUM VARIANCE UNBIASED ESTIMATORS

If we are restricted to the class of unbiased estimators, then the squared bias term in the mean squared error is zero, leaving only the variance term, that is, the mean squared error is the variance of an unbiased estimator, or $E(\hat{\theta}-\theta)^2 = \text{Var}\,\hat{\theta}$. The problem in general, reduces to finding the uniformly minimum variance estimator in the class of all unbiased estimators. This often exists, and we will devote the next chapter to this type of estimator and related topics.

When a uniformly minimum variance unbiased estimator does not exist, or cannot be found, one can restrict the class of eligible estimators still further. One restriction often used is to require that the estimator be a linear function of $X_1, X_2, X_3, \ldots, X_n$

$$\left(\text{e.g.}\quad \hat{\theta} = \sum_{i=1}^{n} a_i X_i + b\right).$$

We then ask for the *uniformly minimum variance linear unbiased estimator.* This can often be found by an algebraic method known as *least squares.* We will not discuss this method in this text.

Remark *Mathematical simplicity was the only justification given for requiring that an estimator be unbiased. One can hardly justify requiring an estimator to be both linear and unbiased on any other grounds.*

The restrictions that an estimator be linear and unbiased have been made so often that many users of statistical methods have the mistaken impression that these restrictions are themselves the *criteria* for *good estimators.* It should, however, be recalled that the criterion which promulgated these restrictions, due to mathematical intractability, was that of minimum mean squared error. Estimator $\hat{\theta}_1$ for θ is, therefore, better than estimator $\hat{\theta}_2$ for θ if

$$E(\hat{\theta}_1-\theta)^2 \leqq E(\hat{\theta}_2-\theta)^2 \qquad \text{all } \theta \in \Theta$$

and

$$E(\hat{\theta}_1-\theta)^2 < E(\hat{\theta}_2-\theta)^2 \qquad \text{some } \theta \in \Theta$$

whether or not $\hat{\theta}_1$ is linear or unbiased.

10-3 Consistency

In this section it is assumed that there is an infinite sequence of random variables $X_1, X_2, X_3, \ldots, X_n, \ldots$ and that for each n there is a family of known distribution functions $\{F_\theta(x_1, x_2, x_3, \ldots, x_n): \theta \in \Theta\}$. For each n there is defined an estimator for θ, $\hat{\theta}_n = \hat{\theta}_n(X_1, X_2, X_3, \ldots, X_n)$, so that there is an infinite sequence of estimators $\hat{\theta}_1, \hat{\theta}_2, \hat{\theta}_3, \ldots, \hat{\theta}_n, \ldots$.

A CONSISTENT SEQUENCE OF ESTIMATORS

If an infinite sequence $\hat{\theta}_1, \hat{\theta}_2, \hat{\theta}_3, \ldots, \hat{\theta}_n, \ldots$ of estimators is such that

$$\hat{\theta}_n \xrightarrow{P} \theta \quad \text{each} \quad \theta \in \Theta$$

then the sequence is said to be a *simple consistent* (or just *consistent*) sequence of estimators. This means that for any $\varepsilon > 0$

$$\lim_{n\to\infty} P[|\,\hat{\theta}_n-\theta\,| > \varepsilon] = 0 \qquad \text{all } \theta \in \Theta$$

where the probability is obtained from the distribution of $\hat{\theta}_n$ for each n (i.e., $F_\theta(x_1, x_2, x_3, \ldots, x_n)$).

Example 10-4

Let $X_1, X_2, X_3, \ldots, X_n, \ldots$ be an infinite sequence of independent identically distributed random variables, each with density

$$f(x) = \frac{1}{\sqrt{(2\pi)}}\, e^{-(x-\theta)^2/2} \qquad \begin{array}{l} -\infty < x < \infty. \\ -\infty < \theta < \infty. \end{array}$$

For the sequence $\hat{\theta}_1, \hat{\theta}_2, \ldots, \hat{\theta}_n, \ldots$, let $\hat{\theta}_n = \bar{X}_n$. Then the law of large numbers ensures that $\hat{\theta}_n \xrightarrow{P} \theta$ for $-\infty < \theta < \infty$. This means that, in general, a sequence of averages is consistent for the mean.

Remark *We often encounter the term "a consistent estimator." However, no estimator is consistent; only a sequence of estimators can be consistent. The term is used to imply that the estimator is defined for arbitrary n, such as $\overline{X}_n$, and that the sequence made up of these estimators is simple consistent.*

A SQUARED ERROR CONSISTENT SEQUENCE OF ESTIMATORS

The infinite sequence $\hat{\theta}_1, \hat{\theta}_2, \hat{\theta}_3, \ldots, \hat{\theta}_n, \ldots$ of estimators is said to be a *squared error consistent* sequence of estimators if

$$\lim_{n \to \infty} E(\hat{\theta}_n - \theta)^2 = 0 \qquad \text{all } \theta \in \Theta$$

where the expectation is with respect to the same distribution as was discussed under simple consistency. That this is a stronger requirement than simple consistency will be demonstrated later. Before deriving some very useful properties of the concepts of consistency, we will demonstrate some of their frequently encountered misconceptions.

MISCONCEPTIONS CONCERNING CONSISTENCY

The definitions of simple consistency and squared error consistency seem to imply that the sequence is in some way converging to the true value of the parameter. Although this may be quite correct, we must be very careful not to read too much into it. For example, we may feel that

$$\textbf{a} \quad E\hat{\theta}_n = \theta \qquad \text{all } n, \text{ all } \theta \in \Theta$$

or

$$\textbf{b} \quad \lim_{n \to \infty} E\hat{\theta}_n = \theta \qquad \text{all } \theta \in \Theta$$

or one of these with the further restriction

$$\textbf{c} \quad E(\hat{\theta}_n - \theta)^2 \leqq M < \infty \qquad \text{all } n, \text{ all } \theta \in \Theta$$

is sufficient for a sequence to be simple (or even squared error) consistent. That this is not the case can easily be seen from the following counter example.

Let each $\hat{\theta}_n$ be such that $P(\hat{\theta}_n = \theta + 1) = 1/2$ and $P(\hat{\theta}_n = \theta - 1) = 1/2$ for all $\theta \in \Theta$. Then $E\hat{\theta}_n = \theta$ for all n and all $\theta \in \Theta$ and $\text{Var } \hat{\theta}_n = E(\hat{\theta}_n - \theta)^2 = 1 < \infty$ for all n and all $\theta \in \Theta$. Thus all the restrictions given above are satisfied. However $P[| \hat{\theta}_n - \theta | > 1/2] = 1$ for all n and all $\theta \in \Theta$. Therefore, the sequence is not simple consistent. Since $E(\hat{\theta}_n - \theta)^2 = 1$ for all n and all $\theta \in \Theta$, the sequence is not squared error consistent either. These restrictions are therefore not sufficient for consistency.

On the other hand we may feel that simple consistency implies certain properties. For example,

$$\textbf{d} \quad E\hat{\theta}_n = \theta \qquad \text{all } n, \text{ all } \theta \in \Theta$$

or

$$\textbf{e} \quad \lim_{n \to \infty} E\hat{\theta}_n = \theta \qquad \text{all } \theta \in \Theta$$

or

$$\textbf{f} \quad E(\hat{\theta}_n - \theta)^2 \leqq M < \infty \qquad \text{all } n, \text{ all } \theta \in \Theta$$

or

$$\text{g} \quad \lim_{n\to\infty} E(\hat{\theta}_n-\theta)^2 = 0 \qquad \text{all } \theta \in \Theta.$$

That none of these represents the necessary properties of a simple consistent sequence of estimators is easily seen from the following example.

For each n let $\hat{\theta}_n$ be such that $P(\hat{\theta}_n = \theta) = (n-1)/n$ and $P(\hat{\theta}_n = \theta+n) = 1/n$ for all $\theta \in \Theta$. Then for any $\varepsilon > 0$ $P(|\hat{\theta}_n-\theta| > \varepsilon) = 1/n$ and so

$$\lim_{n\to\infty} P(|\hat{\theta}_n-\theta| > \varepsilon) = 0$$

and the sequence is simple consistent.

However,

$$E\hat{\theta}_n = \theta\frac{n-1}{n}+(\theta+n)\frac{1}{n} = \theta+1 \neq \theta.$$

This implies that $\lim_{n\to\infty} E\hat{\theta}_n = \theta+1 \neq \theta$. Also

$$E(\hat{\theta}_n-\theta)^2 = n^2\frac{1}{n} = n$$

from which we obtain

$$\lim_{n\to\infty} E(\hat{\theta}_n-\theta)^2 = \infty \neq 0$$

and there is no M such that $E(\hat{\theta}_n-\theta)^2 \leqq M < \infty$ for all n and all $\theta \in \Theta$. Therefore, none of the above properties are satisfied, and thus, they cannot be necessary properties.

SOME PROPERTIES OF CONSISTENCY

We now demonstrate some true properties of consistency.

A If 1. $\lim_{n\to\infty} E\hat{\theta}_n = \theta \quad \text{all } \theta \in \Theta$

and

2. $\lim_{n\to\infty} \text{Var } \hat{\theta}_n = 0 \qquad \text{all } \theta \in \Theta$

then

$$\lim_{n\to\infty} E(\hat{\theta}_n-\theta)^2 = 0,$$

(the sequence is squared error consistent). This is easily seen from the fact that

$$\begin{aligned} E(\hat{\theta}_n-\theta)^2 &= (E\hat{\theta}_n-\theta)^2+E(\hat{\theta}_n-E\hat{\theta}_n)^2 \\ &= (E\hat{\theta}_n-\theta)^2+\text{Var } \hat{\theta}_n. \end{aligned}$$

In the limit both terms on the right-hand side go to zero by assumption and therefore the limit of the left-hand side must be zero.

B If

$$\lim_{n\to\infty} E(\hat{\theta}_n-\theta)^2 = 0 \qquad \text{all } \theta \in \Theta$$

then

$$\lim_{n\to\infty} P(|\,\hat{\theta}_n-\theta\,| > \varepsilon) = 0 \qquad \text{any } \varepsilon > 0$$
$$\text{all } \theta \in \Theta.$$

This can easily be seen by the use of Markov's inequality. For any $\varepsilon > 0$

$$P(|\,\hat{\theta}_n-\theta\,| > \varepsilon) \leqq \frac{E(\hat{\theta}_n-\theta)^2}{\varepsilon^2}.$$

Since the limit of the right-hand side is zero by assumption, the limit of the left-hand side must be zero. This says that squared error consistency implies simple consistency.

C If 1. $\lim_{n\to\infty} P(|\,\hat{\theta}_n-\theta\,| > \varepsilon) = 0 \qquad \text{any } \varepsilon > 0$

$\text{all } \theta \in \Theta$

and

2. $|\,\hat{\theta}_n-\theta\,| \leqq M < \infty \qquad \text{all } \theta \in \Theta$

$\text{all } x_1, x_2, x_3, \ldots, x_n, \ldots$

(which says that $|\,\hat{\theta}_n-\theta\,|$ is uniformly bounded) then

$$\lim_{n\to\infty} E(\hat{\theta}_n-\theta)^2 = 0 \qquad \text{all } \theta \in \Theta.$$

This can be seen from the following. For any $\varepsilon > 0$

$$\begin{aligned} E(\hat{\theta}_n-\theta)^2 &= \int (\hat{\theta}_n-\theta)^2 dF_\theta(x_1, x_2, x_3, \ldots, x_n) \\ &= \int_{|\hat{\theta}_n-\theta|\leqq\varepsilon} (\hat{\theta}_n-\theta)^2 dF_\theta + \int_{|\hat{\theta}_n-\theta|>\varepsilon} (\hat{\theta}_n-\theta)^2 dF_\theta \\ &\qquad\qquad (\text{where } F_\theta = F_\theta(x_1, x_2, x_3, \ldots, x_n)) \\ &\leqq \int_{|\hat{\theta}_n-\theta|\leqq\varepsilon} \varepsilon^2 dF_\theta + \int_{|\hat{\theta}_n-\theta|>\varepsilon} M^2 dF_\theta = \varepsilon^2 \int_{|\hat{\theta}_n-\theta|\leqq\varepsilon} dF_\theta + M^2 \\ &\qquad\qquad \int_{|\theta_n-\theta|>\varepsilon} dF_\theta. \end{aligned}$$

Therefore,

$$E(\hat{\theta}_n-\theta)^2 \leqq \varepsilon^2[1-P(|\,\hat{\theta}_n-\theta\,| > \varepsilon)]+M^2 P(|\,\hat{\theta}_n-\theta\,| > \varepsilon)$$

which implies

$$\lim_{n\to\infty} E(\hat{\theta}_n-\theta)^2 \leqq \varepsilon^2.$$

Since this inequality must hold for every $\varepsilon > 0$, we conclude that

$$\lim_{n\to\infty} E(\hat{\theta}_n - \theta)^2 = 0.$$

This says that if the estimators and the parameter are uniformly bounded, then simple consistency implies squared error consistency.

EXERCISE 10-1

Give an example to illustrate that if the bound is not uniform in n, then consistency may not imply squared error consistency.

10-4 Maximum Likelihood

The principle known as *maximum likelihood* has an intuitive appeal for discrete random variables. The idea is simply the following. Look at the observed values $x_1, x_2, x_3, \ldots, x_n$ of the random variables $X_1, X_2, X_3, \ldots, X_n$ and list the joint probabilities of this event occurring for each value of θ. Then the maximum likelihood estimate $\hat{\theta}_{\text{M.L.}}$ of θ is the value (or the values) of θ for which the probability of this event $\{x_1, x_2, x_3, \ldots, x_n\}$ is the largest.

EXAMPLE 10-5

For each $\theta \in \Theta$ let

$$P(X = \theta) = \tfrac{1}{4}$$
$$P(X = \theta+1) = \tfrac{1}{2}$$
$$P(X = \theta+2) = \tfrac{1}{4}.$$

If x is observed, then what is the maximum likelihood estimate of θ?

$$\begin{aligned} \text{Clearly } P(x) &= \tfrac{1}{4} && \text{for } \theta = x \\ &= \tfrac{1}{2} && \text{for } \theta = x-1 \\ &= \tfrac{1}{4} && \text{for } \theta = x-2 \\ &= 0 && \text{all other } \theta \in \Theta. \end{aligned}$$

Therefore $\hat{\theta}_{\text{M.L.}} = x-1$ is then the maximum likelihood estimate of θ.

Note that in the example above we could have written

$$\hat{\theta}_{\text{M.L.}} = X-1$$

and have had an estimator rather than an estimate. This can usually be done, and, in such cases we call the estimator the *maximum likelihood estimator*.

When $X_1, X_2, X_3, \ldots, X_n$ are all continuous random variables, then each set $\{x_1, x_2, x_3, \ldots, x_n\}$ has zero probability of occurring. Here intuition must be extended in order to apply the maximum likelihood principle, which simply treats the density function as if it were a probability mass function.

EXAMPLE 10-6

Let X be a normal random variable with density function

$$f(x) = \frac{1}{\sqrt{(2\pi)}} e^{-(x-\theta)^2/2} \qquad \begin{array}{l} -\infty < x < \infty \\ -\infty < \theta < \infty. \end{array}$$

Then for an observed value x we obtain

$$f(x) = \frac{1}{\sqrt{(2\pi)}} e^{-(x-\theta)^2/2}.$$

This function is a maximum when $\theta = x$ and, therefore, $\hat{\theta}_{\text{M.L.}} = X$ is the maximum likelihood estimator for θ.

THE GENERAL TREATMENT

The general way to find the maximum likelihood estimator is to consider the function $F_\theta'(x_1, x_2, x_3, \ldots, x_n)$. This is the joint density or mass function when it is a function of the $x_1, x_2, x_3, \ldots, x_n$ for fixed θ. We now consider the x's fixed and let θ be the variable. This is then called the *likelihood function*. The θ which maximizes the likelihood function is called the *maximum likelihood estimator* for θ.

When Θ is discrete, we can list the values of the likelihood function for the different values of θ. We then choose the θ (or one of the values of θ) for which the likelihood is largest.

When Θ is continuous, listing is hopeless. If, however, the likelihood function is continuous and differentiable in θ, then we may use the usual techniques for finding maxima—namely, assuming that θ is real valued, we take a derivative of the likelihood function with respect to θ, set it to zero, and solve for the values of θ. We then complement these with the end points of Θ and then choose from all these the one which affords the largest likelihood as the estimator for θ.

This general procedure is often simplified by noting that log A is a monotonic function of A. We can, therefore, find the same set of θ points by setting the derivative of the logarithm of the likelihood function to zero.

EXAMPLE 10-7

Let $X_1, X_2, X_3, \ldots, X_n$ be such that

$$f(x_1, x_2, x_3, \ldots, x_n) = \frac{1}{(2\pi)^{n/2}} \exp\left[-\frac{1}{2}\sum_{i=1}^{n}(x_i-\theta)^2\right] \qquad \begin{array}{l} -\infty < x_i < \infty \\ \Theta = (-\infty, \infty). \end{array}$$

Now

$$\frac{\partial}{\partial\theta}\ln f(x_1, x_2, x_3, \ldots, x_n) = \frac{\partial}{\partial\theta}[-\ln(2\pi)^{n/2} - \tfrac{1}{2}\sum_{i=1}^{n}(x_i-\theta)^2]$$

$$= \sum_{i=1}^{n}(x_i-\theta).$$

Setting

$$\sum_{i=1}^{n} (x_i - \theta) = 0$$

gives

$$\hat{\theta}_{\text{M.L.}} = \frac{1}{n} \sum_{i=1}^{n} x_i$$

or

$$\hat{\theta}_{\text{M.L.}} = \bar{X}_n$$

as the maximum likelihood estimator.

AN INTERESTING RESULT

Sometimes we are interested in estimating some function of θ, say $g(\theta)$, rather than θ itself. In this case it is interesting to note that if $g(\theta)$ is any function of θ then $\widehat{g(\theta)}_{\text{M.L.}} = g(\hat{\theta}_{\text{M.L.}})$, that is, the maximum likelihood estimator for a function of θ is the same function of the maximum likelihood estimator for θ. This is due to the fact that we are not dealing with probabilities in the likelihood function but rather with an algebraic relation. The likelihood function is maximum for a particular value of θ. If the function had been written in terms of $g(\theta)$ it would be maximum at this same value of θ.

That this property is not necessarily true of expected values was seen in the case of S_n and S_n^2, where the unbiased estimator for σ^2 was not the square of an unbiased estimator for σ.

SOME SPECIAL PROPERTIES OF MAXIMUM LIKELIHOOD ESTIMATORS

If we now consider the very special case where there is an infinite sequence of independent and identically distributed random variables $X_1, X_2, X_3, \ldots, X_n, \ldots$ with Θ an interval and where the likelihood function is twice differentiable in Θ then we obtain the following properties, which we will not derive:

a) The maximum likelihood estimator is squared error consistent.

b) The maximum likelihood estimator is such that

$$\sqrt{n}(\hat{\theta}_{\text{M.L.}} - \theta) \xrightarrow{D} N(0, \sigma^2)$$

where

$$\sigma^2 = -1/\ln E\left[\frac{\partial^2}{\partial\theta^2} \ln F_\theta'\right]$$

and where F_θ' is the common density or mass function for each X.

c) The variance of the maximum likelihood estimator approaches the Cramér–Rao lower bound as $n \to \infty$. (The Cramér–Rao lower bound will be defined in the following chapter.)

These properties are asymptotic properties of the maximum likelihood estimator. Extreme caution should be exercised in using asymptotic properties to judge the characteristics of an estimator for small or even moderate sample sizes.

EXAMPLE 10-8

Let $X_1, X_2, X_3, \ldots, X_n$ be independent and identically distributed random variables with common density function

$$f(x) = \frac{1}{\sqrt{(2\pi)}} e^{-(x-\theta)^2/2} \qquad -\infty < x < \infty \qquad \Theta = (0, .1).$$

We are asked to compare the estimator $\hat{\theta}_1 = \bar{X}_n$ with the constant estimator $\hat{\theta}_2 = 0.05$.

Now

$$E(\hat{\theta}_1 - \theta)^2 = \text{Var } \bar{X}_n = 1/n$$

while

$$E(\hat{\theta}_2 - \theta)^2 = (.05 - \theta)^2 \leqq 0.0025 = \frac{1}{400} \text{all } \theta \in \Theta.$$

From this we conclude that $\hat{\theta}_1$ is squared error consistent (it also satisfies conditions **b** and **c**), while $\hat{\theta}_2$ is not. Asympotically, $\hat{\theta}_1$ has much more desirable properties than does $\hat{\theta}_2$. However for $n < 400$, $\hat{\theta}_2$ has a uniformly smaller mean squared error.

EXERCISE 10-2

Find the maximum likelihood estimator for the situation in Example 10-8 and compare it with $\hat{\theta}_1$ and $\hat{\theta}_2$.

EXERCISE 10-3

Let $X_1, X_2, X_3, \ldots, X_n$ be independent and have common density function

$$f(x) = \frac{1}{\sqrt{(2\pi)}\sigma} e^{-(x^2/2\sigma^2)} \qquad -\infty < x < \infty \qquad \sigma > 0.$$

a) Find $\hat{\sigma}_{\text{M.L.}}$ and, therefore, $\hat{\sigma}^2_{\text{M.L.}}$.

b) Compare $E(\hat{\sigma}^2_{\text{M.L.}} - \sigma^2)^2$ with $E(S_n^2 - \sigma^2)^2$. Which is smaller?

c) Compare the mean squared errors found in b with $E(\phi_n^2 - \sigma^2)^2$ where

$$\phi_n^2 = \frac{n-1}{n+1} S_n^2.$$

10-5 Problems

Problems 1–8 refer to the following estimators:

$$\hat{\theta}_a = X_1 - X_2 \qquad \hat{\theta}_i = |X_1| - |X_2| + |X_3|$$

$$\hat{\theta}_b = \frac{X_1 - X_2}{X_3} \qquad \hat{\theta}_j = \bar{X}_n / S_n^2$$

$$\hat{\theta}_c = X_1^2 - X_2^2 \qquad \hat{\theta}_k = \bar{X}_n / S_n$$

$$\hat{\theta}_d = \frac{X_1^2 - X_2^2}{X_3^2} \qquad \hat{\theta}_l = \bar{X}_n^2 / S_n^2$$

$$\hat{\theta}_e = \sin^2 X_1 + \sin^2 X_2 \qquad \hat{\theta}_m = M_n{}^2$$

$$\hat{\theta}_f = X_1 + X_2 - X_3 \qquad \hat{\theta}_n = \bar{X}_n{}^2/\bar{X}_n{}^2$$

$$\hat{\theta}_g = \frac{X_1{}^2 X_2}{X_3} \qquad \hat{\theta}_o = \bar{X}_n{}^2 - \bar{X}_n{}^2$$

$$\hat{\theta}_h = \frac{1}{n}\sum_{i=1}^{n} |X_i| \qquad \hat{\theta}_p = M_n{}^2/S_n{}^2.$$

1. Which of the estimators are location invariant?
2. Which of the estimators are scale invariant?
3. Which of the estimators are invariant?
4. Which of the estimators are first-order location invariant?
5. Which of the estimators are first-order scale invariant?
6. Which of the estimators are second order scale invariant?
7. Which of the estimators are reflection invariant?
8. Which of the estimators are permutation invariant?
9. If $X_1, X_2, X_3, \ldots, X_n$ are independent and identically distributed with common density function

$$f(x) = \frac{1}{\sqrt{(2\pi)}\sigma} e^{-(x-\mu)^2/2\sigma^2} \qquad -\infty < x < \infty$$

   find the squared bias, variance, and mean squared error of $M_n{}^2$.

10. For the situation of Problem 9, find the squared bias, variance and mean squared error of $S_n{}^2$.
11. Find a linear unbiased estimator for the μ in Problem 9.
12. Let $X_1, X_2, X_3, \ldots, X_n, \ldots$ be independently distributed with the density for X_n given by

$$f(x) = \frac{1}{\Gamma(\alpha\sqrt{n})(\beta/\sqrt{n})^{\beta\sqrt{n}}} x^{\alpha\sqrt{n}-1} e^{-\sqrt{n}x/\beta} \qquad 0 < x < \infty.$$

   a. Is the sequence $\hat{\theta}_n = X_n$ $n = 1, 2, 3, \ldots$ a simple consistent sequence of estimators for $\alpha\beta$?

   b. Is it a squared error consistent sequence of estimators?

13. If $X_1, X_2, X_3, \ldots, X_n, \ldots$ are independent and identically distributed with a gamma distribution, find a squared error consistent sequence of estimators for the mean.
14. If in your answer to Problem 13 you added $1/n$ to $\hat{\theta}_n$ $n = 1, 2, 3, \ldots,$ would you still have a squared error consistent sequence of estimators?
15. Find the maximum likelihood estimator for the parameter in the poisson distribution.
16. Is the maximum likelihood estimator for the parameter β of the uniform distribution unique?
17. If X is a normal random variable with mean μ and variance σ^2 find the maximum likelihood estimators for μ and σ^2.

18. If $X_1, X_2, X_3, \ldots, X_n$ are independent and identically distributed with a common normal distribution with mean μ and variance σ^2, find the maximum likelihood estimator for μ.

19. If in Problem 18 μ is known, find the maximum likelihood estimator for σ^2.

20. If in Problem 18 both μ and σ^2 are unknown, find the maximum likelihood estimator for σ^2.

CHAPTER 11

Uniformly Minimum Variance Unbiased Estimators

We will devote this chapter to methods of finding uniformly minimum variance unbiased estimators. It should be pointed out that, for a particular situation, a uniformly minimum variance unbiased estimator may not exist. When one does exist, however, it is unique. This can be seen from the following.

Let us say that $\hat{\theta}_1$ and $\hat{\theta}_2$ are both uniformly minimum variance unbiased estimators for θ each with variance $m(\theta)$. Define $\hat{\theta}^* = (\hat{\theta}_1 + \hat{\theta}_2)/2$. Clearly, $E\hat{\theta}^* = (\theta+\theta)/2 = \theta$, demonstrating that $\hat{\theta}^*$ is an unbiased estimator. Now

$$\begin{aligned}\text{Var } \hat{\theta}^* &= \tfrac{1}{4}[\text{Var } \hat{\theta}_1 + \text{Var } \hat{\theta}_2 + 2 \text{ Cov}(\hat{\theta}_1, \hat{\theta}_2)] \\ &= \tfrac{1}{4}[2m(\theta) + 2E((\hat{\theta}_1-\theta)(\hat{\theta}_2-\theta))].\end{aligned}$$

The Cauchy–Schwartz inequality for integrals implies that for any functions $g(x)$ and $h(x)$ and non-negative function $f(x)$

$$\int g(x)h(x)f(x)\,dx \leqq \sqrt{\left(\int [g(x)]^2 f(x)\,dx \int [h(x)]^2 f(x)\,dx\right)},$$

with a similar relation holding for discrete variables. Thus,

$$E[(\hat{\theta}_1-\theta)(\hat{\theta}_2-\theta)] \leqq \sqrt{(E(\hat{\theta}_1-\theta)^2 E(\hat{\theta}_2-\theta)^2)} = m(\theta).$$

Therefore

$$\text{Var } \hat{\theta}^* \leqq m(\theta).$$

Since $m(\theta)$ is the uniformly minimum variance we have that $\text{Var } \hat{\theta}^* = m(\theta)$, which occurs only when

$$E((\hat{\theta}_1-\theta)(\hat{\theta}_2-\theta)) = m(\theta).$$

Using this, we obtain

$$\begin{aligned}E(\hat{\theta}_1-\hat{\theta}_2)^2 &= E(\hat{\theta}_1-\theta)^2 + E(\hat{\theta}_2-\theta)^2 - 2E[(\hat{\theta}_1-\theta)(\hat{\theta}_2-\theta)] \\ &= m(\theta) + m(\theta) - 2m(\theta) = 0.\end{aligned}$$

Since $E(\hat{\theta}_1-\hat{\theta}_2)^2$ is the integral of a positive quantity, it can not be zero unless the quantity is zero. Therefore,

$$\hat{\theta}_1(x_1, x_2, x_3, \ldots, x_n) = \hat{\theta}_2(x_1, x_2, x_3, \ldots, x_n)$$

whenever $F_\theta'(x_1, x_2, x_3, \ldots, x_n) \neq 0$. We say that $E(\hat{\theta}_1-\hat{\theta}_2)^2 = 0$ implies that $\hat{\theta}_1 = \hat{\theta}_2$, or the uniformly minimum variance unbiased estimator is unique.

11-1 Sufficient Statistics

One of the more useful properties a statistic can possess is its being sufficient. The objective of this property is to find all the relevant information concerning the parameter θ and use it in a statistic while at the same time eliminating much of the unrelated information. For example, we could not disregard the number of heads in n flips of a coin when estimating the probability of heads. We could, however, disregard the time of day that these heads were recorded without sacrificing any useful information.

DEFINITION OF A SUFFICIENT STATISTIC

Let us say that the random variables $X_1, X_2, X_3, \ldots, X_n$ have a joint distribution which is a member of the family $\{F_\theta(x_1, x_2, x_3, \ldots, x_n): \theta \in \Theta\}$.

We define a sufficient statistic as follows:

Definition A *sufficient statistic* for θ is any statistic $\phi(X_1, X_2, X_3, \ldots, X_n)$ such that

$$F_\theta'(x_1, x_2, x_3, \ldots, x_n \mid \phi) = g(x_1, x_2, x_3, \ldots, x_n) \quad \text{all } \theta \in \Theta$$

where $g(\cdot)$ is a function which does not contain θ and where its range of definition does not contain θ.

EXAMPLE 11-1

Let $\{F_\theta(x_1, x_2, x_3, \ldots, x_n): \theta \in \Theta\}$ be such that F_θ is given by the mass function

$$P_\theta(x_1, x_2, x_3, \ldots, x_n) = \prod_{i=1}^{n} \theta^{x_i}(1-\theta)^{1-x_i} \qquad x_i = 0, 1$$

with $\Theta = (0, 1)$.

Consider the statistic

$$\phi = \sum_{i=1}^{n} X_i.$$

Now

$$P_\theta(x_1, x_2, x_3, \ldots, x_n) = \theta^{\phi}(1-\theta)^{n-\phi}$$

while

$$P_\theta(\phi) = \binom{n}{\phi} \theta^\phi (1-\theta)^{n-\phi}$$

and

$$P_\theta(x_1, x_2, x_3, \ldots, x_n \mid \phi) = \frac{\theta^\phi(1-\theta)^{n-\phi}}{\binom{n}{\phi}\theta^\phi(1-\theta)^{n-\phi}} = \frac{1}{\binom{n}{\phi}}$$

which is independent of θ. Therefore, ϕ is a sufficient statistic for θ.

EXAMPLE 11-2

Let $\{F_\theta(x_1, x_2, x_3, \ldots, x_n): \theta \in \Theta\}$ be such that F_θ is given by the mass function

$$P_\theta(x_1, x_2, x_3, \ldots, x_n) = \prod_{i=1}^{n} \frac{\theta^{x_i}}{x_i!} e^{-\theta} \qquad x_i = 0, 1, 2, \ldots$$

with $\Theta = (0, \infty)$.

Now let

$$\phi = \sum_{1=i}^{n} X_i.$$

As has already been shown in Section 9-2

$$P_\theta(\phi) = \frac{e^{-n\theta}(n\theta)^\phi}{\phi!} \qquad \phi = 0, 1, 2, \ldots.$$

Therefore

$$P_\theta(x_1, x_2, x_3, \ldots, x_n \mid \phi) = \frac{e^{-n\theta}\theta^\phi/(\prod_{i=1}^{n} x_i!)}{\dfrac{e^{-n\theta}(n\theta)^\phi}{\phi!}} = \frac{\phi!}{n^\phi \prod_{i=1}^{n} x_i!}$$

which is independent of θ. Therefore ϕ is a sufficient statistic for θ.

EXAMPLE 11-3

Let $\{F_\theta(x_1, x_2, x_3, \ldots, x_n): \theta \in \Theta\}$ be such that F_θ is given by the density function

$$f_\theta(x_1, x_2, x_3, \ldots, x_n) = \frac{1}{\theta^n} \qquad 0 < \text{all } x_i < \theta$$

$$= 0 \qquad \text{otherwise}$$

with $\Theta = (0, \infty)$.

Let $\phi = \max[X_1, X_2, X_3, \ldots, X_n]$. In order to find the density of ϕ we note that the X_i are independently distributed with the same uniform distribution and that ϕ is less than z if and only if each X_i is less than z. Thus

$$F_\theta(z) = P(\phi < z) = \prod_{i=1}^{n} P(X_i < z) = \prod_{i=1}^{n} \int_0^z \frac{1}{\theta}\, dx_i = \frac{z^n}{\theta^n} \qquad 0 < z < \theta.$$

The density may now be obtained from

$$\frac{\partial}{\partial z} F_\theta(z) = f_\theta(z) = \frac{nz^{n-1}}{\theta^n} \qquad 0 < z < \theta$$

$$= 0 \qquad \text{otherwise.}$$

Or

$$f_\theta(\phi) = \frac{n\phi^{n-1}}{\theta^n} \qquad 0 < \phi < \theta$$

$$= 0 \qquad \text{otherwise.}$$

From this we obtain

$$f(x_1, x_2, x_3, \ldots, x_n \mid \phi) = \frac{1/\theta^n}{n\phi^{n-1}/\theta^n} = \frac{1}{n\phi^{n-1}} \qquad 0 < \text{all } x_i < \phi$$

$$= 0 \qquad \text{otherwise.}$$

The conditional density is therefore independent of θ and ϕ is a sufficient statistic for θ.

NEYMAN FACTORIZATION CRITERION

The definition of sufficiency is often tedious to use in checking sufficiency. The Neyman factorization criterion gives an equivalent definition which is much simpler to use.

The Criterion The statistic ϕ is sufficient for θ if and only if

$$F_\theta'(x_1, x_2, x_3, \ldots, x_n) = g(x_1, x_2, x_3, \ldots, x_n)h_\theta(\phi) \qquad \text{all } \theta \in \Theta$$

where $g(\cdot)$ is not a function of θ and where $h_\theta(\cdot)$ is a function only of θ and $\phi(X_1, X_2, X_3, \ldots, X_n)$.

Proof of Equivalence Assume that ϕ is a sufficient statistic. Then

$$F_\theta'(x_1, x_2, x_3, \ldots, x_n \mid \phi) = g(x_1, x_2, x_3, \ldots, x_n).$$

Now

$$F_\theta'(x_1, x_2, x_3, \ldots, x_n) = F_\theta'(x_1, x_2, x_3, \ldots, x_n \mid \phi)F_\theta'(\phi)$$

$$= g(x_1, x_2, x_3, \ldots, x_n)h_\theta(\phi) \qquad \text{all } \theta \in \Theta$$

where $h_\theta(\phi) = F_\theta'(\phi)$.

For the converse let us assume that

$$F_\theta'(x_1, x_2, x_3, \ldots, x_n) = g(x_1, x_2, x_3, \ldots, x_n)h_\theta(\phi) \qquad \text{all } \theta \in \Theta.$$

In general

$$F_\theta'(x_1, x_2, x_3, \ldots, x_n \mid \phi) = \frac{F_\theta'(x_1, x_2, x_3, \ldots, x_n, \phi)}{F_\theta'(\phi)}.$$

In this case, since ϕ is a function of $x_1, x_2, x_3, \ldots, x_n$, we have

$$F_\theta'(x_1, x_2, x_3, \ldots, x_n, \phi) = F_\theta'(x_1, x_2, x_3, \ldots, x_n)$$

$$= g(x_1, x_2, x_3, \ldots, x_n)h_\theta(\phi).$$

Remark *Since ϕ is a function of $x_1, x_2, x_3, \ldots, x_n$ the probabilities of jointly observing $x_1, x_2, x_3, \ldots, x_n, \phi$ can be non-zero only when the value of ϕ is the value it obtains for the particular set $x_1, x_2, x_3, \ldots, x_n$. We therefore set $F_\theta'(x_1, x_2, x_3, \ldots, x_n, \phi) = F_\theta'(x_1, x_2, x_3, \ldots, x_n)$ with the condition that ϕ has the value determined by $x_1, x_2, x_3, \ldots, x_n$.*

Defining $\Omega(\phi)$ as the set of all those n-tuples $\{x_1, x_2, x_3, \ldots, x_n\}$ such that $\phi(x_1, x_2, x_3, \ldots, x_n) \leqq \phi$, we have

$$F_\theta(\phi) = \int_{\Omega(\phi)} dF_\theta(x_1, x_2, x_3, \ldots, x_n)$$

$$= \int_{\Omega(\phi)} h_\theta(\phi) g(x_1, x_2, x_3, \ldots, x_n) \prod_{i=1}^{n} dx_i$$

(where $\int \Pi_{i=1}^n dx_i$ represents an integral over those x's which are continuous variables and a sum over those x's which are discrete variables)

$$= h_\theta(\phi) \int_{\Omega(\phi)} g(x_1, x_2, x_3, \ldots, x_n) \prod_{i=1}^{n} dx_i.$$

Thus,

$$F_\theta'(\phi) = h_\theta(\phi) G'(\phi)$$

where $G'(\phi)$ is some function which does not depend on θ. Therefore,

$$F_\theta'(x_1, x_2, x_3, \ldots, x_n \mid \phi) = \frac{F_\theta'(x_1, x_2, x_3, \ldots, x_n, \phi)}{F_\theta'(\phi)}$$

$$= \frac{F_\theta'(x_1, x_2, x_3, \ldots, x_n)}{F_\theta'(\phi)}$$

$$= \frac{h_\theta(\phi) g(x_1, x_2, x_3, \ldots, x_n)}{h_\theta(\phi) G'(\phi)}$$

$$= \frac{g(x_1, x_2, x_3, \ldots, x_n)}{G'(\phi)}$$

$$= k(x_1, x_2, x_3, \ldots, x_n) \qquad \text{all } \theta \in \Theta$$

where $k(\cdot)$ is not a function of θ. The proof is now complete.

The Neyman criterion says that ϕ is a sufficient statistic for θ if and only if the density or mass function can be written as a product of two functions. The first function must be independent of θ, and the second function must depend on the random variables only through ϕ.

o

EXAMPLE 11-4

Let $\{F_\theta(x_1, x_2, x_3, \ldots, x_n)\colon \theta \in \Theta\}$ be such that F_θ is given by the density function

$$f_\theta(x_1, x_2, x_3, \ldots, x_n) = \frac{1}{(2\pi)^{n/2}\sigma^n} \exp\left(-\sum_{i=1}^{n} (x_i - \theta)^2/2\sigma^2\right) \qquad -\infty < \text{all } x_i < \infty$$

with $\Theta = (-\infty, \infty)$.

We can rearrange the exponent to obtain

$$f_\theta(x_1, x_2, x_3, \ldots, x_n) = \frac{1}{(2\pi)^{n/2}\sigma^n} \exp\left(-\sum_{i=1}^{n} (x_i - \bar{x}_n + \bar{x}_n - \theta)^2/2\sigma^2\right)$$

$$= \frac{1}{(2\pi)^{n/2}\sigma^n} \exp\left(-\sum_{i=1}^{n} (x_i - \bar{x}_n)^2/2\sigma^2 - n(\bar{x}_n - \theta)^2/2\sigma^2\right)$$

$$= \frac{1}{(2\pi)^{n/2}\sigma^n} \exp\left(-\sum_{i=1}^{n} (x_i - \bar{x}_n)^2/2\sigma^2\right) \exp\left(-n(\bar{x}_n - \theta)^2/2\sigma^2\right).$$

Therefore, $\bar{X}_n$ is a sufficient statistic for θ.

EXAMPLE 11-5

Let $\{F_\theta(x_1, x_2, x_3, \ldots, x_n)\colon \theta \in \Theta\}$ be such that F_θ is given by the density function

$$f_\theta(x_1, x_2, x_3, \ldots, x_n) = \frac{1}{\theta^n} \qquad 0 < \text{all } x_i < \theta$$

$$= 0 \qquad \text{otherwise}$$

with $\Theta = (0, \infty)$.

Let $\phi = \max[X_1, X_2, X_3, \ldots, X_n]$. We can now write the density function as

$$f_\theta(x_1, x_2, x_3, \ldots, x_n) = \frac{1}{\theta^n}\,\delta(\phi < \theta) \qquad 0 < \text{all } x_i < \phi$$

$$\text{otherwise}$$

where

$$\delta(\phi < \theta) = 1 \quad \text{if} \quad \phi < \theta$$

$$= 0 \quad \text{if} \quad \phi \geqq \theta.$$

We have, therefore, factored $f_\theta(x_1, x_2, x_3, \ldots, x_n)$ into a function of the x's (i.e. unity) and a function of θ and ϕ. Thus, ϕ is a sufficient statistic for θ.

A SUFFICINAL ESTIMATOR

We now define a procedure that will be very useful in what follows. Given any estimator $\hat{\theta}$ and any sufficient statistic ϕ we define the *sufficinal estimator* for these by

$$\hat{\theta}^* = E(\hat{\theta} \mid \phi).$$

Since the conditional distribution of the X's given ϕ is independent of θ, $\hat{\theta}^*$ is a function of the observations alone, and is therefore also an estimator, provided, of course, that θ is the only unknown parameter.

From Exercise 4-9 we note that

$$E\hat{\theta}^* = E[E(\hat{\theta} \mid \phi)] = E\hat{\theta}$$

and thus, the sufficinal estimator has the same expected value as the original estimator. This means that if $\hat{\theta}$ is unbiased for θ, then $\hat{\theta}^*$ is unbiased for θ. It should be pointed out that unless $\hat{\theta}^*$ is a one to one function of ϕ it may, itself, not be a sufficient statistic.

The importance of sufficiency lies in the following result.

AN IMPORTANT RESULT

Let $\hat{\theta}$ be any unbiased estimator for θ. For any sufficient statistic ϕ (assuming one exists), form the sufficinal statistic $\hat{\theta}^*$. Now

$$\begin{aligned}\operatorname{Var}\hat{\theta} &= E(\hat{\theta}-\theta)^2 = E(\hat{\theta}-\hat{\theta}^*+\hat{\theta}^*-\theta)^2 \\ &= E(\hat{\theta}^*-\theta)^2+E(\hat{\theta}-\hat{\theta}^*)^2+2E[(\hat{\theta}-\hat{\theta}^*)(\hat{\theta}^*-\theta)].\end{aligned}$$

Looking at the last term, we find that

$$\begin{aligned}E((\hat{\theta}-\hat{\theta}^*)(\hat{\theta}^*-\theta)) &= E\{E[(\hat{\theta}-\hat{\theta}^*)(\hat{\theta}^*-\theta) \mid \phi]\} \\ &= E\{(\hat{\theta}^*-\theta)E[(\hat{\theta}-\hat{\theta}^*) \mid \phi]\} \\ &= E[(\hat{\theta}^*-\theta)(\hat{\theta}^*-\hat{\theta}^*)] = 0.\end{aligned}$$

Thus

$$\operatorname{Var}\hat{\theta} = \operatorname{Var}\hat{\theta}^*+E(\hat{\theta}-\hat{\theta}^*)^2.$$

Since the last term cannot be negative

$$\operatorname{Var}\hat{\theta}^* \leqq \operatorname{Var}\hat{\theta}.$$

Thus, sufficinalization can only reduce the variance of an unbiased estimator.

If $\hat{\theta}$ were the uniformly minimum variance unbiased estimator, then clearly $\operatorname{Var}\hat{\theta}^* = \operatorname{Var}\hat{\theta}$ and $E(\hat{\theta}-\hat{\theta}^*)^2 = 0$. This latter result is an integral of a non-negative function and we can, therefore, say $\hat{\theta} = \hat{\theta}^*$. This means that the uniformly minimum variance unbiased estimator is a sufficinal estimator when both a uniformly minimum variance unbiased estimator and a sufficient statistic exist.

The importance of this result should not be underestimated. The uniformly minimum variance unbiased estimator, if its exists, is a function of any sufficient statistic, if one exists. This means that in order to find the uniformly minimum variance unbiased estimator for θ for the family $\{F_\theta(x_1, x_2, x_3, \ldots, x_n): \theta \in \Theta\}$ one may first find some sufficient statistic for θ say ϕ. Then find the uniformly minimum variance unbiased estimator for θ in the family $\{F_\theta(\phi): \theta \in \Theta\}$ where $F_\theta(\phi)$ is the distribution of ϕ derived from $F_\theta(x_1, x_2, x_3, \ldots, x_n)$. This simplifies the problem greatly by reducing the family of distributions from a family of n-dimensional multivariate distributions to a family of univariate distributions. The uniformly minimum variance unbiased estimator for the second family will also be the uniformly minimum variance unbiased estimator for the first family.

We have already seen that the mean squared error is the sum of a variance term and a bias term. Since $E\hat{\theta}^* = E\hat{\theta}$, the bias term is not altered in the sufficinalization process, while the variance can be no larger. Hence we conclude that the mean squared error can possibly be reduced (but never increased) by the process of sufficinalization.

Given an unbiased estimator and a sufficient statistic, we can form the sufficinal estimator. But how can we be sure that this is the uniformly minimum variance unbiased estimator? This question is answered for many cases in the next section.

11-2 Completeness

Many of the families of distributions which we have studied have the very convenient property of *completeness*, which we now define.

Definition We say that the family of distributions

$$\{F_\theta(x_1, x_2, x_3, \ldots, x_n): \theta \in \Theta\}$$

is *complete* if

$$E\hat{\theta}(X_1, X_2, X_3, \ldots, X_n) = 0 \qquad \text{all } \theta \in \Theta$$

implies that $\hat{\theta} = 0$.

If $\{F_\theta(x_1, x_2, x_3, \ldots, x_n): \theta \in \Theta\}$ is complete then it is easily seen that there is at most one unbiased estimator for θ. If there were two different estimators $\hat{\theta}_1$ and $\hat{\theta}_2$ we could form the estimator $\hat{\theta} = \hat{\theta}_1 - \hat{\theta}_2$ and find $E\hat{\theta} = \theta - \theta = 0$, which would imply that $\hat{\theta}_1 = \hat{\theta}_2$, contradicting that they were different.

Many univariate distributions form a complete family. It is, however, rare to find a multivariate distribution which forms a complete family. If two of the variables (e.g., X_1 and X_2) had the same mean, then $\hat{\theta} = X_1 - X_2$ would be a nonzero estimator with zero expectation. The family then could not be complete. We would not study completeness at all if it were not for the fact that the uniformly minimum variance unbiased estimator must be a function of a sufficient statistic. One can, thus, reduce the problem to one of finding an estimator in the family of univariate distributions of this sufficient statistic.

COMPLETE SUFFICIENT STATISTIC

When we have a family $\{F_\theta(x_1, x_2, x_3, \ldots, x_n): \theta \in \Theta\}$ and a sufficient statistic ϕ for θ then we can form the induced family $\{F_\theta(\phi): \theta \in \Theta\}$ where

$F_\theta(\phi) = \int dF(x_1, x_2, x_3, \ldots, x_n)$ and where the integral is over all $x_1, x_2, x_3, \ldots, x_n$ such that $\phi(x_1, x_2, x_3, \ldots, x_n) \leqq \phi$.

If this induced family is complete we say that ϕ is a *complete sufficient statistic* which will be denoted by ϕ^*. If there is a uniformly minimum variance unbiased estimator we know that it must be a function of every sufficient statistic and, therefore, that it must be a function of ϕ^*. Since ϕ^* has a complete family there is at most one unbiased estimator which is a function of ϕ^*. Therefore, if we have a complete sufficient statistic all one need do is find an unbiased estimator in the distribution of ϕ^* and it will be the *only* candidate for the uniformly minimum variance unbiased estimator.

EXAMPLE 11-6

Consider the family $\{F_\theta(x_1, x_2, x_3, \ldots, x_n): \theta \in \Theta\}$ where F_θ is given by the mass function

$$P_\theta(x_1, x_2, x_3, \ldots, x_n) = \prod_{i=1}^{n} \theta^{x_i}(1-\theta)^{1-x_i} \qquad \text{all } x_i = 0, 1$$

with $\Theta = (0, 1)$.

We have already seen that

$$\phi = \sum_{i=1}^{n} X_i$$

is a sufficient statistic, with distribution

$$P_\theta(\phi) = \binom{n}{\phi} \theta^\phi(1-\theta)^{n-\phi} \qquad \phi = 0, 1, 2, \ldots, n.$$

Let $\hat{\theta}(\phi)$ be any estimator which is a function of ϕ only, and let $E\hat{\theta} = 0$. This means that

$$E\hat{\theta} = \sum_{z=0}^{n} \hat{\theta}(z)\binom{n}{z} \theta^z(1-\theta)^{n-z} = (1-\theta)^n \sum_{z=0}^{n} \hat{\theta}(z)\binom{n}{z}\left(\frac{\theta}{1-\theta}\right)^z$$

$$= (1-\theta)^n \sum_{z=0}^{n} \hat{\theta}(z)\binom{n}{z} \rho^z = 0$$

where $\rho = \dfrac{\theta}{1-\theta}$.

Therefore

$$\sum_{z=0}^{n} \hat{\theta}(z)\binom{n}{z} \rho^z = 0 \qquad 0 < \rho < \infty.$$

Now

$$\frac{\partial^n}{\partial \rho^n}\left[\sum_{z=0}^{n} \hat{\theta}(z)\binom{n}{z} \rho^z\right] = 0$$

implies $n!\hat{\theta}(n) = 0$ or $\hat{\theta}(n) = 0$. Thus

$$\sum_{z=0}^{n-1} \hat{\theta}(z)\binom{n}{z} \rho^z = 0.$$

Further

$$\frac{\partial^{n-1}}{\partial \rho^{n-1}}\left[\sum_{z=0}^{n-1} \hat{\theta}(z)\binom{n}{z} \rho^z\right] = 0$$

implies $n!\hat{\theta}(n-1) = 0$ or $\hat{\theta}(n-1) = 0$. In general

$$\frac{\partial^{n-r}}{\partial p^{n-r}}\left[\sum_{z=0}^{n-r} \hat{\theta}(z)\binom{n}{z}p^z\right] = 0$$

implies $n!/r!\,\hat{\theta}(n-r) = 0$ or $\hat{\theta}(n-r) = 0$, and therefore

$$\hat{\theta}(z) = 0 \qquad z = 0, 1, 2, \dots, n.$$

Thus, ϕ/n is a complete sufficient statistic, and since it is unbiased for θ, it must be the uniformly minimum variance unbiased estimator.

The preceding example demonstrates that the binomial family of distributions is complete.

Exercise 11-1

Show that

$$\hat{\theta} = \frac{n(\bar{X}_n - (\bar{X}_n)^2)}{n-1}$$

is the uniformly minimum variance unbiased estimator for the parameter $\theta(1-\theta)$ in the preceding example.

Example 11-7

Let $\{F_\theta(x_1, x_2, x_3, \dots, x_n)\colon \theta \in \Theta\}$ be such that F_θ is given by the mass function

$$P_\theta(x_1, x_2, x_3, \dots, x_n) = \sum_{i=1}^{n} \frac{e^{-\theta}\theta^{x_i}}{x_i!} \qquad \text{all } x_i = 0, 1, 2, \dots$$

with $\Theta = (0, \infty)$.

We have already seen that

$$\phi = \sum_{i=1}^{n} X_i$$

is a sufficient statistic with mass function

$$P_\theta(\phi) = e^{-n\theta}\frac{(n\theta)^\phi}{\phi!} \qquad \phi = 0, 1, 2, \dots.$$

If $\hat{\theta}(\phi)$ is such that $E\hat{\theta} = 0$, then

$$E\hat{\theta} = \sum_{z=0}^{\infty} \hat{\theta}(z)e^{-n\theta}\frac{(n\theta)^z}{z!} = 0$$

giving

$$\sum_{z=0}^{\infty} \frac{\hat{\theta}(z)}{z!}(n\theta)^z = 0 \qquad 0 < \theta < \infty.$$

This implies that

$$\hat{\theta}(z) = 0 \qquad z = 0, 1, 2, \dots$$

and that ϕ is a complete sufficient statistic.

EXERCISE 11-2
Verify that

$$\sum_{z=0}^{\infty} \frac{\hat{\theta}(z)}{z!}(n\theta)^z = 0 \qquad 0 < \theta < \infty$$

implies that

$$\hat{\theta}(z) = 0 \qquad z = 0, 1, 2, \ldots.$$

The preceding example (setting $n = 1$) demonstrates that the poisson family of distributions is complete.

EXERCISE 11-3
Show that $\bar{X}_n$ is the uniformly minimum variance unbiased estimator for θ in Example 11-7.

EXAMPLE 11-8
Let $\{F_\theta(x_1, x_2, x_3, \ldots, x_n): \theta \in \Theta\}$ be such that F_θ is given by the density function

$$f_\theta(x_1, x_2, x_3, \ldots, x_n) = \frac{1}{\theta^n} \qquad 0 < \text{all } x_i < \theta$$
$$= 0 \qquad \text{otherwise}$$

with $\Theta = (0, \infty)$.

We have already seen that

$$\phi = \max[X_1, X_2, X_3, \ldots, X_n]$$

is a sufficient statistic with density

$$f_\theta(\phi) = \frac{n\phi^{n-1}}{\theta^n} \qquad 0 < \phi < \theta$$
$$= 0 \qquad \text{otherwise.}$$

If $\hat{\theta}(\phi)$ is such that $E\hat{\theta} = 0$ then

$$E\hat{\theta} = \int_0^\theta \hat{\theta}(z)\frac{nz^{n-1}}{\theta^n}\,dz = 0$$

or

$$\int_0^\theta \hat{\theta}(z)z^{n-1}dz = 0 \qquad 0 < \theta < \infty.$$

Remark *Using Lebesgue integration we could conclude immediately that* $\hat{\theta}(z) = 0$, $0 < z < \theta$. *But this type of integration is beyond the prerequisites for this text.*

Consider dividing the function $\hat{\theta}(z)z^{n-1}$ into Riemann rectangles as in Figure 11-1. The integral then becomes

$$\sum_{z=0}^{\theta} \hat{\theta}(z^*)z^{*n-1}\Delta z = 0$$

where the summation is over all the rectangles and where $z \leqq z^* \leqq z+\Delta z$. If this is zero for one value of θ, then it cannot be zero for $\theta+\Delta\theta$ unless $\hat{\theta}(x^*)x^{*n-1} = 0$ where

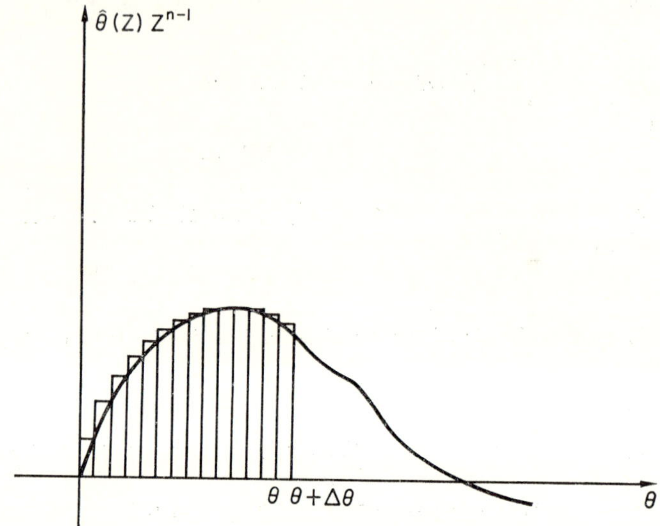

Figure 11-1

$\theta \leqq x^* \leqq \theta + \Delta\theta$. However, this sum must be zero for all values of θ. From this we conclude that $\hat{\theta}(z)z^{n-1}$ is equal to zero somewhere in every z interval no matter how small. This implies that

$$\hat{\theta}(z) = 0 \qquad 0 < z < \theta,$$

and that ϕ is a complete sufficient statistic.

The preceding example (setting $n=1$) demonstrates that the uniform family of distributions is complete.

EXERCISE 11-4

Show that

$$\hat{\theta} = \frac{n+1}{n} \max[X_1, X_2, X_3, \ldots, X_n]$$

is the uniformly minimum variance unbiased estimator for θ in Example 11-8.

EXERCISE 11-5

Assume that $\{F_\theta(x_1, x_2, x_3, \ldots, x_n): \theta \in \Theta\}$ is such that F_θ is given by the density function

$$f_\theta(x_1, x_2, x_3, \ldots, x_n) = \prod_{i=1}^{n} \frac{1}{\Gamma(\alpha)\theta^\alpha} x_i^{\alpha-1} e^{-x_i/\theta} \qquad 0 < \text{all } x_i < \infty, \quad \alpha > 0 \text{ known}$$

$$= 0 \qquad \text{otherwise}$$

with $\Theta = (0, \infty)$.

a. Find a complete sufficient statistic.

b. Find the uniformly minimum variance unbiased estimator for θ.

Hint The Laplace transform

$$\phi\dagger(t) = \int_0^\infty \phi(x)\, e^{-xt} dx$$

is such that at most one function $\phi(x)$ has the transform function $\phi\dagger(t)$.

The preceding exercise demonstrates that the gamma family of distributions (with known α) is complete.

EXERCISE 11-6

Show that

$$\hat{\theta} = \frac{n(\bar{X}_n)^2}{\alpha(n\alpha+1)}$$

is the uniformly minimum variance unbiased estimator for θ^2 in Example 11-8.

It will be of interest for us to consider the following unusual example.

EXAMPLE 11-9

Let $\{F_\theta(x)\colon \theta \in \Theta\}$ be the poisson family given by

$$P_\theta(x) = \frac{e^{-\theta}\theta^x}{x!} \qquad x = 0, 1, 2, \ldots$$

with $\Theta = (0, \infty)$.

X may represent a univariate variable or a sufficient statistic for θ. Let us say that we would like the uniformly minimum variance unbiased estimator for $g(\theta) = e^{-2\theta}$. Clearly $0 < e^{-2\theta} < 1$ for $\theta \in (0, \infty)$.

Consider the estimator

$$\begin{aligned} \widehat{g(\theta)} &= 1 \qquad && x = 0, 2, 4, 6, \ldots \\ &= -1 && x = 1, 3, 5, \ldots. \end{aligned}$$

This is quite a poor estimator for $e^{-2\theta}$.

It is always outside the range of possible values, and about half the time it would be at least a whole unit away. Besides this we note the way it treats all odd numbers alike and all even numbers alike. However

$$\begin{aligned} \widehat{Eg(\theta)} &= e^{-\theta} \sum_{x=0}^{\infty} g(\hat{\theta}) \frac{\theta^x}{x!} = e^{-\theta}\left[1-\theta+\frac{\theta^2}{2!}-\frac{\theta^3}{3!}+\frac{\theta^4}{4!}-\ldots\right] \\ &= = e^{-\theta}e^{-\theta} = e^{-2\theta}. \end{aligned}$$

Since the poisson family is complete, $\widehat{g(\theta)}$ is the only unbiased estimator for $e^{-2\theta}$ and is therefore the uniformly minimum variance unbiased estimator.

This example indicates that there are cases where the uniformly minimum variance unbiased estimator has rather poor properties. Notwithstanding this example, we will continue obtaining methods for finding the uniformly minimum variance unbiased estimator.

11-3 Lower Bounds for the Variance of the Uniformly Minimum Variance Unbiased Estimator

In this section we will be concerned primarily with lower bounds for the variance of the minimum variance unbiased estimator rather than with the estimator itself. If we have a lower bound for the variance of the uniformly minimum variance unbiased estimator and we find an estimator which attains this lower bound, then we have found the uniformly minimum variance unbiased estimator. On the other hand, if we have an estimator whose variance is only slightly higher than the lower bound, we might be willing to accept it in lieu of further effort.

CRAMÉR–RAO LOWER BOUND

The first lower bound that we discuss is known as the *Cramér–Rao lower bound.*

We assume that the family in question is $\{F_\theta(x_1, x_2, x_3, \ldots, x_n): \theta \in \Theta\}$ and make the following regularity assumptions:

1) Θ is an interval on the real line
2) $(\partial/\partial\theta)F_\theta' < \infty$ all $\theta \in \Theta$ and at all $x_1, x_2, x_3, \ldots, x_n$
3) $\dfrac{\partial}{\partial\theta}\displaystyle\int dF_\theta = \int \dfrac{\partial}{\partial\theta}\, dF_\theta$ all $\theta \in \Theta$

(where the integration is over the entire range of the argument of F_θ and the limits do not depend on θ)

4) For any unbiased estimator $\hat{\theta}$: $\dfrac{\partial}{\partial\theta}\displaystyle\int \hat{\theta}\, dF_\theta = \int \dfrac{\partial}{\partial\theta}\hat{\theta}\, dF_\theta = 1$
5) $E\left[\dfrac{\partial}{\partial\theta}\ln F_\theta'\right]^2 > 0$ all $\theta \in \Theta$.

Remark *For the discrete case assumptions three and four require that each P_θ and $\hat{\theta}P_\theta$ be differentiable with respect to θ. In the continuous case, these assumptions require that the derivatives of $f_\theta(x)$ and $\hat{\theta}f_\theta(x)$ with respect to θ exist for all x and that they be at least piecewise continuous.*

For any unbiased estimator $\hat{\theta}$ for θ, we have

$$E\hat{\theta} = \int \hat{\theta}\, dF_\theta = \theta.$$

Assuming that the regularity assumptions hold, we can take a derivative with respect to θ on both sides of this equation and obtain

$$\begin{aligned} 1 &= \frac{\partial}{\partial\theta}\int \hat{\theta}\, dF_\theta = \int \hat{\theta}\frac{\partial}{\partial\theta}\, dF_\theta \\ &= \int \hat{\theta}\frac{1}{dF_\theta}\left(\frac{\partial}{\partial\theta}\, dF_\theta\right) dF_\theta = \int \hat{\theta}\;\frac{1}{F_\theta{}'}\left(\frac{\partial}{\partial\theta}\, F_\theta{}'\right) dF_\theta \\ &= \int \hat{\theta}\left(\frac{\partial}{\partial\theta}\ln F_\theta{}'\right) dF_\theta = E(\hat{\theta}Z) \end{aligned}$$

where

$$Z = \frac{\partial}{\partial\theta}\ln F_\theta{}'.$$

Now

$$1 = \int dF_\theta.$$

Taking a derivative with respect to θ on both sides we obtain

$$0 = \int\left(\frac{\partial}{\partial\theta}\ln F_\theta{}'\right) dF_\theta = EZ.$$

Thus

$$\operatorname{Cov}(\hat{\theta}, Z) = E\hat{\theta}Z - E\hat{\theta}EZ = 1.$$

By the Cauchy–Schwartz inequality, however

$$E[(\hat{\theta}-\theta)Z] \leqq \sqrt{(E(\hat{\theta}-\theta)^2 EZ^2)}$$

or

$$[\operatorname{Cov}(\hat{\theta}, Z)]^2 \leqq \operatorname{Var}\hat{\theta}\operatorname{Var}Z.$$

This implies

$$\operatorname{Var}\hat{\theta} \geqq \frac{[\operatorname{Cov}(\hat{\theta}, Z)]^2}{\operatorname{Var}Z}$$

from which we obtain

$$\operatorname{Var}\hat{\theta} \geqq \frac{1}{\operatorname{Var}Z}.$$

Since $EZ = 0$

$$\operatorname{Var}Z = E\left(\frac{\partial}{\partial\theta}\ln F_\theta{}'\right)^2$$

and we obtain that for any unbiased estimator for θ

$$\operatorname{Var}\hat{\theta} \geqq \frac{1}{E\left(\frac{\partial}{\partial\theta}\ln F_\theta{}'\right)^2}.$$

The right-hand side of this inequality is known as the Cramér–Rao lower bound.

EXAMPLE 11-10

Let $\{F_\theta(x_1, x_2, x_3, \ldots, x_n)\colon \theta \in \Theta\}$ be such that F_θ is given by the mass function

$$P_\theta(x_1, x_2, x_3, \ldots, x_n) = \prod_{i=1}^{n} e^{-\theta} \frac{\theta^{x_i}}{x_i!} \qquad \text{all } x_i = 0, 1, 2, \ldots$$

with $\Theta = (0, \infty)$.

The regularity assumptions hold for this family. We then obtain

$$\ln P_\theta = \sum_{i=1}^{n} (-\theta + x_i \ln \theta - \ln x_i!)$$

$$\frac{\partial}{\partial\theta} \ln P_\theta = -n + \frac{1}{\theta} \sum_{i=1}^{n} x_i = n\left(\frac{\bar{x}_n - \theta}{\theta}\right)$$

$$E\left(\frac{\partial}{\partial\theta} \ln P_\theta\right)^2 = \frac{n^2}{\theta^2} \operatorname{Var} \bar{X}_n = n/\theta.$$

Therefore

$$\operatorname{Var} \hat{\theta} \geqq \theta/n.$$

However, $\operatorname{Var} \bar{X}_n = \theta/n$ thus, $\bar{X}_n$ is the uniformly minimum variance unbiased estimator for θ.

EXAMPLE 11-11

Let $\{F_\theta(x_1, x_2, x_3, \ldots, x_n)\colon \theta \in \Theta\}$ be such that F_θ is given by the density function

$$f_\theta(x_1, x_2, x_3, \ldots, x_n) = \frac{1}{(2\pi)^{n/2}\sigma^n} \exp\left[-\sum_{i=1}^{n} (x_i - \theta)^2/2\sigma^2\right] \qquad -\infty < \text{all } x_i < \infty$$

with $\Theta = (-\infty, \infty)$.

The regularity assumptions hold here, and we obtain

$$\ln f_\theta = -(n/2) \ln 2\pi - n \ln \sigma - \sum_{i=1}^{n} \frac{(x_i - \theta)^2}{2\sigma^2}$$

$$\frac{\partial}{\partial\theta} \ln f_\theta = \sum_{i=1}^{n} \frac{(x_i - \theta)}{\sigma^2} = \frac{(\bar{x}_n - \theta)}{\sigma^2/n}$$

$$E\left(\frac{(\bar{X}_n - \theta)^2}{\sigma^4/n^2}\right) = \frac{n^2}{\sigma^4} \cdot \frac{\sigma^2}{n} = n/\sigma^2$$

giving

$$\operatorname{Var} \hat{\theta} \geqq \sigma^2/n.$$

But $\operatorname{Var} \bar{X}_n = \sigma^2/n$, and therefore $\bar{X}_n$ is the uniformly minimum variance unbiased estimator for θ.

EXAMPLE 11-12

Let $\{F_\theta(x_1, x_2, x_3, \ldots, x_n)\colon \theta \in \Theta\}$ be such that F_θ is given by the density function

$$f_\theta(x_1, x_2, x_3, \ldots, x_n) = \frac{1}{(2\pi)^{n/2}\theta^{n/2}} \exp\left(-\sum_{i=1}^{n} x_i^2/2\theta\right) \qquad -\infty < \text{all } x_i < \infty$$

with $\Theta = (0, \infty)$.

The regularity assumptions are satisfied, and we obtain

$$\ln f_\theta = -(n/2)\ln 2\pi - (n/2)\ln\theta - \sum_{i=1}^{n} x_i^2/2\theta$$

$$\frac{\partial}{\delta\theta}\ln f_\theta = \frac{-n}{2\theta} + \sum_{i=1}^{n} x_i^2/2\theta^2 = \frac{\overline{x_n^2}-\theta}{2\theta^2/n}$$

where $\overline{x_n^2} = \sum_{i=1}^{n} x_i^2/n.$

Since $E\overline{X_n^2} = \theta$ and Var $\overline{X_n^2} = 2\theta^2/n$ we obtain

$$E\left(\frac{\partial}{\partial\theta}\ln f_\theta\right)^2 = \frac{n^2}{4\theta^4}\cdot\frac{2\theta^2}{n} = \frac{n}{2\theta^2}$$

giving

$$\operatorname{Var}\hat{\theta} \geqq 2\theta^2/n.$$

But Var $\overline{X_n^2} = 2\theta^2/n$, and therefore $\overline{X_n^2}$ is the uniformly minimum variance unbiased estimator for θ.

EXAMPLE 11-13

Let $\{F_\theta(x_1, x_2, x_3, \ldots, x_n)\colon \theta \in \Theta\}$ be such that F_θ is given by the density function

$$f_\theta(x_1, x_2, x_3, \ldots, x_n) = \frac{1}{[\Gamma(\alpha)]^n\theta^{n\alpha}}\left(\prod_{i=1}^{n} x_i^{\alpha-1}\right)\exp\left(-\sum_{i=1}^{n} x_i/\theta\right) \qquad 0 < \text{all } x_i < \infty,\ \alpha \text{ fixed}$$

$$= 0 \qquad \text{otherwise}$$

with $\Theta = (0, \infty)$.

The regularity assumptions hold and we obtain

$$\ln f_\theta = -n\ln\Gamma(\alpha) - n\alpha\ln\theta + \sum_{i=1}^{n}(\alpha-1)\ln x_i - \sum_{i=1}^{n} x_i/\theta$$

$$\frac{\partial}{\partial\theta}\ln f_\theta = \frac{-n\alpha}{\theta} + \sum_{i=1}^{n}\frac{x_i}{\theta^2} = \frac{\bar{x}_n - \alpha\theta}{\theta^2/n}.$$

Noting that $E\bar{X}_n = \alpha\theta$ and Var $\bar{X}_n = \alpha\theta^2/n$, we obtain

$$E\left(\frac{\partial}{\partial\theta}\ln f_\theta\right)^2 = \frac{n^2}{\theta^4}\frac{\alpha\theta^2}{n} = n\alpha/\theta^2.$$

Therefore

$$\operatorname{Var}\hat{\theta} \geqq \theta^2/n\alpha.$$

Consider now the unbiased estimator

$$\hat{\theta} = \bar{X}_n/\alpha.$$

Clearly

$$\operatorname{Var}\hat{\theta} = \theta^2/n\alpha$$

and therefore $\hat{\theta}$ is the uniformly minimum variance unbiased estimator for θ.

EXAMPLE 11-14

Let $\{F_\theta(x_1, x_2, x_3, \ldots, x_n)\colon \theta \in \Theta\}$ be such that F_θ is given by the density function

$$f_\theta(x_1, x_2, x_3, \ldots, x_n) = \frac{1}{\theta^n} \qquad 0 < \text{all } x_i < \theta$$
$$= 0 \qquad \text{otherwise}$$

with $\Theta = (0, \infty)$.

Now

$$\frac{\partial}{\partial\theta} \int_0^\theta \int_0^\theta \cdots \int_0^\theta \frac{1}{\theta^n}\, dx_1\, dx_2\, dx_3 \ldots dx_n = 0$$

but

$$\int_0^\theta \int_0^\theta \cdots \int_0^\theta \frac{\partial}{\partial\theta} \frac{1}{\theta^n}\, dx_1\, dx_2\, dx_3 \ldots dx_n = \int_0^\theta \int_0^\theta \cdots \int_0^\theta \frac{-n}{\theta^{n+1}}\, dx_1\, dx_2\, dx_3 \ldots dx_n$$
$$= -n/\theta \neq 0$$

which implies that the regularity conditions do not hold for this family.

The Cramér–Rao lower bound is not always attained even when it exists. When it is attained by the variance of an unbiased estimator, we can find the estimator directly as can be seen in what follows. First, let us derive a very useful fact

$$(EXY)^2 = EX^2EY^2$$

if and only if $X = aY$ for some constant a.

If for random variables X and Y the relation $X = aY$ holds for some constant a, then

$$EXY = aEY^2 = \sqrt{(a^2EY^2EY^2)} = \sqrt{(EX^2EY^2)}$$

or

$$(EXY)^2 = EX^2EY^2.$$

Conversely, if

$$(EXY)^2 = EX^2EY^2$$

then

$$EX^2 - \frac{(EXY)^2}{EY^2} = 0$$

or

$$EX^2 - 2\frac{EXY}{EY^2}EXY + \frac{(EXY)^2}{(EY^2)^2}EY^2 = 0.$$

Letting $a = EXY/EY^2$, we see that this becomes

$$E(X^2 - 2aXY + a^2Y) = E(X - aY)^2 = 0.$$

The last equality implies that $X = aY$. Thus

$$(EXY)^2 = EX^2EY^2$$

if and only if, for some a

$$X = aY.$$

If we now replace X by $\hat{\theta}-\theta$ and Y by Z we obtain

$$E(\hat{\theta}-\theta)Z = \text{Cov}(\hat{\theta}, Z) = \text{Var}\,\hat{\theta}\,\text{Var}\,Z$$

if and only if

$$\hat{\theta}-\theta = a(\theta)Z.$$

Now since $\text{Cov}(\hat{\theta}, Z) = 1$ we obtain

$$\text{Var}\,\hat{\theta} = \frac{1}{\text{Var}\,Z}$$

if and only if

$$\hat{\theta}-\theta = a(\theta)Z.$$

From this we obtain

$$\text{Var}\,\hat{\theta} = a^2(\theta)\,\text{Var}\,Z = \frac{1}{\text{Var}\,Z}$$

which implies

$$a(\theta) = \frac{1}{\text{Var}\,Z}$$

and therefore the Cramér–Rao lower bound is attained by the estimator $\hat{\theta}$ if and only if it can be expressed as

$$\hat{\theta} = \frac{Z}{\text{Var}\,Z}+\theta = \frac{(\partial/\partial\theta)\ln F_\theta'}{E\left(\frac{\partial}{\partial\theta}\ln F_\theta'\right)^2}+\theta.$$

Note that although this combination can always be found, it will not always yield a statistic (i.e. the combination may still contain θ).

Therefore, one procedure for using the Cramér–Rao lower bound is as follows:

1) Check the regularity conditions
2) Form the equation

$$\hat{\theta} = \frac{(\partial/\partial\theta)\ln F_\theta'}{E\left(\frac{\partial}{\partial\theta}\ln F_\theta'\right)^2}+\theta$$

3) Check to see that $\hat{\theta}$ is a statistic. If these are satisfied then $\hat{\theta}$ is the uniformly minimum variance unbiased estimator for θ.

EXAMPLE 11-15

Let $\{F_\theta(x_1, x_2, x_3, \ldots, x_n): \theta \in \Theta\}$ be such that F_θ is given by the density function

$$f_\theta(x_1, x_2, x_3, \ldots, x_n) = \frac{1}{(2\pi)^{n/2}\sigma^n} \exp\left[-\sum_{i=1}^{n} (x_i-\theta)^2/2\sigma^2\right] \qquad -\infty < \text{all } x_i < \infty$$

with $\Theta = (-\infty, \infty)$.

For this density

$$\ln f_\theta = -\frac{n}{2} \ln 2\pi - n \ln \sigma - \sum_{i=1}^{n} \frac{(x_i-\theta)^2}{2\sigma^2}$$

$$Z = \frac{\partial}{\partial\theta} \ln f_\theta = \sum_{i=1}^{n} \frac{(X_i-\theta)}{\sigma^2} = \frac{(\bar{X}_n-\theta)}{\sigma^2/n}$$

$$\text{Var } Z = \frac{n^2}{\sigma^4} \cdot \frac{\sigma^2}{n} = n/\sigma^2$$

giving

$$Z = \bar{X}_n - \theta + \theta = \bar{X}_n$$

which is a statistic and therefore the uniformly minimum variance unbiased estimator.

When the regularity assumptions are not satisfied, as in the case of the uniform distribution, the Cramér–Rao lower bound is not defined. We now derive a lower bound which is harder to work with, but which does not require any cumbersome regularity assumptions.

CHAPMAN–ROBBINS LOWER BOUND

In this subsection we assume that the random variables involved are either discrete or continuous. We write

$$\int F_\theta' \, dZ$$

to mean either

$$\int\int \cdots \int f_\theta(x_1, x_2, x_3, \ldots, x_n) \, dx_1 dx_2 dx_3 \ldots dx_n$$

or

$$\sum_{x_1} \sum_{x_2} \cdots \sum_{x_n} P_\theta(x_1, x_2, x_3, \ldots, x_n)$$

whichever is appropriate. Let us define $Z(\theta)$ to be all those n-tuples $x_1, x_2, x_3, \ldots, x_n$ for which $F_\theta'(x_1, x_2, x_3, \ldots, x_n) > 0$.

Let h be any value such that $Z(\theta+h) \subset Z(\theta)$ with $h \neq 0$. Then

$$\int_{Z(\theta)} F_\theta' \, dZ = 1$$

and

$$\int_{Z(\theta)} F_{\theta+h}' dZ = \int_{Z(\theta+h)} F_{\theta+h}' dZ = 1.$$

If $\hat{\theta}$ is any unbiased estimator for θ, we have

$$\int_{Z(\theta)} \hat{\theta} F_\theta' \, dZ = \theta$$

while

$$\int_{Z(\theta)} \hat{\theta} F'_{\theta+h} dZ = \int_{Z(\theta+h)} \hat{\theta} F'_{\theta+h} dZ = \theta + h.$$

Therefore

$$\int_{Z(\theta)} (\hat{\theta}-\theta)[F'_{\theta+h} - F_\theta'] \, dZ = h$$

or

$$\int_{Z(\theta)} \frac{(\hat{\theta}-\theta)}{h} \frac{(F_\theta')^{1/2}}{(F_\theta')^{1/2}} [F'_{\theta+h} - F_\theta'] \, dZ = 1.$$

Using the Cauchy–Schwartz inequality, we obtain

$$1 \leqq \int_{Z(\theta)} (\hat{\theta}-\theta)^2 F_\theta' \, dZ \int_{Z(\theta)} \left(\frac{F'_{\theta+h} - F_\theta'}{hF_\theta'}\right)^2 F_\theta' \, dZ.$$

Now the first term is Var $\hat{\theta}$, while the second expands to

$$\frac{1}{h^2} \int_{Z(\theta)} \left[\left(\frac{F'_{\theta+h}}{F_\theta'}\right)^2 F_\theta' - 2F'_{\theta+h} + F_\theta'\right] dZ = \frac{1}{h^2} \int_{Z(\theta)} \left(\frac{F'_{\theta+h}}{F_\theta'}\right)^2 F_\theta' \, dZ - \frac{1}{h^2}$$

$$= \frac{1}{h^2}\left[E\left(\frac{F'_{\theta+h}}{F_\theta'}\right)^2 - 1\right].$$

Combining terms we obtain

$$\text{Var } \hat{\theta} \geqq \frac{h^2}{E\left(\frac{F'_{\theta+h}}{F_\theta'}\right)^2 - 1}$$

provided the denominator is not zero. As a matter of fact

$$\text{Var } \hat{\theta} \geqq \sup_h \frac{h^2}{E\left(\frac{F'_{\theta+h}}{F_\theta}\right)^2 - 1}$$

where the Sup is over all h such that $Z(\theta+h) \subset Z(\theta)$.

The restrictions here are so inconsequential that we can reconsider the situation of Example 11-14.

EXAMPLE 11-16

Let $\{F_\theta(x_1, x_2, x_3, \ldots, x_n)\colon \theta \in \Theta\}$ be such that F_θ is given by the density function

$$f_\theta(x_1, x_2, x_3, \ldots, x_n) = \frac{1}{\theta^n} \qquad 0 < \text{all } x_i < \theta$$
$$= 0 \qquad \text{otherwise.}$$

P

Then $Z(\theta+h) \subset Z(\theta)$ wherever $-\theta < h < 0$, and we can therefore write

$$\operatorname{Var}\hat{\theta} \geqq \sup_{-\theta<h<0} \frac{h^2}{E\left(\frac{f_{\theta+h}}{f_\theta}\right)^2 - 1}.$$

Now

$$\frac{f_{\theta+h}}{f_\theta} = \frac{\theta^n}{(\theta+h)^n} \qquad \text{for } 0 < \text{all } x_i < \theta+h$$

and therefore

$$E\left(\frac{f_{\theta+h}}{f_\theta}\right)^2 = \frac{\theta^{2n}}{(\theta+h)^{2n}} \int_0^{\theta+h}\int_0^{\theta+h} \cdots \int_0^{\theta+h} \frac{1}{\theta^n}\, dx_1\, dx_2\, dx_3 \ldots dx_n$$

$$(\text{since } -\theta < h < 0)$$

$$= \frac{\theta^{2n}(\theta+h)^n}{(\theta+h)^{2n}\theta^n} = \frac{\theta^n}{(\theta+h)^n}$$

giving

$$\operatorname{Var}\hat{\theta} \geqq \sup_{-\theta<h<0} \frac{h^2}{\left[\frac{\theta^n}{(\theta+h)^n} - 1\right]} = \sup_{-\theta<h<0} \frac{h^2(\theta+h)^n}{\theta^n - (\theta+h)^n}.$$

Exercise 11-7

Find the Chapman–Robbins lower bound for the preceding example when $n = 2$. Does the variance of the uniformly minimum variance unbiased estimator

$$\hat{\theta} = \frac{n+1}{n} \operatorname{Max}[X_1, X_2, X_3, \ldots, X_n]$$

attain this lower bound?

The Chapman–Robbins lower bound is attained by some unbiased estimator $\hat{\theta}$ if and only if

$$\operatorname{Var}\hat{\theta} = \frac{h_s^2}{E\left(\frac{F'_{\theta+h_s}}{F_\theta'}\right)^2 - 1}$$

where h_s is the allowed value which achieves the supremum. This equality holds if and only if

$$1 = \int_{Z(\theta)} (\hat{\theta}-\theta)^2 F_\theta'\, dZ \int_{Z(\theta)} \left(\frac{F'_{\theta+h_s} - F_\theta'}{h_s F_\theta'}\right)^2 F_\theta'\, dZ$$

which in turn holds if and only if

$$\hat{\theta} - \theta = a(\theta)\left(\frac{F'_{\theta+h_s} - F_\theta'}{h_s F_\theta'}\right).$$

Now

$$E(\hat{\theta}-\theta)^2 = \frac{a^2(\theta)}{h_s^2}\left[E\left(\frac{F'_{\theta+h_s}}{F_\theta'}\right)^2 - 1\right]$$

which implies that

$$a(\theta) = \frac{h_s}{E\left(\frac{F'_{\theta+h_s}}{F_\theta'}\right)^2 - 1}.$$

Therefore, the Chapman–Robbins lower bound is attained if, and only if,

$$\hat{\theta} = \frac{h_s\left(\frac{F'_{\theta+h_s} - F_\theta'}{F_\theta'}\right)}{E\left(\frac{F'_{\theta+h_s}}{F_\theta'}\right)^2 - 1} + \theta$$

is an unbiased estimator for θ.

EXERCISE 11-8

Fill in the details in the preceding derivation.

11-4 Problems

1. Let $\{F_\theta(x_1, x_2, x_3, \ldots, x_n)\colon \theta \in \Theta\}$ be such that F_θ is given by the mass function

$$P_\theta(x_1, x_2, x_3, \ldots, x_n) = \prod_{i=1}^{n} \binom{m}{x_i} \theta^{x_i}(1-\theta)^{m-x_i} \quad \text{all } x_i = 0, 1, 2, \ldots, m$$

with $\Theta = (0, 1)$. Find the distribution of the statistic $\phi = \sum_{i=1}^{n} X_i$ and show that $P_\theta(x_1, x_2, x_3, \ldots, x_n \mid \phi)$ is independent of θ.

2. Let $\{F_\theta(x_1, x_2, x_3, \ldots, x_n)\colon \theta \in \Theta\}$ be such that F_θ is given by the density function

$$f_\theta(x_1, x_2, x_3, \ldots, x_n) = \frac{1}{\theta^n}\prod_{i=1}^{n} e^{-x_i/\theta} \qquad 0 < \text{all } x_i < \infty$$

$$= 0 \qquad \text{otherwise}$$

with $\Theta = (0, \infty)$. Find the distribution of the statistic $\phi = \sum_{i=1}^{n} X_i$ and show that $f_\theta(x_1, x_2, x_3, \ldots, x_n \mid \phi)$ is independent of θ.

3. Let $\{F_\theta(x_1, x_2, x_3, \ldots, x_n)\colon \theta \in \Theta\}$ be such that F_θ is given by the density function

$$f_\theta(x_1, x_2, x_3, \ldots, x_n) = \frac{1}{[\Gamma(\alpha)]^n \theta^{n\alpha}}\prod_{i=1}^{n} x_i^{\alpha-1} e^{-x_i/\theta} \qquad 0 < \text{all } x_i < \infty \qquad \alpha > 0 \text{ fixed}$$

$$= 0 \qquad \text{otherwise}$$

with $\Theta = (0, \infty)$. Find the distribution of the statistic $\phi = \sum_{i=1}^{n} X_i$ and show that $f_\theta(x_1, x_2, x_3, \ldots, x_n \mid \phi)$ is independent of θ.

4. Let $\{F_\theta(x_1, x_2, x_3, \ldots, x_n): \theta \in \Theta\}$ be such that F_θ is given by the density function

$$f_\theta(x_1, x_2, x_3, \ldots, x_n) = \frac{1}{(2\pi)^{n/2}\sigma^n} \prod_{i=1}^{n} e^{-(x_i-\theta)^2/2\sigma^2} \qquad -\infty < \text{all } x_i < \infty$$
$$\sigma > 0 \text{ fixed}$$

with $\Theta = (-\infty, \infty)$. Find the distribution of the statistic $\phi = \sum_{i=1}^{n} X_i$ and show that $f_\theta(x_1, x_2, x_3, \ldots, x_n \mid \phi)$ is independent of θ.

5. For the family of distributions and the estimator of Problem 1, show that the Neyman factorization criterion holds.

6. For the family of distributions and the estimator of Problem 2, show that the Neyman factorization criterion holds.

7. For the family of distributions and the estimator of Problem 3, show that the Neyman factorization criterion holds.

8. For the family of distributions and the estimator of Example 11-1, show that the Neyman factorization criterion holds.

9. For the family of distributions and the estimator of Example 11-2, show that the Neyman factorization criterion holds.

10. Let $\{F_\theta(x_1, x_2, x_3, \ldots, x_n): \theta \in \Theta\}$ be such that F_θ is given by the density function

$$f_\theta(x_1, x_2, x_3, \ldots, x_n) = \frac{1}{[\Gamma(\theta)]^n \beta^{n\theta}} \prod_{i=1}^{n} x_i^{\theta-1} e^{-x_i/\beta} \qquad 0 < \text{all } x_i < \infty$$
$$\beta > 0$$
$$= 0 \qquad \text{otherwise}$$

with $\Theta = (0, \infty)$. Find a sufficient statistic for θ.

11. Let $\{F_\theta(x_1, x_2, x_3, \ldots, x_n): \theta \in \Theta\}$ be such that F_θ is given by the density function

$$f_\theta(x_1, x_2, x_3, \ldots, x_n) = \left[\frac{\Gamma(\alpha+\beta)}{\Gamma(\alpha)\Gamma(\beta)}\right]^n \prod_{i=1}^{n} x_i^{\alpha-1}(1-x_i)^{\beta-1} \qquad 0 < \text{all } x_i < 1$$
$$\alpha, \beta > 0$$
$$= 0 \qquad \text{otherwise}$$

with $\theta = \alpha$ and $\Theta = (0, \infty)$. Find a sufficient statistic for θ.

12. For the family of distributions given in Problem 11 but with $\theta = \beta$, find a sufficient statistic for θ.

13. Find the uniformly minimum variance unbiased estimator for the parameter of Problem 1.

14. Find the uniformly minimum variance unbiased estimator for the parameter of Problem 2.

15. Find the uniformly minimum variance unbiased estimator for the parameter of Problem 3.

16. Find the Cramér–Rao lower bound for the family of distributions and the parameter of Problem 1.

17. Find the Cramér–Rao lower bound for the family of distributions and the parameter of Problem 2.

18. Find the uniformly minimum variance unbiased estimator for the parameter in Problem 1 by using the procedure of Section 11-3.

19. Find the uniformly minimum variance unbiased estimator for the parameter in Problem 2 by using the procedure of Section 11-3.

20. Find the Chapman–Robbins lower bound when $n = 2$ for the parameter in Problem 2.

CHAPTER 12

Hypothesis Testing

In Chapter Eight we defined both a test function and a randomized test function. In practice, as we shall see, randomized test functions need be considered only when the random variables involved are discrete. In order to better picture test functions, we will first discuss what are known as critical regions.

CRITICAL REGIONS

Generally, a nonrandomized test of the hypothesis $\theta \in \Theta_0$ vs. the alternate hypothesis that $\theta \in \Theta_1 = \Theta - \Theta_0$, says *reject* when the observed value $x_1, x_2, x_3, \ldots, x_n$ lies in some prescribed region $C_\delta{}^r$. This region $C_\delta{}^r$ is known as the *critical region* for the test δ and contains all those values of $X_1, X_2, X_3, \ldots, X_n$ for which $\delta =$ *reject*. The complement of $C_\delta{}^r$ or all those values of $X_1, X_2, X_3, \ldots, X_n$ for which $\delta =$ *accept* is called the *acceptance region*. For randomized tests there is a third region, usually on the boundary between the rejection region and the acceptance region, which contains all of those values of $X_1, X_2, X_3, \ldots, X_n$ for which we must perform a bernoulli trial to determine whether or not to reject. This we will call the *randomization region*. It is possible to have several of these randomization regions, each with its own probability of rejection. In practice, however, one or two is usually sufficient. We will use $C_\delta{}^a$ for the acceptance region, and $C_\delta{}^\lambda$ for a randomization region which has probability λ of rejecting.

The mechanism of a test $\delta(X_1, X_2, X_3, \ldots, X_n)$ should now be clear. We first record the regions $C_\delta{}^r$ and $C_\delta{}^a$ (and $C_\delta{}^\lambda$ when necessary). Then we observe $x_1, x_2, x_3, \ldots, x_n$ and note which region it is in; if in $C_\delta{}^r$, we reject the hypothesis, and if in $C_\delta{}^a$, we accept the hypothesis. If we are dealing with a randomized test $\delta(X_1, X_2, X_3, \ldots, X_n, Z)$ and $x_1, x_2, x_3, \ldots, x_n$ is in $C_\delta{}^\lambda$, we perform a bernoulli trial, like spinning a dial, with probability λ for success, and reject the hypothesis for a success and accept otherwise.

THE POWER FUNCTION

The *power function* for a particular test function δ (randomized or not) is defined by

$$\beta_\delta(\theta) = P_\theta(\delta = \textit{reject}) \qquad \text{all } \theta \in \Theta.$$

This is the probability that the test function will say *reject* when θ is the true value of the parameter. It is a function of θ for a particular test function δ. Mathematically we can write this as

$$\beta_\delta(\theta) = P_\theta(\delta = reject \mid x_1, x_2, x_3, \ldots, x_n \in C_\delta{}^r)P_\theta(x_1, x_2, x_3, \ldots, x_n \in C_\delta{}^r)$$
$$+ \sum_i P_\theta(\delta = reject \mid x_1, x_2, x_3, \ldots, x_n \in C_\delta{}^{\lambda}\text{i})P_\theta$$
$$(x_1, x_2, x_3, \ldots, x_n \in C_\delta{}^{\lambda_i})$$
$$= 1 \cdot P_\theta(x_1, x_2, x_3, \ldots, x_n \in C_\delta{}^r) + \sum_i \lambda_i P_\theta(x_1, x_2, x_3, \ldots, x_n \in C_\delta{}^{\lambda_i})$$
$$= \int_{C_\delta{}^r} dF_\theta + \sum_i \lambda_i \int_{C_\delta{}^{\lambda_i}} dF_\theta .$$

This power function, which is a function of θ, is very important in hypothesis testing.

When $\theta \in \Theta_0$ the value of the power function is said to be the *size* of test δ at θ. When $\theta \in \Theta_0$, a "good" test function should not say *reject*. It is unreasonable to require that a test function be correct all the time, but we can require that the frequency of its being incorrect be small, that is, require that the size of a good test be small. When $\theta \in \Theta_0$ and $\delta = reject$, we say that we have committed *an error of the first kind*. The probability of this error of the first kind is the size of the test.

When $\theta \in \Theta_1$, we would prefer that a test function say *reject*. When $\theta \in \Theta_1$ and $\delta = accept$, we say that we have committed *an error of the second kind*. A good test function should have a small probability of committing an error of the second kind. Since

$$P_\theta(\delta = accept) = 1 - P_\theta(\delta = reject),$$

the probability of an error of the second kind is simply $1 - \beta_\delta(\theta)$ for all $\theta \in \Theta_1$. This will be small when $\beta_\delta(\theta)$ is large. Therefore, $\beta_\delta(\theta)$ should be small when $\theta \in \Theta_0$ and large when $\theta \in \Theta_1$. We still have the problem of deciding what we mean by "small" and "large". The usual attitude toward this problem is to require that the size of a good test be no larger than some preassigned small value α, and with this restriction, to obtain a test function which has the uniformly largest power function. To this end we define a *uniformly most powerful test* δ^*, for given α, as any test such that:

$$\beta_{\delta^*}(\theta) \leqq \alpha \text{ (fixed)} \qquad \text{all } \theta \in \Theta_0$$

and

$$\beta_{\delta^*}(\theta) \leqq \beta_\delta(\theta) \qquad \text{all } \theta \in \Theta_1$$

where δ is any other test such that $\beta_\delta(\theta) \leqq \alpha$ all $\theta \in \Theta_\theta$.

In the following section we investigate a situation in which most powerful tests exist.

12-1 The Simple vs. Simple Hypothesis Testing Problem and the Neyman–Pearson Fundamental Lemma

In this section we consider the very special case where the null and alternate hypotheses can be written

$$H_0\colon \theta = \theta_0$$

and

$$H_1\colon \theta = \theta_1.$$

NEYMAN–PEARSON FUNDAMENTAL LEMMA

For this simple vs. simple hypothesis testing situation

A) There exists a test function δ and a constant k such that

a $\beta_\delta(\theta_0) = \alpha$

b $\delta = \textit{reject} \text{ if } \dfrac{F'_{\theta_1}(x_1, x_2, x_3, \ldots, x_n)}{F'_{\theta_0}(x_1, x_2, x_3, \ldots, x_n)} > k$

$$= \textit{accept} \text{ if } \frac{F'_{\theta_1}(x_1, x_2, x_3, \ldots, x_n)}{F'_{\theta_0}(x_1, x_2, x_3, \ldots, x_n)} < k.$$

B) These conditions are necessary and sufficient for the test function δ to be a most powerful size α test.

Proof of A Let k be the smallest number for which $P_0(C_\delta^r) \leqq \alpha$, where of course, C_δ^r is the region of values of $X_1, X_2, X_3, \ldots, X_n$ for which $(F'_{\theta_1}/F'_{\theta_0}) > k$ and where P_0 is the probability mass function assuming $\theta = \theta_0$.

For discrete random variables, we may find a k such that $P_0((F'_{\theta_1}/F'_{\theta_0}) > k) < \alpha$ but that $P_0((F'_{\theta_1}/F'_{\theta_0}) \geqq k) > \alpha$. In this case we define C_δ^a as the region of values for $X_1, X_2, X_3, \ldots, X_n$ such that $(F'_{\theta_1}/F'_{\theta_0}) < k$ and C_δ^λ as the region for which $F'_{\theta_1}/F'_{\theta_0} = k$ where

$$\lambda = \frac{\alpha - P_0(C_\delta^r)}{1 - P_0(C_\delta^r) - P_0(C_\delta^a)}.$$

Thus,

$$\delta = \textit{reject} \quad \text{if} \quad F'_{\theta_1} > kF'_{\theta_0}$$

$$\delta = \textit{accept} \quad \text{if} \quad F'_{\theta_1} < kF'_{\theta_0}$$

$$\delta = \textit{reject} \text{ with probability } \lambda$$

if

$$F'_{\theta_1} = kF'_{\theta_0}.$$

This is a randomized test when $\lambda \neq 0$, where Z is the bernoulli trial with probability λ for rejection. Now

$$\begin{aligned}\beta_\delta(\theta_0) &= P_0(\delta = \textit{reject})\\ &= P_0(\delta = \textit{reject} \mid C_\delta^r)P_0(C_\delta^r)+P_0(\delta = \textit{reject} \mid C_\delta^\lambda)P_0(C_\delta^\lambda)\\ &= P_0(C_\delta^r)+\lambda P_0(C_\delta^\lambda) = P_0(C_\delta^r)+\frac{\alpha - P_0(C_\delta^r)}{1-P_0(C_\delta^r)-P_0(C_\delta^a)}P_0(C_\delta^\lambda)\\ &= \alpha\end{aligned}$$

since $P_0(C_\delta^\lambda) = 1-P_0(C_\delta^r)-P_0(C_\delta^a)$, and thus A is proved.

Proof of B Assume that δ satisfies **a** and **b**, but that δ^* does not. Then

$$\begin{aligned}\beta_\delta(\theta_0)-\beta_{\delta^*}(\theta_0) &= P_0(C_\delta^r)+\lambda P_0(C_\delta^\lambda)-P_0(C_{\delta^*}^r)-\sum_i \lambda_i P_0(C_{\delta^*}^{\lambda_i})\\ &= P_0(C_\delta^r \cap [C_{\delta^*}^a \cup_i C_{\delta^*}^{\lambda_i}])-P_0(C_{\delta^*}^r \cap [C_\delta^a \cup C_\delta^\lambda])\\ &\quad +\lambda P_0(C_\delta^\lambda)-\sum_i \lambda_i[P_0(C_{\delta^*}^{\lambda_i} \cap C_\delta^r)+P_0(C_{\delta^*}^{\lambda_i} \cap C_\delta^a)\\ &\quad +P_0(C_{\delta^*}^{\lambda_i} \cap C_\delta^\lambda)]\\ &= P_0(C_\delta^r \cap C_{\delta^*}^a)-P_0(C_{\delta^*}^r \cap C_\delta^a)+\lambda P_0(C_\delta^\lambda)\\ &\quad -P_0(C_{\delta^*}^r \cap C_\delta^\lambda)+\sum_i(1-\lambda_i)P_0(C_{\delta^*}^{\lambda_i} \cap C_\delta^r)\\ &\quad -\sum_i \lambda_i P_0(C_{\delta^*}^{\lambda_i} \cap C_\delta^a)-\sum_i \lambda_i P_0(C_{\delta^*}^{\lambda_i} \cap C_\delta^\lambda).\end{aligned}$$

Similarly, if $\theta = \theta_1$,

$$\begin{aligned}\beta_\delta(\theta_1)-\beta_{\delta^*}(\theta_1) &= P_1(C_\delta^r \cap C_{\delta^*}^a)-P_1(C_{\delta^*}^r \cap C_\delta^a)+\lambda P_1(C_\delta^\lambda)\\ &\quad -P_1(C_{\delta^*}^r \cap C_\delta^\lambda)+\sum_i(1-\lambda_i)P_1(C_{\delta^*}^{\lambda_i} \cap C_\delta^r)\\ &\quad -\sum_i \lambda_i P_1(C_{\delta^*}^{\lambda_i} \cap C_\delta^a)-\sum_i \lambda_i P_1(C_{\delta^*}^{\lambda_i} \cap C_\delta^\lambda).\end{aligned}$$

Now for subsets of C_δ^r

$$P_1(A) = \int_A dF_{\theta_1} > k\int_A dF_{\theta_0} = kP_0(A)$$

while for subsets of C_δ^a

$$P_1(A) = \int_A dF_{\theta_1} < k\int_A dF_{\theta_0} = kP_0(A)$$

and for subsets of C_δ^λ

$$P_1(A) = \int_A dF_{\theta_1} = k\int_A dF_{\theta_0} = kP_0(A).$$

Using these we find that

$$\beta_\delta(\theta_1)-\beta_{\delta^*}(\theta_1) \geqq k[\beta_\delta(\theta_0)-\beta_{\delta^*}(\theta_0)].$$

Now since $\beta_{\delta^*}(\theta_0) \leqq \alpha$ while $\beta_\delta(\theta_0) = \alpha$, we obtain that $\beta_\delta(\theta_1) \geqq \beta_{\delta^*}(\theta_1)$. The equality can hold only if δ^* satisfies both conditions **a** and **b**. This completes the proof of B.

EXERCISES 12-1

The details were omitted in the proof above. Put in these details.

It is interesting to note that

$$\delta^* = \textit{reject} \text{ with probability } \alpha \text{ for all } x_1, x_2, x_3, \ldots, x_n,$$

is a size α test with power α. The derivation above indicates that since this test function does not satisfy conditions **a** and **b**, we have $\beta_\delta(\theta_1) > \beta_{\delta^*}(\theta_1) = \alpha$, i.e., the power of a most powerful test is strictly greater than its size.

EXAMPLE 12-1

Consider the situation in which we have two machines manufacturing a particular computer part. Machine A routinely produces 5% defectives, and Machine B produces 10% defectives. One box has failed to be labeled as to which machine it came from. By sampling ten items from the box (assuming there is a large enough number of parts in the box for the sampling to be considered *with replacement*), we must decide which machine manufactured the parts. Let us say that if the machine was A, then we would not like to have more than a 1-in-20 chance of mistakenly labeling the parts "B". Thus we need an $\alpha = 0.05$ test of the hypothesis

$$H_0 \colon \theta = 0.05$$

vs.

$$H_1 \colon \theta = 0.10$$

where θ is the proportion of defectives.

If we let x be the number of defectives found in the sample of ten items then

$$P(x) = \binom{10}{x} \theta^x (1-\theta)^{10-x} \qquad x = 0, 1, 2, \ldots, 10.$$

Thus

$$\frac{P_1(x)}{P_0(x)} = \frac{\binom{10}{x}(.10)^x(.90)^{10-x}}{\binom{10}{x}(.05)^x(.95)^{10-x}} = 2^x \left(\frac{18}{19}\right)^{10-x}.$$

Using a standard binomial table with $n = 10$ and $\theta \doteq 0.05$ we find that the smallest c for which $P_0(x > c) \leqq 0.05$ is $c = 2$. Thus, the smallest k for which

$$P_0\left[\frac{P_1(x)}{P_0(x)} > k\right] \leqq 0.05$$

is $k = 2.59$ to two decimal places, with

$$P_0\left[\frac{P_1(x)}{P_0(x)} > k\right] = P_0(x > c) = 0{,}0116$$

and

$$P_0\left[\frac{P_1(x)}{P_0(x)} = k\right] = P_0(x = c) = 0.0746.$$

Therefore, a most powerful size .05 test of the hypothesis is given by

$$\delta = reject \quad \text{if} \quad 2^x\left(\frac{18}{19}\right)^{10-x} > 2.59$$

(or equivalently if $x > 2$)

$$= accept \quad \text{if} \quad 2^x\left(\frac{18}{19}\right)^{10-x} < 2.59$$

(or equivalently if $x < 2$)

$$= reject \quad \text{with probability} \quad \lambda = \frac{.05 - .0116}{.0746} = 0.515$$

$$\text{if} \quad 2^x\left(\frac{18}{19}\right)^{10-x} = 2.59$$

(or equivalently if $x = 2$).

When Θ_0 and Θ_1 have more than one element, we are in a composite hypothesis testing situation. In this case there may not be a test which is a most powerful test for each $\theta \in \Theta_0$ vs. each $\theta \in \Theta_1$. However, in the special case where the family of distributions involved admits a monotone likelihood ratio, then there does exist a uniformly most powerful test function.

MONOTONE LIKELIHOOD RATIO

If $\{F_\theta(x_1, x_2, x_3, \dots, x_n) : \theta \in \Theta\}$ is such that for any values θ_0 and θ_1 with $\theta_0 < \theta_1$ there exists a statistic $\phi(X_1, X_2, X_3, \dots, X_n)$ such that $F'_{\theta_1}/F'_{\theta_0}$ is a strictly increasing function of $\phi(X_1, X_2, X_3, \dots, X_n)$, then we say that the family has a *monotone likelihood ratio.* Note that the statistic $\phi(X_1, X_2, X_3, \dots, X_n)$ must not depend on which values of θ are chosen for θ_0 and θ_1. When a family has a monotone likelihood ratio the hypothesis testing problems

A) H_0: $\theta \leqq \theta_0$ (fixed)

vs.

H_1: $\theta > \theta_0$

and

B) H_0: $\theta \geqq \theta_0$ (fixed)

vs.

H_1: $\theta < \theta_0$

admit test functions which are *uniformly most powerful* for any fixed size,

Let us first consider the simple hypothesis testing problem

$$H_0: \ \theta = \theta_0$$

vs.

$$H_1: \ \theta = \theta_1 > \theta_0.$$

From the fundamental lemma it is necessary and sufficient for a most powerful size α test to be of the form

$$\delta = \textit{reject} \qquad \text{if} \qquad \frac{F'_{\theta_1}}{F'_{\theta_0}} > k$$

$$= \textit{accept} \qquad \text{if} \qquad \frac{F'_{\theta_1}}{F'_{\theta_0}} < k.$$

Since $F'_{\theta_1}/F'_{\theta_0}$ is a strictly increasing function of ϕ, there is a unique constant c such that $\phi > c$ or $\phi < c$ whenever $F'_{\theta_1}/F'_{\theta_0} > k$ or $F'_{\theta_1}/F'_{\theta_0} < k$. See, for example, Figure 12-1. Thus, a necessary and sufficient condition for δ to be a most powerful test is that it be of the form

C) $\delta = \textit{reject} \qquad \text{if} \qquad \phi > c$

$\quad = \textit{accept} \qquad \text{if} \qquad \phi < c.$

We have defined k as the smallest number such that $P_0(F'_{\theta_1}/F'_{\theta_0}) > k) \leqq \alpha$. Thus, c is the smallest number such that $P_0(\phi > c) \leqq \alpha$. Now if the hypothesis to be tested were

$$H_0: \ \theta = \theta_0$$

vs.

$$H_2: \ \theta = \theta_2 > \theta_0 \qquad \theta_2 \neq \theta_1$$

a most powerful test would be similar to the one obtained for θ_1 with $(F'_{\theta_1}/F'_{\theta_0}) > k^*$ and $(F'_{\theta_1}/F'_{\theta_0}) < k^*$ replacing $(F'_{\theta_1}/F'_{\theta_0}) > k$ and $(F'_{\theta_1}/F'_{\theta_0}) < k$. Since $(F'_{\theta_1}/F'_{\theta_0})$ is a strictly increasing function of ϕ there is a c^* to be used

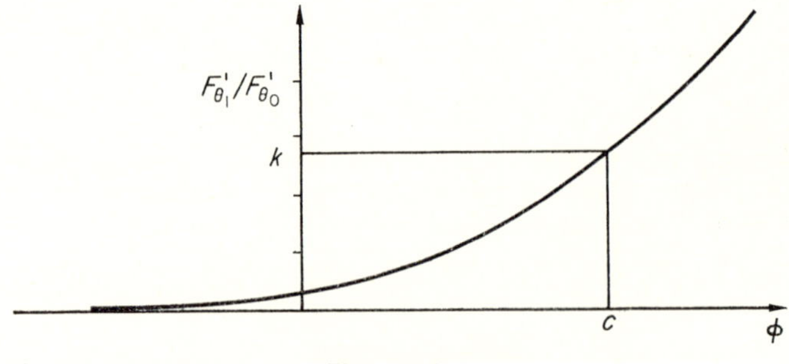

Figure 12-1

in condition **C** corresponding to k^*. However, c^* is defined as the smallest number such that $P_0(\phi > c^*) \leqq \alpha$ which is the definition for c. Thus $c^* = c$ and condition **C** is necessary and sufficient for a size α test to be most powerful for the alternate hypothesis H_2. Since this same test is most powerful for all θ values in Θ_1 we say that it is a uniformly most powerful size α test for the problem

$$H_0: \ \theta = \theta_0$$

vs.

$$H_1: \ \theta > \theta_0.$$

Now consider the hypothesis

$$H_0: \ \theta = \theta_3 < \theta_0$$

vs.

$$H_1: \ \theta = \theta_0.$$

Let δ satisfy conditions **C** and let $\beta_\delta(\theta_0) = \alpha$. Since δ is in the form used in the fundamental lemma with $\theta_3 < \theta_0$, $\beta_\delta(\theta_3) < \beta_\delta(\theta_0)$ or $\beta_\delta(\theta_3) < \alpha$. This says that any test, satisfying condition **C** and for which $\beta_\delta(\theta_0) = \alpha$, has size no larger than α for $\theta \leqq \theta_0$. Thus, any test satisfying these conditions is a *uniformly most powerful size* α *test* of

$$H_0: \ \theta \leqq \theta_0$$

vs.

$$H_1: \ \theta > \theta_0.$$

We have now to prove the result for

$$H_0: \ \theta \geqq \theta_0$$

vs.

$$H_1: \ \theta < \theta_0.$$

If we set $\phi^*(x_1, x_2, x_3, \ldots, x_n) = -\phi(x_1, x_2, x_3, \ldots, x_n)$ then there is a strictly decreasing likelihood ratio in ϕ^*. The proof is then similar to the above, with ϕ^* replacing ϕ.

EXERCISE 12-2

Prove that any test satisfying

1) $\beta_\delta(\theta_0) = \alpha$

2) $\delta = \textit{reject}$ if $\phi^* > c$

$\quad = \textit{accept}$ if $\phi^* < c$

with c the smallest number such that $P_0(\phi^* > c) \leqq \alpha$ is a uniformly most powerful size α test for the hypothesis

$$H_0: \theta \geqq \theta_0$$

vs.

$$H_1: \theta < \theta_0.$$

The type of test most often used as a uniformly most powerful size α test is the following randomized test:

$$\begin{aligned}\delta &= \textit{reject} \quad \text{if} \quad \phi > c \\ &= \textit{accept} \quad \text{if} \quad \phi < c \\ &= \textit{reject} \text{ with probability } \lambda \\ &\qquad\qquad \text{if} \quad \phi = c\end{aligned}$$

where

$$\begin{aligned}\lambda &= 0 & &\text{if } P_0[\phi = c] = 0 \\ &= \frac{\alpha - P_0[\phi > c]}{P[\phi = c]} & &\text{otherwise}\end{aligned}$$

and where c is the smallest number such that $P_0[\phi > c] \leqq \alpha$. This is clearly a uniformly most powerful size α test.

Although the requirement of a strictly increasing likelihood ratio seems to be very restrictive, we can show that many of the families of distributions considered in statistics have this property. It is easily shown that if F_θ is such that

$$F_\theta'(x_1, x_2, x_3, \ldots, x_n) = A(\theta)B(x_1, x_2, x_3, \ldots, x_n) \exp[C(\theta)\phi(x_1, x_2, x_3, \ldots, x_n)]$$

where $C(\theta)$ is a strictly increasing function of θ then the family has a monotone likelihood ratio in $\phi(x_1, x_2, x_3, \ldots, x_n)$. This can be seen by forming

$$\frac{F'_{\theta_1}}{F'_{\theta_0}} = \frac{A(\theta_1)}{A(\theta_0)} \exp\{[c(\theta_1) - c(\theta_0)]\phi(x_1, x_2, x_3, \ldots, x_n)\}$$

which is a strictly increasing function of $\phi(x_1, x_2, x_3, \ldots, x_n)$, when $\theta_1 > \theta_0$. Distributions which can be put into the exponential form given above are said to belong to the *exponential family* of distributions.

Example 12-2

Let us say that there are n independent observations from a normal distribution with unknown mean θ and known variance σ^2. We are to test the hypothesis that the mean is less than or equal to a given quantity θ_0 against the alternative that it is greater than θ_0. Here $\{F_\theta(x_1, x_2, x_3, \ldots, x_n) \colon \theta \in \Theta\}$ is such that

$$\begin{aligned}f_\theta(x_1, x_2, x_3, \ldots, x_n) &= \frac{1}{(2\pi)^{n/2}\sigma^n} \exp\left(-\sum_{i=1}^n (x_i - \theta)^2/2\sigma^2\right) \qquad -\infty < \text{all } x_i < \infty \\ &= \frac{1}{(2\pi)^{n/2}\sigma^n} \exp\left(-\sum_{i=1}^n x_i^2/2\sigma^2\right) e^{-n\theta^2/2\sigma^2} e^{n\theta x_n/\sigma^2}.\end{aligned}$$

Thus this family has a monotone likelihood ratio in $\bar{X}_n$.

In order to find the smallest number c for which $P_0(\bar{x}_n > c) \leqq \alpha$ all we need do is note that $\bar{X}_n$ has a normal distribution with mean θ_0 (under H_0) and variance σ^2/n. Therefore, c is the solution of

$$\Phi\left(\frac{\sqrt{(n)}(c-\theta_0)}{\sigma}\right) = 1-\alpha$$

where $\Phi(\cdot)$ is the cumulative distribution function for the standard normal. The uniformly most powerful size α test of the hypothesis is then given by

$$\begin{aligned} \delta &= \textit{reject} \qquad \text{if} \quad \bar{x}_n \geqq c \\ &= \textit{accept} \qquad \text{if} \quad \bar{x}_n < c. \end{aligned}$$

Remark *The position of the equal sign is irrelevant since* $P_0[\bar{x}_n = c] = 0$.

Example 12-3

A coin is flipped n times with unknown probability θ for heads. The hypothesis to be tested is

$$H_0\colon\ \theta \leqq \theta_0 \text{ (fixed)}$$

vs.

$$H_1\colon \theta > \theta_0.$$

If we let $X_i = 1$ if a head occurs on the ith flip and $X_i = 0$ otherwise then $\{F_\theta(x_1, x_2, x_3, \ldots, x_n)\colon \theta \in \Theta\}$ is such that

$$\begin{aligned} P_\theta(x_1, x_2, x_3, \ldots, x_n) &= \prod_{i=1}^{n} \theta^{x_i}(1-\theta)^{1-x_i} \qquad \text{all } x_i = 0, 1 \\ &= \exp_\theta\left(\sum_{i=1}^{n} x_i\right) \exp_{(1-\theta)}\left(n - \sum_{i=1}^{n} x_i\right) \\ &= \exp_{(1-\theta)^n[(\theta)/1-\theta]} \sum_{i=1}^{n} x_i \\ &= (1-\theta)^n \exp\left\{\ln\left(\frac{\theta}{1-\theta}\right)\right] \sum_{i=1}^{n} x_i\right\}. \end{aligned}$$

Therefore there is a monotone likelihood ratio in $\Sigma_{i=1}^n X_i$. Under H_0, $\Sigma_{i=1}^n X_i$ has a binomial distribution with parameters n and θ_0. Thus, c is the smallest integer such that

$$\sum_{v=c+1}^{n} \binom{n}{v} \theta_0{}^v (1-\theta_0)^{n-v} \leqq \alpha$$

and

$$\lambda = \frac{\alpha - \displaystyle\sum_{v=c+1}^{n} \binom{n}{v} \theta_0{}^v (1-\theta_0)^{n-v}}{\binom{n}{c} \theta_{0c}(1-\theta_0)^{n-c}}.$$

Using this, we obtain the following uniformly most powerful size α test of the given hypothesis:

$$\begin{aligned} \delta &= \textit{reject} \quad \text{if} \quad \sum_{i=1}^{n} x_i > c \\ &= \textit{accept} \quad \text{if} \quad \sum_{i=1}^{n} x_i < c \\ &= \textit{reject} \text{ with probability } \lambda \\ &\qquad\qquad \text{if} \quad \sum_{i=1}^{n} x_i = c. \end{aligned}$$

EXAMPLE 12-4

Let us say that there are n independent observations from a normal distribution with known mean μ and unknown variance θ. The hypothesis to be treated is

$$H_0\colon \theta \leqq \theta_0 \text{ (fixed)}$$

vs.

$$H_1\colon \theta > \theta_0.$$

Here $\{F_\theta(x_1, x_2, x_3, \ldots, x_n)\colon \theta \in \Theta\}$ is such that

$$f_\theta(x_1, x_2, x_3, \ldots, x_n) = \frac{1}{(2\pi\theta)^{n/2}} \exp\left(- \sum_{i=1}^{n} (x_i-\mu)^2/2\theta \right) \qquad -\infty < \text{all } x_i < \infty$$

indicating that there is a monotone likelihood ratio in

$$\sum_{i=1}^{n} (X_i-\mu)^2.$$

Since X_i has a normal distribution with mean μ and variance θ, $Z_i = X_i-\mu$ has a normal distribution with mean zero and variance θ, and

$$\sum_{i=1}^{n} Z_i^2 = \sum_{i=1}^{n} (X_i-\mu)^2$$

has a chi-squared distribution with n degrees of freedom and parameter θ. Thus if

$$\phi(X_1, X_2, X_3, \ldots, X_n) = \sum_{i=1}^{n} (X_i-\mu)^2$$

we have

$$\begin{aligned} f_{\theta_0}(\phi) &= \frac{1}{\Gamma(n/2)2^{n/2}\theta_0^{n/2}} \phi^{n/2-1} e^{-\phi/2\theta} \qquad 0 < \phi < \infty \\ &= 0 \qquad\qquad \text{otherwise.} \end{aligned}$$

One can now find the value of c such that

$$P_0(\phi > c) = 1-F_{\theta_0}(c) = \alpha.$$

Using this, we see that the uniformly most powerful size α test is given by

$$\begin{aligned} \delta &= \textit{reject} \quad \text{if} \quad \sum_{i=1}^{n} (x_i-\mu)^2 \geqq c \\ &= \textit{accept} \quad \text{if} \quad \sum_{i=1}^{n} (x_i-\mu)^2 < c. \end{aligned}$$

Let us now consider a numerical example.

EXAMPLE 12-5

Let X_1, X_2, X_3, X_4, X_5 be independent and identically distributed with a poisson distribution with unknown mean θ. We are to find a uniformly most powerful size 0.05 test of the hypothesis that $\theta \leqq 1$ against the alternative that $\theta > 1$.

$$P_\theta(x_1, x_2, x_3, x_4, x_5) = \prod_{i=1}^{5} e^{-\theta} \frac{\theta^{x_i}}{x_i!} \qquad \text{all } x_i = 0, 1, 2, \ldots$$

$$= e^{-5\theta} \left(\frac{1}{\prod_{i=1}^{5} x_i!} \right) \exp(\ln \theta) \sum_{i=1}^{5} x_i$$

indicating that there is a monotone likelihood ratio in

$$\sum_{i=1}^{5} X_i.$$

But

$$\phi(X_1, X_2, X_3, X_4, X_5) = \sum_{i=1}^{5} X_i$$

has a poisson distribution with parameter 5θ. We now define c to be the smallest integer such that

$$P_0(\phi > c) = \sum_{\phi=c+1}^{\infty} e^{-5} \frac{5^\phi}{\phi!} \leqq 0.05.$$

Using a table of the poisson distribution we find that $c = 9$ with $P_0(\phi > 9) = 0.0317$ and $P_0(\phi = 9) = 0.0363$. Thus

$$\lambda = \frac{.05 - .0317}{.0363} = 0.504.$$

A uniformly most powerful size 0.05 test of the hypothesis is then given by

$$\begin{aligned} \delta &= \textit{reject} \quad \text{if } \sum_{i=1}^{5} x_i > 9 \\ &= \textit{accept} \quad \text{if } \sum_{i=1}^{5} x_i < 9 \\ &= \textit{reject} \text{ with probability } 0.504 \\ &\qquad \text{if } \sum_{i=1}^{5} x_i = 9. \end{aligned}$$

12-2 Two Sided Hypotheses

In the previous section we saw how to test hypotheses of the form

$$H_0: \quad \theta = \theta_0$$

vs.

$$H_1: \quad \theta > \theta_0$$

Q

or

$$H_1^*: \quad \theta < \theta_0.$$

But what about the alternate hypothesis

$$H_1^{**}: \theta \neq \theta_0$$

which is known as a *two sided hypothesis*? Clearly H_1^{**} is true only if either H_1 or H_1^* is true. That is, $\theta \neq \theta_0 \Leftrightarrow \theta > \theta_0$ or $\theta < \theta_0$. If a uniformly most powerful test were to exist for H_0 vs. H_1^{**} it would have to be at least as powerful as a uniformly most powerful test of H_0 vs. H_1. If this were the case, it would have to be a uniformly most powerful test of H_0 vs. H_1. As we have seen, however, the power function is a nondecreasing function. This would imply that this test could not be uniformly most powerful for H_0 vs. H_1^*. Thus, it could not have been uniformly most powerful for H_0 vs. H_1^{**} proving that a uniformly most powerful test for a two-sided alternative cannot exist.

Since a uniformly most powerful test does not exist, we will discuss a method for obtaining "reasonable" tests.

Let us recall that when there was a monotone likelihood ratio in the statistic $\phi = \phi(X_1, X_2, X_3, \ldots, X_n)$, then for H_0 vs. H_1 there was a uniformly most powerful test of any size. For a particular size α_1 we define $c(\alpha_1)$ to be the smallest number such that $P_0(\phi > c(\alpha_1)) \leqq \alpha_1$ and set

$$\begin{aligned} \lambda_1 &= \frac{\alpha_1 - P_0[\phi > c(\alpha_1)]}{P_0[\phi = c(\alpha_1)]} && \text{when it exists} \\ &= 0 && \text{otherwise.} \end{aligned}$$

Then

$$\begin{aligned} \delta_1 &= \textit{reject} && \text{if} \quad \phi > c(\alpha_1) \\ &= \textit{accept} && \text{if} \quad \phi < c(\alpha_1) \\ &= \textit{reject} \text{ with probability } \lambda_1 && \text{if} \quad \phi = c(\alpha_1) \end{aligned}$$

is a uniformly most powerful size α_1 test. On the other hand, to test H_0 vs. H_1^* we define (for any size α_2) $c(\alpha_2)$ to be the smallest number such that $P_0[\phi^* = -\phi > c(\alpha_2)] \leqq \alpha_2$. This is equivalent to finding the largest number $c^*(\alpha_2)$ such that $P_0[\phi < c^*(\alpha_2)] \leqq \alpha_2$ [where $c^*(\alpha_2) = -c(\alpha_2)$]. Then we can set

$$\begin{aligned} \lambda_2 &= \frac{\alpha_2 - P_0[\phi < c^*(\alpha_2)]}{P_0[\phi = c^*(\alpha_2)]} && \text{when it exists} \\ &= 0 && \text{otherwise} \end{aligned}$$

and obtain

$$\begin{aligned}\delta_2 &= \textit{reject} && \text{if} \quad \phi < c^*(\alpha_2)\\ &= \textit{accept} && \text{if} \quad \phi > c^*(\alpha_2)\\ &= \textit{reject} \text{ with probability } \lambda_2 && \text{if} \quad \phi = c^*(\alpha_2)\end{aligned}$$

as a uniformly most powerful size α_2 test of H_0 vs. $H_1{}^*$.

Since $P_0[\phi < c^*(\alpha_2)] \leqq \alpha_2$ and $P_0[\phi < c(\alpha_1)] \geqq 1-\alpha_1$ we see that $c^*(\alpha_2) < c(\alpha_1)$ if $\alpha_2 < 1-\alpha_1$. This means that if $\alpha_2 < 1-\alpha_1$ there will be no overlapping of the rejection regions for δ_1 and δ_2. That is, there are no values of ϕ for which both δ_1 and δ_2 say *reject*. This suggests the test function

$$\begin{aligned}\delta &= \textit{reject} && \text{if } \delta_1 = \textit{reject} \quad \text{or} \quad \delta_2 = \textit{reject}\\ &= \textit{accept} && \text{otherwise.}\end{aligned}$$

Now

$$P_0(\delta = \textit{reject}) = P_0(\delta_1 = \textit{reject}) + P_0(\delta_2 = \textit{reject})$$

(assuming that $\alpha_2 < 1-\alpha_1$) which says that if $\alpha_1+\alpha_2 = \alpha$ then δ is a size α test. Consider now the power function

$$P_\theta(\delta = \textit{reject}) = P_\theta(\delta_1 = \textit{reject}) + P_\theta(\delta_2 = \textit{reject}).$$

As θ increases above θ_0 $P_\theta(\delta_1 = \textit{reject})$ increases above α_1 but $P_\theta(\delta_2 = \textit{reject})$ decreases below α_2. If both α_1 and α_2 are small then it seems reasonable to expect that $P_\theta(\delta_1 = \textit{reject})$ will increase more rapidly than $P_\theta(\delta_2 = \textit{reject})$ will decrease. On the other hand, when θ decreases below θ_0 $P_\theta(\delta_2 = \textit{reject})$ will increase while $P_\theta(\delta_1 = \textit{reject})$ will decrease. Again when α_1 and α_2 are small it is reasonable to expect that $P_\theta(\delta_2 = \textit{reject})$ will increase more rapidly than $P_\theta(\delta_1 = \textit{reject})$ will decrease.

Exercise 12-3

Show that if the family of distributions admits a monotone likelihood function, then the power of the reasonable test is at least as large as $\alpha_1+\alpha_2$.

Although any values for α_1 and α_2 such that $\alpha_1+\alpha_2 = \alpha$ will cause δ to be a size α test, the most often used values are $\alpha_1 = \alpha_2 = \alpha/2$. The test is then called an *equal tails test*.

Example 12-6

We have a sample of n independent observations from a normal distribution with unknown mean θ and variance σ^2. We are required to find a size α test of the hypothesis that the mean is θ_0 with the alternative hypothesis that it is not θ_0.

Here

$$f_\theta(x_1, x_2, x_3, \dots, x_n) = \frac{1}{(2\pi)^{n/2}\sigma^n} \exp\left(-\sum_{i=1}^{n} (x_i-\theta)^2/2\sigma^2\right) \qquad -\infty < \text{all } x_i < \infty.$$

In Example 12-2 we found δ_1 to be given by

$$\begin{aligned} \delta_1 &= \textit{reject} \qquad \text{if} \quad \bar{x}_n \geqq c(\alpha_1) \\ &= \textit{accept} \qquad \text{if} \quad \bar{x}_n < c(\alpha_1) \end{aligned}$$

with $c(\alpha_1)$ satisfying the equation

$$\phi\left(\frac{\sqrt{(n)}[c(\alpha_1)-\theta_0]}{\sigma}\right) = 1-\alpha_1$$

and $\phi(\cdot)$ the standard normal distribution function. Similarly we can find

$$\begin{aligned} \delta_2 &= \textit{reject} \qquad \text{if} \quad \bar{x}_n \leqq c^*(\alpha_2) \\ &= \textit{accept} \qquad \text{if} \quad \bar{x}_n > c^*(\alpha_2) \end{aligned}$$

with $c^*(\alpha_2)$ defined by the equation

$$\phi\left(\frac{\sqrt{(n)}(c^*(\alpha_2)-\theta_0)}{\sigma}\right) = \alpha_2.$$

Thus

$$\begin{aligned} \delta &= \textit{accept} \qquad \text{if} \quad c^*\left(\frac{\alpha}{2}\right) < \bar{x}_n < c\left(\frac{\alpha}{2}\right) \\ &= \textit{reject} \qquad \text{otherwise} \end{aligned}$$

is the equal tailed size α test of the above hypothesis.

Other two-sided hypotheses can be treated in a similar manner. For example, we may have to test the hypothesis

A) H_0: $\theta_1 \leqq \theta \leqq \theta_2$ (θ_1, θ_2 fixed)

vs.

H_1: $\theta < \theta_1$ or $\theta > \theta_2$

or the hypothesis

B) H_0: $\theta \leqq \theta_1$ or $\theta \geqq \theta_2$ (θ_1, θ_2 fixed)

vs.

H_1: $\theta_1 < \theta < \theta_2$.

EXERCISE 12-4

Find a size α test for situation **A)** above by using the uniformly most powerful tests of

$$H_0: \theta \geqq \theta_1$$

vs.

$$H_1: \theta < \theta_1$$

and

$$H_0\colon \theta \leqq \theta_2$$

vs.

$$H_1\colon \theta > \theta_2.$$

EXERCISE 12-5

Find a size α test for situation **B** above by using the uniformly most powerful tests of

$$H_0\colon \theta \leqq \theta_1$$

vs.

$$H_1\colon \theta > \theta_1$$

and

$$H_0\colon \theta \geqq \theta_2$$

vs.

$$H_1\colon \theta < \theta_2.$$

12-3 Interpretation of a Test of Hypothesis

Before completing this chapter on hypothesis testing it will be appropriate to discuss the meaning of the results of these tests.

When the null hypothesis is true, then the probability that $\delta = reject$ is at most α, which presumably is small (perhaps 0.05). Thus, if the random variables are observed, and if δ is calculated with the result that $\delta = reject$, one can conclude one of two things:

1) The hypothesis is true and an event with small probability has occurred. Using a frequency interpretation, we know that this should occur with frequency approximately α (1-in-20 for $\alpha = 0.05$).

2) The null hypothesis is not true and the alternative hypothesis is such that it made $\delta = reject$ quite likely.

When $\delta = reject$, the most useful interpretation is the second, and this is the interpretation usually quoted. But we should be cautious here since it may not be the correct interpretation.

When the observations are such that $\delta = accept$, then the interpretation is even more dubious. We give two possible interpretations:

1) The null hypothesis is true, and an event of high probability $(1-\alpha)$ has occurred.

2) The alternate hypothesis is true and an event of probability γ has occurred.

For a uniformly most powerful test, γ must be less than $1-\alpha$, but not necessarily much less. Thus, when $\delta = accept$ no conclusion can be reached and we see that, at best, a test has some significance only when $\delta = reject$.

12-4 Problems

1. Assume $\{F_\theta(x): \theta \in \Theta\}$ is such that

$$f_\theta(x) = \frac{1}{\sqrt{(2\pi)}} e^{-(x-\theta)^2/2} \qquad -\infty < x < \infty$$

with

$$H_0: \theta = -1$$

vs.

$$H_1: \theta = +1$$

and

$$\delta = \begin{cases} \textit{reject} & \text{if } x \geqq 0.96 \\ \textit{accept} & \text{otherwise.} \end{cases}$$

Find the size of the test δ.

2. For the situation of Problem 1 find the power.
3. Is the test function a most powerful test (of size found in Problem 1)?
4. If in Problem 1 the alternate hypothesis had been $H_1: \theta > -1$, what would be the size of the test?
5. For the situation of Problem 4 find the value of the power function at $\theta = 0$. At $\theta = +1$.
6. Assume $\{F_\theta(x): \theta \in \Theta\}$ is such that

$$f_\theta(x) = \frac{1}{\sqrt{(2\pi)}} e^{-(x-\theta)^2/2} \qquad -\infty < x < \infty$$

with $\Theta = (-\infty, \infty)$. Find the equal tails size .05 test of the hypothesis

$$H_0: \theta = -1$$

vs.

$$H_1: \theta \neq -1.$$

7. Sketch the power function for the situation in Problem 6.
8. The number of counts recorded by a geiger counter which has been placed near a radioactive source is assumed to be a random variable with a poisson distribution. Assume that we will record the counts per second for ten consecutive seconds. With these data, we are required to test the hypothesis that the mean of the poisson distribution is 0.8 against the alternative that it is 1.0. Find a uniformly most powerful size 0.05 test for this situation.
9. If we were to accept, rather than randomize, what would be the size of the test in Problem 2?
10. What is the power of the test found in Problem 8?
11. What would be the power of the test suggested in Problem 9?
12. If, in Problem 8 the alternate hypothesis was H_1: *the mean is greater than* 0.8, find a uniformly most powerful size 0.05 test.
13. If, in Problem 8 the alternate hypothesis was H_1: *the mean is not* 0.8, find an equal tails size 0.05 test of the hypothesis.

14. Assume $\{F_\theta(x_1, x_2, x_3, \ldots, x_n): \theta \in \Theta\}$ is such that

$$f_\theta(x_1, x_2, x_3, \ldots, x_n) = \theta^n \prod_{i=1}^{n} e^{-\theta x_i} \qquad 0 < \text{all } x_i < \infty$$

with $\Theta = (1, \infty)$. Find a uniformly most powerful size α test of the hypothesis

$$H_0: \theta \leqq \theta_0 \quad (\theta_0 \text{ fixed})$$

vs.

$$H_1: \theta > \theta_0.$$

15. For the family of distributions in Problem 14 find an equal tails test of the hypothesis

$$H_0: \theta = \theta_0 \quad (\theta_0 \text{ fixed})$$

vs.

$$H_1: \theta \neq \theta_0.$$

16. Assume $\{F_\theta(x): \theta \in \Theta\}$ is such that

$$f_\theta(x) = \theta e^{-\theta x} \qquad 0 < x < \infty$$

with $\Theta = (0, \infty)$. Find the most powerful size 0.05 test of the hypothesis

$$H_0: \theta = 1$$

vs.

$$H_1: \theta = 2.$$

17. Find the uniformly most powerful size 0.05 test if, in Problem 16 the hypothesis had been

$$H_0: \theta = 1$$

vs.

$$H_1: \theta > 1.$$

18. Find the uniformly most powerful size .05 test if, in Problem 16 the hypothesis had been

$$H_0: \theta = 2$$

vs.

$$H_1: \theta < 2.$$

19. Find the equal tails size 0.05 test if in Problem 16 the hypothesis had been

$$H_0: 1 \leqq \theta \leqq 2$$

$$H_1: \theta < 1 \quad \text{or} \quad \theta > 2.$$

20. Find a *reasonable* size 0.05 test if in Problem 16 the hypothesis had been

$$H_0: \theta \leqq 1 \quad \text{or} \quad \theta \geqq 2$$

vs.

$$H_1: 1 < \theta < 2.$$

CHAPTER 13

Confidence Intervals

In Chapter eight $\Omega_L(X_1, X_2, X_3, \dots, X_n)$, the lower confidence bound, and $\Omega_U(X_1, X_2, X_3, \dots, X_n)$, the upper confidence bound were defined. A confidence interval was then defined by the double inequality

$$\Omega_L(X_1, X_2, X_3, \dots, X_n) \leqq \theta \leqq \Omega_U(X_1, X_2, X_3, \dots, X_n).$$

In this chapter we will not only consider criteria for *good* confidence intervals, but we will also consider three essentially different methods for obtaining confidence intervals.

13-1 Neyman-Pearson Type Confidence Intervals

Assume that the family of distributions $\{F_\theta(x_1, x_2, x_3, \dots, x_n): \theta \in \Theta\}$ is known. For a particular $\theta \in \Theta$ the distribution $F_\theta(x_1, x_2, x_3, \dots, x_n)$ is completely specified, and the probability of any set of n-tuples can be determined (ignoring algebraic difficulties). In particular, the probability of the set of all n-tuples such that $\Omega_L(X_1, X_2, X_3, \dots, X_n) \leqq \theta \leqq \Omega_U(X_1, X_2, X_3, \dots, X_n)$ can be found. This probability, namely

$$P_\theta[\Omega_L(X_1, X_2, X_3, \dots, X_n) \leqq \theta \leqq \Omega_U(X_1, X_2, X_3, \dots, X_n)] = \beta(\theta)$$

is known as the *confidence* of the interval at θ (fixed). The minimum of these confidences for a particular interval for all θ is known as the *confidence of the confidence interval*; that is, the interval

$$\Omega_L(X_1, X_2, X_3, \dots, X_n) \leqq \theta \leqq \Omega_U(X_1, X_2, X_3, \dots, X_n)$$

has *confidence* β if

$$\beta = \min_{\theta \varepsilon \Theta} \beta(\theta).$$

INTERPRETATION OF NEYMAN–PEARSON TYPE CONFIDENCE INTERVALS

For any fixed $\theta \in \Theta$, the statistics $\Omega_L(X_1, X_2, X_3, \dots, X_n)$ and $\Omega_U(X_1, X_2, X_3, \dots, X_n)$ form a random interval of real numbers. If we were to observe the values $x_1, x_2, x_3, \ldots, x_n$ from the distribution $F_\theta(x_1, x_2, x_3, \ldots, x_n)$ then either the interval

$$[\Omega_L(x_1, x_2, x_3, \ldots, x_n), \Omega_U(x_1, x_2, x_3, \ldots, x_n)]$$

contains θ or it does not. If another n-tuple were observed from this same distribution, then once again the interval would or would not contain θ. If this were repeated many times the *relative frequency* that the interval would contain θ should approach $\beta(\theta)$. If, however, each time an n-tuple were repeated, the values of $\theta \in \Theta$ were not necessarily the same, then the limit, assuming that it exists, would depend on the frequency with which each θ occurred. This limit, however, could not be smaller than β. For example, if, in N repetitions, θ_i occurred N_i times $i = 1, 2, 3, \ldots r \leqq N$ then the *relative frequency* that the appropriate value of θ was in the recorded interval would approach

$$\sum_{i=1}^{r} \frac{N_i}{N} \beta(\theta_i)$$

which is no smaller than β.

Therefore, we can say that if we were to use the confidence interval $\Omega_L(X_1, X_2, X_3, \ldots, X_n) \leqq \theta \leqq \Omega_U(X_1, X_2, X_3, \ldots, X_n)$ repeatedly, the limiting frequency and, therefore, the probability of being correct is not less than β, regardless of the true value of $\theta \in \Theta$.

It is important to note that $X_1, X_2, X_3, \ldots, X_n$ are random variables while θ is an unknown parameter. The probability statement, therefore, refers to sets of possible observations on $X_1, X_2, X_3, \ldots, X_n$ and does not say anything about θ for fixed or observed $x_1, x_2, x_3, \ldots, x_n$. It is not correct to interpret the numerical confidence interval

$$\Omega_L(x_1, x_2, x_3, \ldots, x_n) \leqq \theta \leqq \Omega_U(x_1, x_2, x_3, \ldots, \beta_n)$$

to mean that the frequency that θ will lie between the given numbers will, in some sense, approach β.

EXAMPLE 13-1

Let $\{F_\theta(x_1, x_2, x_3, \ldots, x_n) : \theta \in \Theta\}$ be such that F_θ is given by the density function

$$f_\theta(x_1, x_2, x_3, \ldots, x_n) = \frac{1}{(2\pi)^{n/2}\sigma^n} \exp\left(-\sum_{i=1}^{n} (x_i - \theta)^2/2\sigma^2\right) \qquad -\infty < \text{all } x_i < \infty$$

$$\sigma^2 \text{ known}$$

with $\Theta = (-\infty, \infty)$. Consider the confidence interval $\bar{X}_n - 1.96\,\sigma/\sqrt{(n)} \leqq \theta \leqq \bar{X}_n + 1.96\,\sigma/\sqrt{(n)}$. In order for the interval to contain θ, $\bar{X}_n$ must have a value such that

$\theta - 1.96\,\sigma/\sqrt{(n)} \leqq \bar{X}_n \leqq \theta + 1.96\,\sigma/\sqrt{(n)}$. The distribution of $\bar{X}_n$ is normal with mean θ and variance σ^2/n. Thus

$$\beta(\theta) = P_\theta(\theta - 1.96\sigma/\sqrt{(n)} \leqq \bar{X}_n \leqq \theta + 1.96\sigma/\sqrt{(n)}) = P_\theta\left(-1.96 \leqq \frac{\sqrt{(n)}(\bar{X}_n - \theta)}{\sigma} \leqq 1.96\right)$$
$$= 0.95$$

for any $\theta \in \Theta$ and therefore $\beta = .95$. One can thus state that $\bar{X}_n - 1.96\,\sigma/\sqrt{n} \leqq \theta \leqq \bar{X}_n + 1.96\,\sigma/\sqrt{n}$ with confidence .95 whatever the value of θ might be.

CRITERIA FOR GOOD CONFIDENCE INTERVALS: SYMMETRIC CONFIDENCE INTERVALS

As in estimation and hypothesis testing, criteria for good confidence intervals must depend on the particular situation. For example, an engineer is often required to estimate the "true" value of a parameter θ. After this is done, he may be required to give a value α and present his result as $\hat{\theta} \pm \alpha$ where $\hat{\theta}$ is his estimate of θ and where the interval $(\hat{\theta} - \alpha, \hat{\theta} + \alpha)$ is "almost certain" to contain θ. This can be interpreted as the confidence interval $\hat{\theta} - \alpha \leqq \theta \leqq \hat{\theta} + \alpha$. By "almost certain" he may mean "with high confidence". How near unity the confidence should be, varies from situation to situation. The confidence of the interval can often easily be found since $\hat{\theta} - \alpha \leqq \theta \leqq \hat{\theta} + \alpha$ is equivalent to $\theta - \alpha \leqq \hat{\theta} \leqq \theta + \alpha$. Example 13-1 demonstrates this type of confidence interval with $\hat{\theta} = \bar{X}_n$ and $\alpha = 1.96\,\sigma/\sqrt{n}$.

SHORTEST LENGTH CONFIDENCE INTERVALS

A confidence interval can always be given in the form $-\infty < \theta < \infty$, which will have confidence one. The obvious objection to this interval is its infinite length. The shorter the length of an interval of given confidence, the more informative is the interval. If we could find, from among all those intervals with confidence at least β (fixed), one which had the shortest length for all $\theta \in \Theta$, then this would be called a *shortest length confidence interval of confidence* β. Shortest length confidence intervals can often be found. Consider the univariate situation where $\{F_\theta(x)\colon \theta \in \Theta\}$ is such that $f_\theta(x)$ is a continuous unimodal (monotonically decreasing on both sides of the maximum value) density function with θ, a location parameter.

Remark *A location parameter θ is such that $F_\theta(x) = F_{\theta+a}(x+a)$ for any θ and $\theta + a \in \Theta$. For different values of θ, the distribution function (and the density function) has the same shape but is translated along the coordinate axis.*

For this situation, it will be possible to find two values l and u such that

a. $f_\theta(\theta - l) = f_\theta(\theta + u)$

and

b. $\int_{\theta-l}^{\theta+u} f_\theta(x)\,dx = \beta$ (β fixed).

Condition **b** says that

$$P_\theta(\theta-l \leqq X < \theta+u) = F_\theta(\theta+u)-F_\theta(\theta-l) = \beta.$$

This probability is constant for all $\theta \in \Theta$ when θ is a location parameter. Therefore, the interval $X-u \leqq \theta \leqq X+l$ (obtained from the interval $\theta-l \leqq X \leqq \theta+u$) is a β confidence interval. That it has the shortest length among all β confidence intervals can be seen from Figure 13-1. If one were to

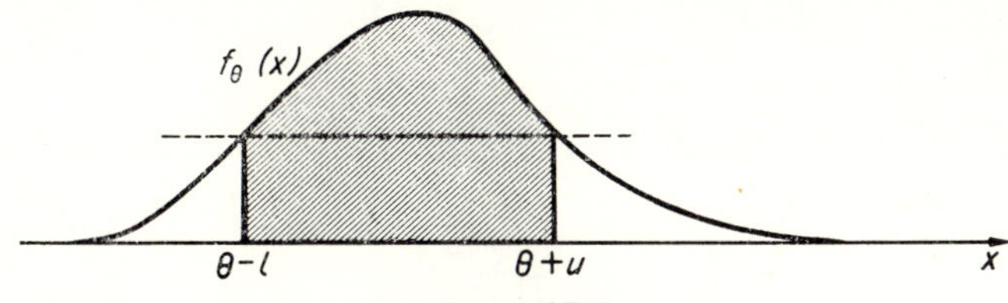

Figure 13-1

move the interval to the left or to the right the area under the curve would decrease (since the effective height of the area eliminated would be greater than the effective height of the area included). In order to increase the area (confidence) to β, we must therefore lengthen the interval indicating that the first interval was shorter.

EXAMPLE 13-2

Let $\{F_\theta(x): \theta \in \Theta\}$ be such that F_θ is given by the density function

$$f_\theta(x) = \frac{1}{\sqrt{(2\pi)}\sigma} e^{-(x-\theta)^2/2\sigma^2} \qquad \begin{array}{l} -\infty < x < \infty \\ \sigma > 0 \text{ known} \end{array}$$

with $\Theta = (-\infty, \infty)$. Here θ is a location parameter and $f_\theta(x)$ is unimodal and continuous. Since this density is symmetric about θ, we have for any value α

a. $f_\theta(\theta-\alpha) = f_\theta(\theta+\alpha)$ all $\theta \in \Theta$

and

b. $\int_{\theta-\alpha}^{\theta+\alpha} f_\theta(x)\,dx = \Phi(\alpha/\sigma)-\Phi(-\alpha/\sigma) = 2\Phi(\alpha/\sigma)-1$

where $\Phi(\cdot)$ is the cumulative distribution function of the standard normal. Thus, if α is such that

$$\Phi(\alpha/\sigma) = \frac{1+\beta}{2}$$

then $X-\alpha \leqq \theta \leqq X+\alpha$ is the shortest length β confidence interval for θ.

The assumption of continuity of $f_\theta(x)$ can be relaxed. For example, the density function may be truncated (i.e., cut off) at a point which depends on θ. In this case, $\theta - l$ (or $\theta + u$) may be replaced by the truncation point when the density function at the truncation point is above $f_\theta(\theta - l)$ (or $f_\theta(\theta + u)$).

UNIFORMLY MOST ACCURATE CONFIDENCE INTERVALS

The confidence $\beta(\theta)$ can be considered a special case of the function

$$\beta(\theta, \theta^*) = P_\theta[\Omega_L(X_1, X_2, X_3, \dots, X_n) \leqq \theta^* \leqq \Omega_U(X_1, X_2, X_3, \dots, X_n)].$$

This function is the probability of the confidence interval covering a value θ^* when θ is the true value. When $\theta^* = \theta$, $\beta(\theta, \theta) = \beta(\theta)$. The confidence of an interval is then

$$\beta = \min_{\theta \varepsilon \Theta} \beta(\theta, \theta).$$

The *inaccuracy* of an interval can be defined as the probability with which the interval covers the wrong θ value. That is, the *inaccuracy* of the interval about θ (call it I) at θ^* is defined to be $\beta_I(\theta, \theta^*)$, where the subscript is being used to distinguish the inaccuracies of different intervals. A β confidence interval I is said to be *uniformly more accurate* than a β confidence interval I^* if

$$\beta_I(\theta, \theta^*) \leqq \beta_I{}^*(\theta, \theta^*)$$

whenever $\theta, \theta^* \in \Theta$ and $\theta^* \neq \theta$. If I is uniformly more accurate than any other β confidence interval it is said to be a *uniformly most accurate* confidence interval. In general, uniformly most accurate confidence intervals do not exist. However, they do exist in a limited sense, to be discussed in the following subsection.

ONE SIDED UNIFORMLY MOST ACCURATE CONFIDENCE INTERVALS

Often we are interested in only one bound, either $\Omega_L(X_1, X_2, X_3, \dots, X_n)$ or $\Omega_U(X_1, X_2, X_3, \dots, X_n)$. For the present, let us say that we are interested in the upper bound $\Omega_U(X_1, X_2, X_3, \dots, X_n)$; that is, the intervals of interest are $\theta \leqq \Omega_U(X_1, X_2, X_3, \dots, X_n)$. In this situation, inaccuracy only has meaning for values of $\theta^* > \theta$, since if $\theta^* < \theta$, then $\Omega_U(X_1, X_2, X_3, \dots, X_n)$ must be greater than θ^* whenever it is greater than θ. Therefore, we define the concept of *uniformly more accurate one-sided* confidence intervals by replacing the condition $\theta^* \neq \theta$ in the definition of *uniformly most accurate* confidence intervals by the condition $\theta^* > \theta$. For this case, as we will see, there often exist uniformly most accurate confidence intervals.

Consider the hypothesis testing situation where the family of distributions $\{F_\theta(x_1, x_2, x_3, \ldots, x_n) : \theta \in \Theta\}$ is such that there is a monotone likelihood ratio in $\phi = \phi(x_1, x_2, x_3, \ldots, x_n)$. The hypothesis

$$H_0 : \theta \leqq \theta_0$$

vs.

$$H_1 : \theta > \theta_0$$

admits a uniformly most powerful size α test of the form

$$\begin{aligned} \delta &= \textit{reject} \qquad \text{if} \quad \phi \geqq c \\ &= \textit{accept} \qquad \text{if} \quad \phi < c \end{aligned}$$

where we have assumed that randomization is not necessary.

Remark *Randomization could be used here as was done in hypothesis testing; we will not use it. In the discrete case, therefore, it will not be possible to obtain β confidence intervals for every β. In this case, a value near β which can be realized will be substituted.*

For this test we have shown that

$$P_\theta(\delta = \textit{reject}) > P_\theta(\delta^* = \textit{reject}) \qquad \text{all } \theta > \theta_0$$

where δ^* is any size α test not of this form. What was not shown however, is that

$$P_\theta(\delta = \textit{reject}) < P_\theta(\delta^* = \textit{reject}) \qquad \text{all } \theta < \theta_0.$$

This can be seen by noting that for $\theta < \theta_0$ the inequalities in the proof are reversed.

EXERCISE 13-1

Prove the above statements.

Let us now assume that the set of n-tuples for which $\phi \geqq c$ is equivalent to the set of n-tuples for which $\theta_0 \leqq \Omega_U[\phi(X_1, X_2, X_3, \ldots, X_n)]$. If I represents the interval $\theta_0 \leqq \Omega_U[\phi(X_1, X_2, X_3, \ldots, X_n)]$ while I^* represents any other interval $\theta_0 \leqq \Omega_U{}^*(X_1, X_2, X_3, \ldots, X_n)$ such that $P_0[\theta_0 \leqq \Omega_U{}^*(X_1, X_2, X_3, \ldots, X_n)] = \alpha$ then the result proved in Exercise 13-1 indicates that

$$\begin{aligned} \beta_I(\theta, \theta_0) &= P_\theta\{\theta_0 \leqq \Omega_U[\phi(X_1, X_2, X_3, \ldots, X_n)]\} \\ &< P_\theta[\theta_0 \leqq \Omega_U{}^*(X_1\ \ X_2, X_3, \ldots, X_n)] = \beta_{I^*}(\theta, \theta_0) \end{aligned}$$

for all $\theta_0 > \theta$. Therefore $\theta \leqq \Omega_U[\phi(X_1, X_2, X_3, \ldots, X_n)]$ is a uniformly most accurate one sided β confidence interval. This is called *lower β confidence interval.* A similar result holds for *upper β confidence intervals* of the form $\Omega_L(X_1, X_2, X_3, \ldots, X_n) \leqq \theta$.

EXAMPLE 13-3

Assume $\{F_\theta(x_1, x_2, x_3, \ldots, x_n)\colon \theta \in \Theta\}$ is such that F_θ is given by the density function

$$f_\theta(x_1, x_2, x_3, \ldots, x_n) = \frac{1}{(2\pi)^{n/2}\sigma^n} \exp\left[-\sum_{i=1}^{n} (x_i - \theta)^2/2\sigma^2\right] \qquad -\infty < \text{all } x_i < \infty$$

σ^2 known

with $\Theta = (-\infty, \infty)$. The uniformly most powerful size α test of the hypothesis

$$H_0\colon \theta \leqq \theta_0 \quad (\theta_0 \text{ fixed})$$

vs.

$$H_1\colon \theta > \theta_0$$

is given by

$$\delta = \begin{cases} \textit{reject} & \text{if } \bar{X}_n > \theta_0 + \alpha \\ \textit{accept} & \text{if } \bar{X}_n \leqq \theta_0 + \alpha \end{cases}$$

where α is determined by the equation $\phi(\sqrt{(n)}\alpha/\sigma) = 1-\alpha$. Since α does not depend on θ_0

$$P_\theta(\bar{X}_n \leqq \theta + \alpha) = P_\theta(\bar{X}_n - \alpha \leqq \theta) = 1 - \alpha.$$

Thus if we set $\alpha = 1-\beta$, we find that $\Omega_L(X_1, X_2, X_3, \ldots, X_n) = \bar{X}_n - \alpha \leqq \theta$ is a uniformly most accurate upper β confidence interval for θ.

REASONABLY ACCURATE CONFIDENCE INTERVALS

If $\Omega_L(X_1, X_2, X_3, \ldots, X_n) \leqq \theta$ is a uniformly most accurate upper confidence interval of confidence β_U and $\theta \leqq \Omega_U(X_1, X_2, X_3, \ldots, X_n)$ is a uniformly most accurate lower confidence interval of confidence β_L, then $\Omega_L(X_1, X_2, X_3, \ldots, X_n) \leqq \theta \leqq \Omega_U(X_1, X_2, X_3, \ldots, X_n)$ is a *reasonably accurate confidence interval* of confidence $\beta = \beta_L + \beta_U - 1$ (assuming $\beta_L + \beta_U > 1$). This can be seen from the fact that the set of $x_1, x_2, x_3, \ldots, x_n$ for which $\Omega_L(X_1, X_2, X_3, \ldots, X_n) > \theta$ is disjoint from the set of $x_1, x_2, x_3, \ldots, x_n$ for which $\theta > \Omega_U(X_1, X_2, X_3, \ldots, X_n)$ when $\beta_L + \beta_U > 1$. [This is the same as saying that $\phi > c(1-\beta_U)$ and $\phi < c(1-\beta_L)$ are disjoint for the two sided hypothesis testing problem when $(1-\beta_U)+(1-\beta_L) < 1$.] Thus

$$\begin{aligned} P_\theta[\Omega_L(X_1, X_2, X_3, \ldots, X_n) > \theta \quad &\text{or} \quad \theta > \Omega_U(X_1, X_2, X_3, \ldots, X_n)] \\ &= (1-\beta_U)+(1-\beta_L) = 2-\beta_L-\beta_U. \end{aligned}$$

Now since

$$\begin{aligned} P_\theta[\Omega_L(X_1, X_2, X_3, \ldots, X_n) &\leqq \theta \leqq \Omega_U(X_1, X_2, X_3, \ldots, X_n)] \\ &= 1 - P_\theta[\Omega_L(X_1, X_2, X_3, \ldots, X_n) > \theta \quad \text{or} \\ &\qquad \theta > \Omega_U(X_1, X_2, X_3, \ldots, X_n)] \end{aligned}$$

we have

$$P_\theta[\Omega_L(X_1, X_2, X_3, \ldots, X_n) \leqq \theta \leqq \Omega_U(X_1, X_2, X_3, \ldots, X_n)] = \beta_L + \beta_U - 1.$$

It is often assumed that

$$\beta_L = \beta_U = (1+\beta)/2.$$

EXERCISE 13-2

For the family of distributions of Example 13-3, find a reasonably accurate β confidence interval for θ. Use $\beta_L = \beta_U = (1+\beta)/2$.

13-2 Bayes Type "Confidence Intervals"

Assume, as usual, that there is a family of distributions $\{F_\theta(x_1, x_2, x_3, \ldots, x_n): \theta \in \Theta\}$. Assume further that the "true" value of θ is obtained through a random mechanism with distribution function $F(\theta)$; that is, θ is the outcome of a random variable Θ with probability distribution function $F(\theta)$. Since Θ is a random variable we can construct intervals $\alpha \leqq \theta \leqq b$ where a and b are constants and interpret the confidence of this interval to be the probability under $F(\theta)$ that the random variable will be in the interval (a, b). This ignores the fact that we will observe the random variables $X_1, X_2, X_3, \ldots, X_n$. One can, however, find the probability that Θ will be in the interval (a, b) given that $x_1, x_2, x_3, \ldots, x_n$ are the observed values of $X_1, X_2, X_3, \ldots, X_n$. The distribution necessary for this is the conditional distribution $F(\theta \mid x_1, x_2, x_3, \ldots, x_n)$.

Given the family of distributions and $F(\theta)$ we note that

$$F'(\theta \mid x_1, x_2, x_3, \ldots, x_n) = \frac{F_\theta'(x_1, x_2, x_3, \ldots, x_n)F'(\theta)}{\int_\Theta F_\theta'(x_1, x_2, x_3, \ldots, x_n)dF(\theta)}$$

from which the conditional probabilities of intervals can be found. Any interval $\theta_L = \Omega_L(x_1, x_2, x_3, \ldots, x_n) \leqq \theta \leqq \Omega_U(x_1, x_2, x_3, \ldots, x_n) = \theta_U$ such that

$$F(\theta_U \mid x_1, x_2, x_3, \ldots, x_n) - F(\theta_L \mid x_1, x_2, x_3, \ldots, x_n) = \beta$$

is a β "confidence interval".

Symmetric confidence intervals about any estimate are then easily found. For example, if $\hat{\theta} = \hat{\theta}(x_1, x_2, x_3, \ldots, x_n)$ is any estimate of θ, then a symmetric confidence interval about it is

$$\hat{\theta} - \alpha \leqq \theta \leqq \hat{\theta} + \alpha$$

where α is determined by the equation

$$F(\hat{\theta} + \alpha \mid x_1, x_2, x_3, \ldots, x_n) - F(\hat{\theta} - \alpha \mid x_1, x_2, x_3, \ldots, x_n) = \beta.$$

EXAMPLE 13-4

Assume $\{F_\theta(x): \theta \in \Theta\}$ is such that F_θ is given by the density function

$$f_\theta(x) = \frac{1}{\sqrt{(2\pi)}} e^{-(x-\theta)^2/2} \qquad -\infty < x < \infty$$

with $\Theta = (-\infty, \infty)$ and where

$$f(\theta) = \frac{1}{\sqrt{(2\pi)}} e^{-\theta^2/2} \qquad -\infty < \theta < \infty.$$

Now since

$$\frac{1}{2\pi}\int_{-\infty}^{\infty} e^{-(x-\theta)^2/2}\, e^{-\theta^2/2} d\theta = \frac{1}{2\pi}\int_{-\infty}^{\infty} e^{-[\sqrt{2}\theta-(x)/\sqrt{2}]^2/2} e^{-x^2/4} d\theta$$

$$= \frac{1}{\sqrt{(\pi)}}\int_{-\infty}^{\infty} e^{-[\theta-(x/2)]^2} d\theta \left[\frac{1}{2\sqrt{(\pi)}} e^{-x^2/4}\right] = \frac{1}{2\sqrt{(\pi)}} e^{-x^2/4}$$

we obtain

$$f(\theta \mid x) = \frac{\frac{1}{2\pi} e^{-(x-\theta)^2/2} e^{-\theta^2/2}}{\frac{1}{2\pi}\int_{-\infty}^{\infty} e^{-(x-\theta)^2/2} e^{-\theta^2/2} d\theta}$$

$$= \frac{\frac{1}{2\pi} e^{-[\theta-(x/2)]^2} e^{-x^2/4}}{\frac{1}{2\sqrt{(\pi)}} e^{-x^2/4}}$$

$$= \frac{1}{\sqrt{(\pi)}} e^{-(\theta-x/2)^2} \qquad -\infty < \theta < \infty.$$

The uniformly minimum variance unbiased estimator for θ, for this family, is known to be $\hat{\theta}(X) = X$. Using $\hat{\theta}(x) = x$ as the estimate of θ, let us find the symmetric β confidence interval about $\hat{\theta}$.

Clearly since $F(\theta \mid x)$ is a normal distribution with mean $x/2$ and variance $1/2$ we have

$$F(x+\alpha \mid x) - F(x-\alpha \mid x) = \Phi(x/\sqrt{(2)}+\alpha\sqrt{(2)}) - \Phi(x/\sqrt{2}-\alpha\sqrt{2}) = \beta$$

as the defining equation for α. Then $x-\alpha \leqq \theta \leqq x+\alpha$ is the required β confidence interval.

The shortest length confidence interval is also easily obtained for the Bayes approach since the conditional density function for θ is known.

Example 13-5

For the situation of Example 13-4 $F(\theta \mid x)$ is a normal distribution with mean $x/2$ and variance $1/2$. The shortest length confidence interval in this distribution is $x/2-b \leqq \theta \leqq x/2+b$ where b is obtained from the equation $2\Phi(b\sqrt{(2)})-1 = \beta$.

WHEN θ IS NOT A RANDOM VARIABLE

If θ is not a random variable then, as we have already seen, there is no probabilistic interpretation to the numerical interval $\Omega_L(x_1, x_2, x_3, \ldots, x_n) \leqq \theta \leqq \Omega_U(x_1, x_2, x_3, \ldots, x_n)$. The Bayes approach is not really applicable here. It is still used, however, as a *method* for obtaining *intervals*. Its interpretation must, of course, be dubious.

In this method we assume some distribution function $F(\theta)$ for Θ. Three criteria for choosing $F(\theta)$ are worthy of mention.

1) Choose the $F(\theta)$ which one feels best describes the way in which the parameters are occurring in practice.

2) Use a distribution which allows the mathematical manipulations to be most easily accomplished.

3) Assume $F(\theta)$ to be a uniform distribution over the range Θ.

The first proposal is sensible when there is information concerning how the true values are occurring in practice. The second proposal is expedient. The third proposal assumes that lack of knowledge of the true value of θ can be interpreted to mean that no interval of θ values has more probability of occurring than any other interval of θ values of the same length. This proposal is the oldest of the three, and is usually what is meant by the *classical Bayes* approach to confidence intervals.

When the range of θ values is infinite, it would seem that the classical Bayes approach might not be applicable. However, from the following it is seen that the normalizing factor can be made to cancel, yielding a unique result. That is, for a finite range on θ

$$F'(\theta \mid x_1, x_2, x_3, \ldots, x_n) = \frac{KF'(x_1, x_2, x_3, \ldots, x_n \mid \theta)}{K\int F'(x_1, x_2, x_3, \ldots, x_n \mid \theta)d\theta}$$

$$(\text{where } 1/K = \int d\theta)$$

$$= \frac{F'(x_1, x_2, x_3, \ldots, x_n \mid \theta)}{\int F'(x_1, x_2, x_3, \ldots, x_n \mid \theta)d\theta}.$$

and, therefore, we can define

$$F'(\theta \mid x_1, x_2, x_3, \ldots, x_n) = \frac{F'(x_1, x_2, x_3, \ldots, x_n \mid \theta)}{\int F'(x_1, x_2, x_3, \ldots, x_n \mid \theta)d\theta}$$

for an infinite range on θ.

EXAMPLE 13-6

For the family of distributions in Example 13-3, we find that the classical Bayes conditional distribution function is given by the density function

$$f(\theta \mid x) = \frac{\dfrac{1}{\sqrt{(2\pi)}}e^{-(x-\theta)^2/2}}{\dfrac{1}{\sqrt{(2\pi)}}\displaystyle\int_{-\infty}^{\infty} e^{-(x-\theta)^2/2}d\theta} = \frac{1}{\sqrt{(2\pi)}}e^{-(x-\theta)^2/2}.$$

EXERCISE 13-3

For the situation of Example 13-6, find:

a) The symmetric β confidence interval about the uniformly minimum variance unbiased estimator, and

b) The shortest length β confidence interval, and compare the two.

When θ is not a random variable, a Bayes confidence interval for any assumed distribution $F(\theta)$, has the following important feature. If the interval $\Omega_L(x_1, x_2, x_3, \ldots, x_n) \leqq \theta \leqq \Omega_U(x_1, x_2, x_3, \ldots, x_n)$ were used many times, and if the values of θ which occurred did actually occur as if they came from the distribution $F(\theta)$, then the interval would have the properties derived by the Bayes method. That is, it would have a relative frequency of being correct approximately β and it would be of shortest average length (assuming that we are dealing with a shortest length confidence interval). Thus no other interval or procedure could yield a shorter average length β confidence interval for this set of θ values. Since no interval could be *better* for this particular set of θ values, no interval can be *uniformly better* than this interval. Therefore no confidence interval can be *uniformly better* than a Bayes confidence interval. This property is known as the *admissibility* of Bayes confidence intervals.

13-3 Fiducial Type "Confidence Intervals"

The fiducial approach to confidence intervals is a mathematical procedure which yields intervals and corresponding numbers which are called *fiducial intervals* and *fiducial confidences.* There is, however, no general attempt to justify these fiducial intervals or fiducial confidences in terms of probability or frequency, since this can only be done for particular situations. The method will be presented here along with examples to illustrate its utility in the particular situations where frequency interpretations can be given.

Assume that $\{F_\theta(x_1, x_2, x_3, \ldots, x_n): \theta \in \Theta\}$ is such that there is a sufficient statistic $\phi = \phi(X_1, X_2, X_3, \ldots, X_n)$ which has a distribution function

$$F_\theta(\phi) = F(\phi, \theta).$$

Assume further that Θ is an interval on the real line and that either

a) $F(\phi, \theta)$ is a nondecreasing differentiable function of θ with $F(\phi, \theta)$ approaching zero as θ approaches its minimum value and $F(\phi, \theta)$ approaching one as θ approaches its maximum value, or

b) $F(\phi, \theta)$ is a nonincreasing differentiable function of θ with $F(\phi, \theta)$ approaching zero as θ approaches its maximum value and $F(\phi, \theta)$ approaching one as θ approaches its minimum value.

Then define the *fiducial density* for θ by

$$f_\phi(\theta) = \frac{\partial F(\phi, \theta)}{\partial \theta} \qquad \text{for condition } \mathbf{a}.$$

$$= -\frac{\partial F(\phi, \theta)}{\partial \theta} \qquad \text{for condition } \mathbf{b}.$$

EXAMPLE 13-7

Let $\{F_\theta(x_1, x_2, x_3, \ldots, x_n): \theta \in \Theta\}$ be such that F_θ is given by the density function

$$f_\theta(x_1, x_2, x_3, \ldots, x_n) = \frac{1}{(2\pi)^{n/2}\sigma^n} \exp\left(-\sum_{i=1}^{n} (x_i - \theta)^2/2\sigma^2 \right) \qquad -\infty < \text{all } x_i < \infty$$

with $\Theta = (-\infty, \infty)$. We know that $\phi = \bar{X}_n$ is a sufficient statistic for this family with

$$f(\phi, \theta) = f_\theta(\phi) = \frac{\sqrt{n}}{\sqrt{(2\pi)}\sigma} e^{-n(\phi-\theta)^2/2\sigma^2} \qquad -\infty < \phi < \infty.$$

Since

$$F(\phi, \theta) = \frac{\sqrt{n}}{\sqrt{(2\pi)}\sigma} \int_{-\infty}^{\phi} e^{-n(x-\theta)^2/2\sigma^2} dx$$

satisfies conditions **b**, the fiducial density is given by

$$f_\phi(\theta) = -\frac{\partial}{\partial \theta} F(\phi, \theta) = \frac{-\sqrt{n}}{\sqrt{(2\pi)}\sigma} \int_{-\infty}^{\phi} \frac{n(x-\theta)}{\sigma^2} e^{-n(x-\theta)^2/2\sigma^2} dx$$

$$= \frac{-2\sqrt{n}}{\sqrt{(2\pi)}\sigma} \int_{-\infty}^{\frac{\sqrt{n}(\phi-\theta)}{\sqrt{2}\sigma}} y e^{-y^2} dy$$

$$\left(\text{where } \quad y = \frac{\sqrt{n}(x-\theta)}{\sqrt{2}\sigma} \right)$$

$$= \frac{-2\sqrt{n}}{\sqrt{(2\pi)}\sigma} \int_{-\infty}^{0} y e^{-y^2} dy - \frac{2\sqrt{n}}{\sqrt{(2\pi)}\sigma} \int_{0}^{\frac{\sqrt{n}(\phi-\theta)}{\sqrt{2}\sigma}} y e^{-y^2} dy$$

$$= \frac{-2\sqrt{n}}{\sqrt{(2\pi)}\sigma} \int_{0}^{\frac{\sqrt{n}(\phi-\theta)}{\sqrt{2}\sigma}} y e^{-y^2} dy + \frac{2\sqrt{n}}{\sqrt{(2\pi)}\sigma} \int_{0}^{\infty} y e^{-y^2} dy$$

$$= \frac{-\sqrt{n}}{\sqrt{(2\pi)}\sigma} \int_0^{\frac{n(\phi-\theta)^2}{2\sigma^2}} e^{-z} dz + \frac{\sqrt{n}}{\sqrt{(2\pi)}\sigma} \int_0^{\infty} e^{-z} dz$$

$$\text{(where } z = y^2\text{)}$$

$$= \frac{\sqrt{n}}{\sqrt{(2\pi)}\sigma} e^{-n(\phi-\theta)^2/2\sigma^2} \qquad -\infty < \theta < \infty.$$

Thus θ has a normal fiducial distribution with mean ϕ and variance σ^2/n.

Once $f_\phi(\theta)$ is obtained we can proceed as in the Bayes confidence interval situation, finding fiducial confidence intervals of various types; symmetric, shortest length, etc. The confidence or "probability" covered by the interval in the fiducial distribution is called the *fiducial confidence*. This fiducial confidence, however, may not have any frequency interpretation. It does have a frequency interpretation for the following situations.

WHEN θ IS A LOCATION PARAMETER

We have already defined a location parameter by the equation

$$F_\theta(\phi) = F_{\theta+\alpha}(\phi+\alpha)$$

where θ and $\theta+\alpha$ are in Θ. This, however, implies that

$$F(\phi, \theta) = g(\phi-\theta)$$

where $g(\cdot)$ is some function. If we now assume that the ranges of ϕ and θ are the same intervals on the real line, it is easily seen that condition **b** must hold, and therefore that

$$f_\theta(\phi) = \frac{\partial}{\partial\phi} F(\phi, \theta) = \frac{\partial}{\partial\phi} g(\phi-\theta)$$

while

$$f_\phi(\theta) = -\frac{\partial}{\partial\theta} F(\phi, \theta) = -\frac{\partial}{\partial\theta} g(\phi-\theta).$$

This, however, implies that

$$f_\phi(\theta) = f_\theta(\phi).$$

Thus the fiducial density function for a location parameter (assuming the range of values of ϕ and θ are the same) is identical to the probability density function $f_\theta(\phi)$. This result was seen in Example 13-7.

It is interesting to note that in the case of a location parameter where $f_\phi(\theta) = f_\theta(\phi)$, we have

$$\int_\Theta f_\theta(\phi)\, d\theta = \int_\Theta f_\phi(\theta)\, d\theta = 1.$$

Therefore, the classical Bayes conditional density function is given by

$$f_\phi(\theta) = \frac{K f_\theta(\phi)}{K \int_\Theta f_\theta(\phi)\, d\theta} = f_\theta(\phi).$$

Thus, for this situation the classical Bayes density is identical to the fiducial density.

WHEN θ IS A SCALE PARAMETER

We define θ to be a scale parameter for a family of distributions if

$$F(\phi, \theta) = g(\phi\theta)$$

for some function $g(\cdot)$. Here we assume that the range of ϕ and θ is the same interval on the positive real line.

For this case condition **a** holds, and

$$f_\theta(\phi) = \frac{\partial}{\partial\phi} F(\phi, \theta) = \frac{\partial}{\partial\phi} g(\phi\theta)$$

while

$$f_\phi(\theta) = \frac{\partial}{\partial\theta} F(\phi, \theta) = \frac{\partial}{\theta\partial} g(\phi\theta).$$

This implies that $f_\phi(\theta)$ has the same form as $f_\theta(\phi)$ with ϕ replacing θ and θ replacing ϕ.

EXAMPLE 13-8

Assume $\{F_\theta(x)\colon \theta \in \Theta\}$ is such that $F_\theta(x)$ has the density function

$$\begin{aligned} f_\theta(x) &= \theta e^{-\theta x} && 0 < x < \infty \\ &= 0 && \text{otherwise} \end{aligned}$$

with $\Theta = (0, \infty)$. Then

$$F_\theta(x) = \theta \int_0^x e^{-\theta t} dt = 1 - e^{-\theta x}$$

illustrating that θ is a scale parameter and that

$$\begin{aligned} f_x(\theta) &= x e^{-x\theta} && 0 < \theta < \infty \\ &= 0 && \text{otherwise.} \end{aligned}$$

All of the examples considered so far have been such that the sufficient statistic ϕ was continuous. The requirement for the use of fiducial approach is that Θ be an interval without any restrictions on the form of the variable ϕ. To illustrate that ϕ may be discrete, consider the following and final example.

EXAMPLE 13-9

Assume $\{F_\theta(x)\colon \theta \in \Theta\}$ is such that F_θ is given by the mass function

$$P_\theta(x) = e^{-\theta}\frac{\theta^x}{x!} \qquad x = 0, 1, 2, \ldots$$

with $\Theta = (0, \infty)$. Thus

$$F(\phi, \theta) = \sum_{x=0}^{\phi} \frac{e^{-\theta}\theta^x}{x!}$$

which satisfies condition **b**. Therefore

$$f_\phi(\theta) = -\frac{\partial}{\partial\theta}F(\phi, \theta) = \sum_{x=0}^{\phi} e^{-\theta}\frac{\theta^x}{x!} - \sum_{x=0}^{\phi} xe^{-\theta}\frac{\theta^{x-1}}{x!}$$

$$= \sum_{x=0}^{\phi} e^{-\theta}\frac{\theta^x}{x!} - \sum_{x=1}^{\phi} e^{-\theta}\frac{\theta^{x-1}}{(x-1)!} = \frac{e^{-\theta}\theta^\phi}{\phi!}$$

Thus

$$f_\phi(\theta) = \frac{1}{\Gamma(\phi+1)}\theta^\phi e^{-\theta} \qquad 0 < \theta < \infty$$

with ϕ a fixed integer

$$= 0 \qquad \text{otherwise.}$$

13-4 Problems

1. Assume that X_1, X_2, X_3, X_4, X_5 are independent and identically distributed with a normal distribution with unknown mean θ and variance two. Find the symmetric 0.9 confidence interval for θ of the form $\bar{X}_5 - \alpha \leqq \theta \leqq \bar{X}_5 + \alpha$.
2. Assume that X has an exponential distribution with unknown mean θ and with $\Theta = (0, \infty)$. Is there a symmetric 0.95 confidence interval for θ of the form $X - \alpha \leqq \theta \leqq X + \alpha$? If so, find it. If not, why?
3. For the situation of Problem 2 is there a 0.95 confidence interval for θ of the form $0 \leqq \theta \leqq X + \alpha$? If so, find it. If not, why?
4. For the situation of Problem 2 find the 0.95 confidence interval for θ of the form $0 \leqq \theta \leqq X/\alpha$.
5. Show that the interval of Problem 4 is the shortest length 0.95 confidence interval for θ.
6. Assume that X_1, X_2, X_3, X_4, X_5 are independent and identically distributed with an exponential distribution with unknown mean θ and with $\Theta = (0, \infty)$. Find the 0.95 confidence interval for θ of the form $0 \leqq \theta \leqq \bar{X}_5/\alpha$. *Hint*: Use a table of the incomplete gamma function such as *Tables of the Incomplete Γ-Function*, Karl Pearson, Cambridge University Press 1951.

7. Assume that X_1, X_2, X_3, X_4, X_5 are independent and identically distributed with a poisson distribution with unknown mean θ and with $\Theta = (1, 2)$. Find a uniformly most accurate upper confidence interval for θ with confidence approximately 0.8 but not smaller.
8. For the situation of Problem 7, find a uniformly most accurate lower confidence interval for θ with confidence approximately 0.8 but not smaller.
9. For the situation of Problem 7, find a reasonably accurate two sided confidence interval for θ with confidence approximately 0.6 but not smaller.
10. For the situation of Problem 1, find a reasonably accurate 0.9 confidence interval for θ (use $\beta_L = \beta_U = 0.95$). Compare your answer with the answer to Problem 1.
11. Assume that X is normally distributed with unknown mean θ and known variance σ^2. Assume further that θ is a random variable with an exponential distribution with unit mean. Find the conditional distribution of θ, given X, and use it to find the shortest length Bayes confidence interval for θ.
12. Assume that X has a binomial distribution with known parameter n and unknown parameter θ. Assume further that θ is a random variable with a beta distribution. Find the conditional distribution of θ, given x.
13. Assume that X has the distribution described in Problem 12. Find the classical Bayes form for the conditional distribution of θ, given X.
14. Assume that $X_1, X_2, X_3, \ldots, X_n$ are independent and identically distributed with a poisson distribution with unknown mean θ. Assume further that θ is a random variable with an exponential distribution with unit mean. Find the conditional distribution of θ given $x_1, x_2, x_3, \ldots, x_n$.
15. Assume that X has the distribution described in Problem 14. Find the classical Bayes form for the conditional distribution of θ, given $x_1, x_2, x_3, \ldots, x_n$.
16. For the situation of Problem 1, find the shortest length classical Bayes 0.9 confidence interval for θ.
17. Assume that $X_1, X_2, X_3, \ldots, X_n$ are independent and identically distributed with a uniform distribution on $(0, \theta)$ with θ unknown and with $\Theta = (0, \infty)$. Find the fiducial distribution for θ. Show how this can be used to obtain a fiducial β confidence interval for θ.
18. Assume that X has a binomial distribution with known parameter n and unknown parameter θ and with $\Theta = (0, 1)$. Find the fiducial distribution for θ.
19. Assume that $X_1, X_2, X_3, \ldots, X_n$ are independent and identically distributed with an exponential distribution with unknown mean θ and with $\Theta = (0, \infty)$. Find the fiducial distribution for θ.
20. For the situation of Problem 1, find the shortest length fiducial 0.9 confidence interval for θ.

APPENDIX A

Summary of Distributions and Their Properties

A-1 The Binomial Distribution

If there are m independent trials each with probability p for success, and if x equals the number of successes, then:

PROBABILITY MASS FUNCTION

$$P(x) = \binom{m}{x} p^x (1-p)^{m-x} \qquad x = 0, 1, 2, 3, \ldots, m$$

$$0 < p < 1$$

$$m = 1, 2, 3, \ldots$$

MOMENTS

$$\text{Mean} = EX = mp$$

$$\text{Variance} = \text{Var } X = mp(1-p)$$

MOMENT GENERATING FUNCTION

$$m(t) = E\,e^{tX} = (pe^t + 1 - p)^m.$$

If $X_1, X_2, X_3, \ldots, X_n$ are independently distributed with binomial distributions with parameters $m_1, m_2, m_3, \ldots, m_n$ and each with the same parameter p, then:

MASS FUNCTION OF THE SUM

with

$$\phi = \sum_{i=1}^{n} x_i$$

$$P(\phi) = \left(\sum_{i=1}^{n} m_i\right) p^{\phi} \exp_{(1-p)}\left(\sum_{i=1}^{n} m_i - \phi\right) \qquad \phi = 0, 1, 2, \ldots \sum_{i=1}^{n} m_i$$

$$0 < p < 1.$$

SUFFICIENT STATISTIC

$$\phi = \sum_{i=1}^{n} X_i \quad \text{is sufficient for } p.$$

If $m_1 = m_2 = m_3 = \ldots m_n = m$ then:

CRAMÉR–RAO LOWER BOUND

The Cramér–Rao lower bound for the variance of any unbiased estimator for p is $p(1-p)/n$, which is attained by

$$\overline{X}_n = \frac{1}{mn} \sum_{i=1}^{n} X_i.$$

UNIFORMLY MINIMUM VARIANCE UNBIASED ESTIMATOR

$$\overline{X}_n = \frac{1}{mn} \sum_{i=1}^{n} X_i$$

is the uniformly minimum variance unbiased estimator for p.

COMPLETENESS

The binomial distributions are complete.

MONOTONE LIKELIHOOD RATIO

There is a monotone likelihood ratio in

$$\phi = \sum_{i=1}^{n} X_i,$$

A-2 The Poisson Distribution

MASS FUNCTION

$$P(x) = \frac{e^{-\lambda}\lambda^x}{x!} \qquad x = 0, 1, 2, \ldots$$

$$0 < \lambda < \infty$$

MOMENTS

$$\text{Mean} = EX = \lambda$$

$$\text{Variance} = \text{Var}\, X = \lambda$$

MOMENT GENERATING FUNCTION

$$m(t) = E\, e^{tX} = e^{-\lambda} e^{\lambda e^t}.$$

If $X_1, X_2, X_3, \ldots, X_n$ are independently distributed with poisson distributions with means $\lambda_1, \lambda_2, \lambda_3, \ldots, \lambda_n$, then:

MASS FUNCTION OF THE SUM

with

$$\phi = \sum_{i=1}^{n} X_i$$

$$P(\phi) = \frac{\exp\left(-\sum_{i=1}^{n} \lambda_i\right)\left(\sum_{i=1}^{n} \lambda_i\right)^{\phi}}{\phi!} \qquad \phi = 0, 1, 2, \ldots$$

$$0 < \lambda < Œ.$$

If $\lambda_1 = \lambda_2 = \lambda_3 = \ldots = \lambda_n = \lambda$ then:

SUFFICIENT STATISTIC

$$\phi = \sum_{i=1}^{n} X_i$$

is sufficient for λ.

CRAMÉR–RAO LOWER BOUND

The Cramér–Rao lower bound for the variance of any unbiased estimator for λ is λ/n, which is attained by

$$\bar{X}n = \frac{1}{n}\sum_{i=1}^{n} X_i.$$

UNIFORMLY MINIMUM VARIANCE UNBIASED ESTIMATOR

$$\bar{X}_n = \frac{1}{n}\sum_{i=1}^{n} X_i$$

is the uniformly minimum variance unbiased estimator for λ.

COMPLETENESS

The poisson distributions are complete.

MONOTONE LIKELIHOOD RATIO

There is a monotone likelihood ratio in $\phi = \sum_{i=1}^{n} X_i$.

FIDUCIAL DENSITY

If

$$P(x) = \frac{e^{-\lambda}\lambda^{x}}{x!} \qquad x = 0, 1, 2, \ldots$$

then

$$f_x(\lambda) = \frac{1}{\Gamma(x+1)} \lambda^x e^{-\lambda} \qquad 0 < \lambda < \infty$$

$$= 0 \qquad \text{otherwise.}$$

A-3 The Negative Binomial Distribution

Independent trials with probability p for success. If x is the number of failures before the rth success, then:

MASS FUNCTION

$$P(x) = \binom{r+x-1}{r-1} p^r (1-p)^x \qquad x = 0, 1, 2, \ldots$$

$$0 < p < 1$$

$$r = 1, 2, 3, \ldots$$

MOMENTS

$$\text{Mean} = EX = \frac{r(1-p)}{p}$$

$$\text{Variance} = \text{Var } X = \frac{r(1-p)}{p^2}$$

MOMENT GENERATING FUNCTION

$$m(t) = E\, e^{tX} = \frac{p^r}{[1-(1-p)e^t]^r}.$$

A-4 The Hypergeometric Distribution

There are r objects selected (without replacement) from n objects, m of which have a specified property. If x is the number of objects selected which have the specified property, then:

PROBABILITY MASS FUNCTION

$$P(x) = \frac{\binom{m}{x}\binom{n-m}{r-x}}{\binom{n}{r}} \qquad x = \max(0, r-n+m), \max(0, r-n+m) + 1, \ldots, \min(m, r)$$

MOMENTS

$$\text{Mean} = EX = mr/n$$

$$\text{Variance} = \text{Var } X = \frac{m(n-m)r(n-r)}{n^2(n-1)}.$$

A-5 The Discrete Uniform Distribution

PROBABILITY MASS FUNCTION

$$P(x) = \frac{1}{n} \qquad x = 1, 2, 3, \ldots, n$$

MOMENTS

$$\text{Mean} = EX = \frac{(n+1)}{2}$$

$$\text{Variance} = \text{Var } X = \frac{(n^2-1)}{12}$$

MOMENT GENERATING FUNCTION

$$m(t) = E\,e^{tX} = \frac{1}{n}\left(\frac{e^{nt}-1}{1-e^{-t}}\right).$$

A-6 The Uniform Distribution

PROBABILITY DENSITY FUNCTION

$$f(x) = \frac{1}{\beta-\alpha} \qquad \alpha < x < \beta$$

$$-\infty < \alpha < \beta < \infty$$

$$= 0 \qquad \text{otherwise}$$

MOMENTS

$$\text{Mean} = EX = \frac{(\beta+\alpha)}{2}$$

$$\text{Variance} = \text{Var } X = \frac{(\beta-\alpha)^2}{12}$$

MOMENT GENERATING FUNCTION

$$m(t) = E\, e^{tX} = \frac{e^{t\beta}-e^{t\alpha}}{t(\beta-\alpha)}.$$

If $X_1, X_2, X_3, \ldots, X_n$ are independent and identically distributed with uniform distribution with $\alpha = 0$, then:

SUFFICIENT STATISTIC

$\phi = \mathrm{Max}\{X_1, X_2, X_3, \ldots, X_n\}$ is a sufficient statistic for β.

CRAMÉR–RAO LOWER BOUND

The regularity assumptions do not hold.

UNIFORMLY MINIMUM VARIANCE UNBIASED ESTIMATOR

$$\phi = \frac{n+1}{n}\,\mathrm{Max}\{X_1, X_2, X_3, \ldots, X_n\}$$

is the uniformly minimum variance unbiased estimator of β with variance $1/(n+2)$.

COMPLETENESS

$\phi = \mathrm{Max}\{X_1, X_2, X_3, \ldots, X_n\}$ is a complete sufficient statistic for β.

A-7 The Normal Distribution

PROBABILITY DENSITY FUNCTION

$$f(x) = \frac{1}{\sqrt{(2\pi)}\sigma}\, e^{-(x-\mu)^2/2\sigma^2} \qquad -\infty < x < \infty$$

$$-\infty < \mu < \infty$$

$$0 < \sigma$$

MOMENTS

Mean $= EX = \mu$

Variance $= \mathrm{Var}\, X = \sigma^2$

Odd central moments $= EX^{2r+1} = 0$

Even central moments $= EX^{2r} = 1\cdot 3\cdot 5 \ldots (2r-1)\sigma^{2r}$

MOMENT GENERATING FUNCTION

$$m(t) = E\, e^{tX} = e^{t\mu + t^2\sigma^2/2}.$$

If $X_1, X_2, X_3, \ldots, X_n$ are independent and identically distributed with a normal distribution with mean μ and variance σ^2, then:

DISTRIBUTION OF THE AVERAGE

$$\overline{X}_n = \frac{1}{n}\sum_{i=1}^{n} X_i$$

has a normal distribution with mean μ and variance σ^2/n.

SUFFICIENT STATISTIC

$$\overline{X}_n = \frac{1}{n}\sum_{i=1} X_i$$

is a sufficient statistic for μ.

CRAMÉR–RAO LOWER BOUND

The Cramér–Rao lower bound for the variance of any unbiased estimator for μ is σ^2/n, which is attained by

$$\overline{X}_n = \frac{1}{n}\sum_{i=1}^{n} X_i.$$

UNIFORMLY MINIMUM VARIANCE UNBIASED ESTIMATOR

$$\overline{X}_n = \frac{1}{n}\sum_{i=1}^{n} X_i$$

is the minimum variance unbiased estimator for μ.

MONOTONE LIKELIHOOD RATIO

There is a monotone likelihood ratio in

$$\phi = \sum_{i=1}^{n} X_i.$$

CLASSICAL BAYES DENSITY

If

$$\phi = \overline{X}_n = \frac{1}{n}\sum_{i=1}^{n} X_i,$$

then

$$f_\phi(\mu) = \frac{\sqrt{n}}{\sqrt{(2\pi)}\sigma}\, e^{-n(\mu-\phi)^2/2\sigma^2} \qquad -\infty < \mu < \infty$$

FIDUCIAL DENSITY

The same as the classical Bayes density.

SPECIAL PROPERTY

If X has a normal distribution with mean μ and variance σ^2, then

$$P(a \leqq X \leqq b) = \Phi\left(\frac{b-\mu}{\sigma}\right) - \Phi\left(\frac{a-\mu}{\sigma}\right)$$

where $\Phi(\cdot)$ is the standard normal distribution function.

A-8 The Gamma Distribution

PROBABILITY DENSITY FUNCTION

$$f(x) = \frac{1}{\Gamma(\alpha)\beta^\alpha} x^{\alpha-1} e^{-x/\beta} \qquad 0 < x < \infty$$

$$\alpha, \beta > 0$$

$$= 0 \qquad \text{otherwise}$$

PROPERTIES OF THE GAMMA FUNCTION

$$\Gamma(\alpha) = \int_0^\infty x^{\alpha-1} e^{-x} dx$$

$$= 2 \int_0^\infty x^{2\alpha-1} e^{-x^2} dx$$

$$= (\alpha-1)\Gamma(\alpha-1)$$

$$\Gamma(n) = (n-1)!$$

$$\Gamma(\tfrac{1}{2}) = \sqrt{\pi}$$

MOMENTS

$$\text{Mean} = EX = \alpha\beta$$

$$\text{Variance} = \text{Var } X = \alpha\beta^2$$

$$\text{Higher moments} = EX^r = \beta^r \frac{\Gamma(\alpha+r)}{\Gamma(\alpha)}$$

MOMENT GENERATING FUNCTION

$$m(t) = E\, e^{tX} = (1-\beta t)^{-\alpha}$$

SPECIAL CASES

Exponential distribution:

$$\alpha = 1 \quad \text{and} \quad \beta = 1/\lambda$$

Chi-squared distribution:

$$\alpha = n/2 \quad \text{and} \quad \beta = 2\sigma^2.$$

If $X_1, X_2, X_3, \ldots, X_n$ are independently distributed with parameters $\alpha_1, \alpha_2, \alpha_3, \ldots, \alpha_n$ and the same parameter β, then

$$\phi = \sum_{i=1}^{n} X_i$$

has a gamma distribution with parameters

$$\sum_{i=1}^{n} \alpha_i \quad \text{and} \quad \beta.$$

SUFFICIENT STATISTIC

$$\phi = \sum_{i=1}^{n} X_i$$

is a sufficient statistic for β.

If $\alpha_1 = \alpha_2 = \alpha_3 = \ldots = \alpha_n = \alpha$ (known), then:

CRAMÉR–RAO LOWER BOUND

The Cramér–Rao lower bound for the variance of any unbiased estimator for β is given by $\beta^2/n\alpha$, which is attained by

$$\phi = \sum_{i=1}^{n} X_i/n\alpha \quad (\alpha \text{ known}).$$

COMPLETENESS

$$\phi = \sum_{i=1}^{n} X_i$$

is a complete sufficient statistic for β.

MONOTONE LIKELIHOOD RATIO

There is a monotone likelihood ratio in

$$\phi = \sum_{i=1}^{n} X_i \quad (\text{with } \alpha \text{ known}).$$

A-9 The Chi-Squared Distribution

A special case of the gamma distribution with $\alpha = n/2$ and $\beta = 2\sigma^2$.

PROBABILITY DENSITY FUNCTION

$$f(x) = \frac{1}{\Gamma(n/2)2^{n/2}\sigma^n} x^{n/2-1} e^{-x/2\sigma^2} \qquad 0 < x < \infty$$

$$n = 1, 2, 3, \ldots$$

$$0 < \sigma$$

$$= 0 \qquad \text{otherwise}$$

MOMENTS

$$\text{Mean} = EX = n\sigma^2$$

$$\text{Variance} = \text{Var } X = 2n\sigma^4.$$

SPECIAL PROPERTY

If $X_1, X_2, X_3, \ldots, X_n$ are independent and identically distributed with a normal distribution with mean μ and variance σ^2 then

$$\phi = \sum_{i=1}^{n} (X_i - \mu)^2$$

has a chi-squared distribution with n degrees of freedom and parameter σ^2, while

$$S_n^2 = \frac{1}{n-1} \sum_{i=1}^{n} (X_i - \bar{X}_n)^2 \quad \left(\text{with } \bar{X}_n = \frac{1}{n} \sum_{i=1}^{n} X_i\right)$$

has a chi-squared distribution with $n-1$ degrees of freedom and parameter $\sigma^2/(n-1)$.

A-10 The Chi Distribution

PROBABILITY DENSITY FUNCTION

$$f(x) = \frac{1}{\Gamma(n/2)2^{n/2-1}\sigma^n} x^{n-1} e^{-x^2/2\sigma^2} \qquad 0 < x < \infty$$

$$n = 1, 2, 3, \ldots$$

$$0 < \sigma$$

$$= 0 \qquad \text{otherwise}$$

MOMENTS

$$\text{Mean} = EX = \frac{\sqrt{(2)}\sigma\Gamma[(n+1)/2]}{\Gamma(n/2)}$$

$$\text{Variance} = n\sigma^2 - 2\sigma^2 \left\{\frac{\Gamma[(n+1)/2]}{\Gamma(n/2)}\right\}^2$$

$$\text{Higher moments} = EX^r = \frac{\Gamma[(n+r)/2]2^{r/2}\sigma^r}{\Gamma(n/2)}.$$

SPECIAL PROPERTY

If X has a chi-squared distribution then $Y = \sqrt{X}$ has a chi distribution.

A-11 The Beta Distribution

PROBABILITY DENSITY FUNCTION

$$f(x) = \frac{\Gamma(\alpha+\beta)}{\Gamma(\alpha)\Gamma(\beta)} x^{\alpha-1}(1-x)^{\beta-1} \qquad 0 < x < 1$$

$$\alpha, \beta > 0$$

$$= 0 \qquad \text{otherwise}$$

MOMENTS

$$\text{Mean} = EX = \frac{\alpha}{\alpha+\beta}$$

$$\text{Variance} = \text{Var } X = \frac{\alpha\beta}{(\alpha+\beta)^2(\alpha+\beta+1)}$$

$$\text{Higher Moments} = EX^r = \frac{\Gamma(\alpha+\beta)\Gamma(\alpha+r)}{\Gamma(\alpha)\Gamma(\alpha+\beta+r)}.$$

A-12 The F Distribution

PROBABILITY DENSITY FUNCTION

$$f(x) = \frac{\Gamma[(m+n)/2]}{\Gamma(m/2)\Gamma(n/2)} \left(\frac{m}{n}\right)^{m/2} x^{m/2-1} \left(1+\frac{m}{n}x\right)^{-(m+n/2)}$$

$$0 < x < \text{Œ}$$

$$m = 1, 2, 3, \ldots$$

$$n = 1, 2, 3, \ldots$$

$$= 0 \qquad \text{otherwise}$$

MOMENTS

$$\text{Mean} = EX = \frac{n}{n-2} \qquad (n > 2)$$

$$\text{Variance} = \text{Var } X = \frac{2n^2(m+n-2)}{m(n-2)^2(n-4)} \qquad (n > 4)$$

$$\text{Higher Moments} = EX^r = \frac{\Gamma(m/2+r)\Gamma(n/2-r)}{\Gamma(m/2)\Gamma(n/2)} \left(\frac{n}{m}\right)^r \qquad (r < n/2).$$

SPECIAL PROPERTY

If X has a chi-squared distribution with m degrees of freedom and parameter σ^2 while Y has a chi-squared distribution with n degrees of freedom and parameter σ^2 independent of X, then $F = nX/mY$ has an F distribution with m and n degrees of freedom.

A-13 The t Distribution

PROBABILITY DENSITY FUNCTION

$$f(x) = \frac{\Gamma[(n+1)/2]}{\sqrt{(n\pi)}\Gamma(n/2)} \left(1+\frac{x^2}{n}\right)^{-[(n+1)/2]} \qquad -\infty < x < \infty$$

$$n = 1, 2, 3, \ldots$$

MOMENTS

$$\text{Mean} = EX = 0 \qquad (n > 1)$$

$$\text{Variance} = \text{Var } X = \frac{n}{n-2} \qquad (n < 2)$$

$$\text{Odd moments} = EX^{2r+1} = 0 \qquad (0 \leqq r < n/2-1)$$

$$\text{Even moments} = EX^{2r} = \frac{\Gamma(r+1/2)\Gamma(n/2-r)}{\sqrt{(\pi)}\Gamma(n/2)} n^r \qquad (0 \leqq r < n/2).$$

SPECIAL CASE: CAUCHY DISTRIBUTION

When $n = 1$ the t distribution is called the *cauchy distribution*. For this distribution the mean does not exist. The odd moments do not exist and the even moments are infinite.

SPECIAL PROPERTY

If $X_1, X_2, X_3, \ldots, X_n$ are independently and identically distributed with a normal distribution with mean μ and variance σ^2 then

$$t = \frac{\sqrt{(n)}(\bar{X}_n - \mu)}{S_n}, \quad \text{where} \quad \bar{X}_n = \frac{1}{n}\sum_{i=1}^{n} X_i$$

and

$$S_n = \sqrt{\left(\frac{1}{n-1}\sum_{i=1}^{n}(X_i - \bar{X}_n)^2\right)},$$

has a t distribution with $n-1$ degrees of freedom (i.e., $n-1$ replaces n in the above density function).

A-14 The Trinomial Distribution

n independent trials with probability p for obtaining property A and probability q for obtaining property B ($p+q < 1$). If x is the number of times property A is obtained, and if y is the number of times property B is obtained, then:

JOINT PROBABILITY MASS FUNCTION

$$P(x, y) = \binom{n}{n}\binom{n-x}{y} p^x q^y (1-p-q)^{n-x-y} \qquad 0 \leqq x+y \leqq n$$

$$p, q > 0$$

$$0 < p+q < 1$$

$$n = 1, 2, 3, \ldots$$

MOMENTS

$$EX = np$$

$$\text{Var } X = np(1-p)$$

$$EY = nq$$

$$\text{Var } Y = nq(1-q)$$

$$\text{Cov }(X,Y) = -npq$$

MOMENT GENERATING FUNCTION

$$m(t_1, t_2) = E\, e^{t_1 X + t_2 Y} = (1-p-q+pe^{t_1}+qe^{t_2})^n$$

MARGINAL MASS FUNCTIONS

$$P(x) = \binom{n}{x} p^x (1-p)^{n-x} \qquad x = 0, 1, 2, \ldots, n \quad 0 < p < 1 \quad n = 1, 2, 3, \ldots$$

$$P(y) = \binom{n}{x} q^y (1-q)^{n-y} \qquad y = 0, 1, 2, \ldots, n \quad 0 < q < 1 \quad n = 1, 2, 3, \ldots$$

CONDITIONAL MASS FUNCTIONS

$$P(x \mid y) = \binom{n-y}{x} \left(\frac{1-q}{p}\right)^x \left(\frac{1-p-q}{1-q}\right)^{n-y-x} \qquad 0 \leqq x \leqq n-y \quad p, q > 0 \quad 0 < p+q < 1 \quad n = 1, 2, 3, \ldots$$

$$P(y \mid x) = \binom{n-x}{y} \left(\frac{q}{1-p}\right)^y \left(\frac{1-p-q}{1-p}\right)^{n-x-y} \qquad 0 \leqq y \leqq n-x \quad p, q > 0 \quad 0 < p+q < 1 \quad n = 1, 2, 3, \ldots.$$

A-15 The Bivariate Normal Distribution

JOINT PROBABILITY DENSITY FUNCTION

$$f(x, y) = \frac{1}{2\pi\sigma_x\sigma_y\sqrt{(1-\rho^2)}} \exp\left\{-\frac{1}{2(1-\rho^2)}\left[\left(\frac{x-\mu_x}{\sigma_x}\right)^2 - 2\rho\left(\frac{x-\mu_x}{\sigma_x}\right)\left(\frac{y-\mu_y}{\sigma_y}\right) + \left(\frac{y-\mu_y}{\sigma_y}\right)^2\right]\right\}$$

$$-\infty < x < \infty \quad -\infty < y < \infty \quad -\infty < \mu_x < \infty \quad -\infty < \mu_y < \infty \quad \sigma_x, \sigma_y > 0$$

MOMENTS

$$EX = \mu_x$$
$$\text{Var } X = \sigma_x^{\ 2}$$
$$EY = \mu_y$$
$$\text{Var } Y = \sigma_y^{\ 2}$$
$$\text{Cov}\,(X,Y) = \rho\sigma_x\sigma_y$$

MOMENT GENERATING FUNCTION

$$m(t_1, t_2) = E\exp(t_1X+t_2Y)$$
$$= \exp(t_1\mu_x+t_2\mu_y+t_1^{\ 2}\sigma_x^{\ 2}/2+\rho t_1t_2\sigma_x\sigma_y+t_2\sigma_y^{\ 2}/2)$$

MARGINAL DENSITY FUNCTIONS

$$f(x) = \frac{1}{\sqrt{(2\pi)}\sigma_x}\exp[-(x-\mu_x)^2/2\sigma_x^{\ 2})] \qquad -\infty < x < \infty$$
$$-\infty < \mu_x < \infty$$
$$0 < \sigma_x$$

$$f(y) = \frac{1}{\sqrt{(2\pi)}\sigma_y}\exp[-(y-\mu_y)^2/2\sigma_y^{\ 2}] \qquad -\infty < y < \infty$$
$$-\infty < \mu_y < \infty$$
$$0 < {}_y$$

CONDITIONAL DENSITY FUNCTIONS

$$f(x \mid y) = \frac{1}{\sqrt{(2\pi)}\sigma_x\sqrt{(1-\rho^2)}}\exp\left\{-\frac{1}{2\sigma_x^{\ 2}(1-\rho^2)}\right.$$
$$\left.\left[\left\{x-\left(\mu_x+\frac{\rho\sigma_x}{\sigma_y}(y-\mu_y)\right)\right\}^2\right]\right\} \qquad -\infty < x < \infty$$
$$-\infty < \mu_x < \infty$$
$$0 < \sigma_x$$

$$f(y \mid x) = \frac{1}{\sqrt{(2\pi)}\sigma_y\sqrt{(1-\rho^2)}}\exp\left\{-\frac{1}{2\sigma_y^{\ 2}(1-\rho^2)}\right.$$
$$\left.\left[\left\{y-\left(\mu_y+\frac{\rho\sigma_y}{\sigma_x}(x-\mu_x)\right)\right\}^2\right]\right\} \qquad -\infty < y < \infty$$
$$-\infty < \mu_y < \infty$$
$$0 < \sigma_y.$$

SPECIAL PROPERTY

X and Y are independent if and only if $\rho = 0$.

A-16 Properties of Probabilities in General

If A can occur in $n(A)$ mutually exclusive and equally likely ways, then

$$P(A) = n(A)/n$$

where n is the total number of mutually exclusive and equally likely ways possible.

NUMBER OF WAYS

n^r is the number of r-tuples which can be made from n objects with replacement. $n!/[(n-r)!]$ is the number of r-tuples which can be made from n objects without replacement. $\binom{n}{r}$ is the number of r-object combinations which can be made from n objects.

ODDS

If the odds on A occurring are $n(A)$: $n(\bar{A})$, then

$$P(A) = \frac{n(A)}{n(A)+n(\bar{A})}.$$

DISJOINT EVENTS

If A_i $i = 1, 2, 3, \ldots$ are mutually disjoint, then

$$P(\bigcup_i A_i) = \sum_i P(A_i).$$

INDEPENDENCE

If $X_1, X_2, X_3, \ldots, X_n$ are independent, then

$$F(x_1, x_2, x_3, \ldots, x_n) = F(x_1)F(x_2)F(x_3) \ldots F(x_n)$$
$$P(x_1, x_2, x_3, \ldots, x_n) = P(x_1)P(x_2)P(x_3) \ldots P(x_n)$$
$$f(x_1, x_2, x_3, \ldots, x_n) = f(x_1)f(x_2)f(x_3) \ldots f(x_n)$$
$$F'(x_1, x_2, x_3, \ldots, x_n) = F'(x_1)F'(x_2)F'(x_3) \ldots F'(x_n)$$
$$F(x_i \mid x_v, x_\mu, x_\rho, \ldots) = F(x_i)$$
$$\text{Var } \Sigma X_i = \sum_i \text{Var } X_i$$
$$m_{\sum_i x_i}(t) = \Pi_i m_{x_i}(t).$$

CONDITIONAL PROBABILITIES

$$F'(x, y) = F'(x \mid y)F'(y)$$

$$F'(x, y, z) = F'(x \mid y, z)F'(y, z) = F'(x \mid y, z)F'(y \mid z)F'(z)$$

Bayes Formula:

$$dF(y \mid x) = \frac{dF(x \mid y)dF(y)}{\int_y dF(x \mid y)dF(y)}.$$

MARKOV'S INEQUALITY

If $g(x)$ is symmetric about zero and is nondecreasing on $(0, \infty)$, then

$$P(|X| > \varepsilon) \leqq \frac{Eg(X)}{g(\varepsilon)}.$$

APPENDIX B

Tables

The values for the following tables were calculated by the author on an IBM 7040 computer at the Virginia Polytechnic Institute and compared with existing tables.

TABLE B-1

The Binomial Mass Function

$$P(x) = \binom{n}{x} p^x(1-p)^{n-x}$$

n	x	.05	.10	.15	.20	p .25	.30	.35	.40	.45	.50
1	0	.9500	.9000	.8500	.8000	.7500	.7000	.6500	.6000	.5500	.5000
	1	.0500	.1000	.1500	.2000	.2500	.3000	.3500	.4000	.4500	.5000
2	0	.9025	.8100	.7225	.6400	.5625	.4900	.4225	.3600	.3025	.2500
	1	.0950	.1800	.2550	.3200	.3750	.4200	.4550	.4800	.4950	.5000
	2	.0025	.0100	.0225	.0400	.0625	.0900	.1225	.1600	.2025	.2500
3	0	.8574	.7290	.6141	.5120	.4219	.3430	.2746	.2160	.1664	.1250
	1	.1354	.2430	.3251	.3840	.4219	.4410	.4436	.4320	.4084	.3750
	2	.0071	.0270	.0574	.0960	.1406	.1890	.2389	.2880	.3341	.3750
	3	.0001	.0010	.0034	.0080	.0156	.0270	.0429	.0640	.0911	.1250
4	0	.8145	.6561	.5220	.4096	.3164	.2401	.1785	.1296	.0915	.0625
	1	.1715	.2916	.3685	.4096	.4219	.4116	.3845	.3456	.2995	.2500
	2	.0135	.0486	.0975	.1536	.2109	.2646	.3105	.3456	.3675	.3750
	3	.0005	.0036	.0115	.0256	.0469	.0756	.1115	.1536	.2005	.2500
	4	.0000	.0001	.0005	.0016	.0039	.0081	.0150	.0256	.0410	.0625
5	0	.7738	.5905	.4437	.3277	.2373	.1681	.1160	.0778	.0503	.0312
	1	.2036	.3280	.3915	.4096	.3955	.3602	.3124	.2592	.2059	.1562
	2	.0214	.0729	.1382	.2048	.2637	.3087	.3364	.3456	.3369	.3125
	3	.0011	.0081	.0244	.0512	.0879	.1323	.1811	.2304	.2757	.3125
	4	.0000	.0004	.0022	.0064	.0146	.0284	.0488	.0768	.1128	.1562
	5	.0000	.0000	.0001	.0003	.0010	.0024	.0053	.0102	.0185	.0312

TABLE B-1 (Cont.)

n	x	.05	.10	.15	.20	.25	.30	.35	.40	.45	.50
6	0	.7351	.5314	.3771	.2621	.1780	.1176	.0754	.0467	.0277	.0156
	1	.2321	.3543	.3993	.3932	.3560	.3025	.2437	.1866	.1359	.0938
	2	.0305	.0984	.1762	.2458	.2966	.3241	.3280	.3110	.2780	.2344
	3	.0021	.0146	.0415	.8019	.1318	.1852	.2355	.2765	.3032	.3125
	4	.0001	.0012	.0055	.0154	.0330	.0595	.0951	.1382	.1861	.2344
	5	.0000	.0001	.0004	.0015	.0044	.0102	.0205	.0369	.0609	.0938
	6	.0000	.0000	.0000	.0001	.0002	.0007	.0018	.0041	.0083	.0156
7	0	.6983	.4783	.3206	.2097	.1335	.0824	.0490	.0280	.0152	.0078
	1	.2573	.3720	.3960	.3670	.3115	.2471	.1848	.1306	.0872	.0547
	2	.0406	.1240	.2097	.2753	.3115	.3177	.2985	.2613	.2140	.1641
	3	.0036	.0230	.0617	.1147	.1730	.2269	.2679	.2903	.2918	.2734
	4	.0002	.0026	.0109	.0287	.0577	.0972	.1442	.1935	.2388	.2734
	5	.0000	.0002	.0012	.0043	.0115	.0250	.0466	.0774	.1172	.1641
	6	.0000	.0000	.0001	.0004	.0013	.0036	.0084	.0172	.0320	.0547
	7	.0000	.0000	.0000	.0000	.0001	.0002	.0006	.0016	.0037	.0078
8	0	.6634	.4305	.2725	.1678	.1001	.0576	.0319	.0168	.0084	.0039
	1	.2793	.3826	.3847	.3355	.2670	.1977	.1373	.0896	.0548	.0312
	2	.0515	.1488	.2376	.2936	.3115	.2965	.2587	.2090	.1569	.1094
	3	.0054	.0331	.0839	.1468	.2076	.2541	.2786	.2787	.2568	.2188
	4	.0004	.0046	.0185	.0459	.0865	.1361	.1875	.2322	.2627	.2734
	5	.0000	.0004	.0026	.0092	.0231	.0467	.0808	.1239	.1719	.2188
	6	.0000	.0000	.0002	.0011	.0038	.0100	.0217	.0413	.0703	.1094
	7	.0000	.0000	.0000	.0001	.0004	.0012	.0033	.0079	.0164	.0312
	8	.0000	.0000	.0000	.0000	.0000	.0001	.0002	.0007	.0017	.0039
9	0	.6302	.3874	.2316	.1342	.0751	.0404	.0207	.0101	.0046	.0020
	1	.2985	.3874	.3679	.3020	.2253	.1556	.1004	.0605	.0339	.0176
	2	.0629	.1722	.2597	.3020	.3003	.2668	.2162	.1612	.1110	.0703
	3	.0077	.0446	.1069	.1762	.2336	.2668	.2716	.2508	.2119	.1641
	4	.0006	.0074	.0283	.0661	.1168	.1715	.2194	.2508	.2600	.2461
	5	.0000	.0008	.0050	.0165	.0389	.0735	.1181	.1672	.2128	.2461
	6	.0000	.0001	.0006	.0028	.0087	.0210	.0424	.0742	.1160	.1641
	7	.0000	.0000	.0000	.0003	.0012	.0039	.0098	.0212	.0407	.0703
	8	.0000	.0000	.0000	.0000	.0001	.0004	.0013	.0035	.0083	.0176
	9	.0000	.0000	.0000	.0000	.0000	.0000	.0001	.0003	.0008	.0020
10	0	.5987	.3487	.1969	.1074	.0563	.0282	.0135	.0060	.0025	.0010
	1	.3151	.3874	.3474	.2684	.1877	.1211	.0725	.0403	.0207	.0098

(Column heading: p)

TABLE B-1 (Cont.)

n	x	.05	.10	.15	.20	.25	.30	.35	.40	.45	.50
10	2	.0746	.1937	.2759	.3020	.2816	.2335	.1757	.1209	.0763	.0439
	3	.0105	.0574	.1298	.2013	.2503	.2668	.2522	.2150	.1665	.1172
	4	.0010	.0112	.0401	.0881	.1460	.2001	.2377	.2508	.2384	.2051
	5	.0001	.0015	.0085	.0264	.0584	.1029	.1536	.2007	.2340	.2461
	6	.0000	.0001	.0012	.0055	.0162	.0368	.0689	.1115	.1596	.2051
	7	.0000	.0000	.0001	.0008	.0031	.0090	.0212	.0425	.0746	.1172
	8	.0000	.0000	.0000	.0001	.0004	.0014	.0043	.0106	.0229	.0439
	9	.0000	.0000	.0000	.0000	.0000	.0001	.0005	.0016	.0042	.0098
	10	.0000	.0000	.0000	.0000	.0000	.0000	.0000	.0001	.0003	.0010

TABLE B-2

The Poisson Mass Function

$$P(x) = \frac{e^{-\lambda}\lambda^x}{x!}$$

x	0.1	0.2	0.3	0.4	0.5	0.6	0.7	0.8	0.9	1.0
0	.9048	.8187	.7408	.6703	.6065	.5488	.4966	.4493	.4066	.3679
1	.0905	.1637	.2222	.2681	.3033	.3298	.3476	.3595	.3659	.3679
2	.0045	.0164	.0333	.0536	.0758	.0988	.1217	.1438	.1647	.1839
3	.0002	.0011	.0033	.0072	.0126	.0198	.0284	.0383	.0494	.0613
4	.0000	.0001	.0002	.0007	.0016	.0030	.0050	.0077	.0111	.0153
5	.0000	.0000	.0000	.0001	.0002	.0004	.0007	.0012	.0020	.0031
6	.0000	.0000	.0000	.0000	.0000	.0000	.0001	.0002	.0003	.0005
7	.0000	.0000	.0000	.0000	.0000	.0000	.0000	.0000	.0000	.0001

x	1.1	1.2	1.3	1.4	1.5	1.6	1.7	1.8	1.9	2.0
0	.3329	.3012	.2725	.2466	.2231	.2019	.1827	.1653	.1496	.1353
1	.3662	.3614	.3543	.3452	.3347	.3230	.3106	.2975	.2842	.2707
2	.2014	.2169	.2303	.2417	.2510	.2584	.2640	.2678	.2700	.2707
3	.0738	.0867	.0998	.1128	.1253	.1378	.1496	.1607	.1710	.1804
4	.0203	.0260	.0324	.0395	.0471	.0551	.0636	.0723	.0812	.0902

TABLE B-2 (Cont.)

					λ					
x	1.1	1.2	1.3	1.4	1.5	1.6	1.7	1.8	1.9	2.0
5	.0045	.0062	.0084	.0111	.0141	.0176	.0216	.0260	.0309	.0361
6	.0008	.0012	.0018	.0026	.0035	.0047	.0061	.0078	.0098	.0120
7	.0001	.0002	.0003	.0005	.0008	.0011	.0015	.0020	.0027	.0034
8	.0000	.0000	.0001	.0001	.0001	.0002	.0003	.0005	.0006	.0009
9	.0000	.0000	.0000	.0000	.0000	.0000	.0001	.0001	.0001	.0002

					λ					
x	2.1	2.2	2.3	2.4	2.5	2.6	2.7	2.8	2.9	3.0
0	.1225	.1108	.1003	.0907	.0821	.0743	.0672	.0608	.0550	.0498
1	.2572	.2438	.2306	.2177	.2052	.1931	.1815	.1703	.1596	.1494
2	.2700	.2681	.2652	.2613	.2565	.2510	.2450	.2384	.2314	.2240
3	.1890	.1966	.2033	.2090	.2138	.2176	.2205	.2225	.2237	.2240
4	.0992	.1082	.1169	.1254	.1336	.1414	.1488	.1557	.1622	.1680
5	.0417	.0476	.0538	.0602	.0668	.0735	.0804	.0872	.0940	.1008
6	.0146	.0174	.0206	.0241	.0278	.0319	.0362	.0407	.0455	.0504
7	.0044	.0055	.0068	.0083	.0099	.0118	.0139	.0163	.0188	.0216
8	.0011	.0015	.0019	.0025	.0031	.0038	.0047	.0057	.0068	.0081
9	.0003	.0004	.0005	.0007	.0009	.0011	.0014	.0018	.0022	.0027
10	.0001	.0001	.0001	.0002	.0002	.0003	.0004	.0005	.0006	.0008
11	.0000	.0000	.0000	.0000	.0000	.0001	.0001	.0001	.0002	.0002
12	.0000	.0000	.0000	.0000	.0000	.0000	.0000	.0000	.0000	.0001

					λ					
x	3.1	3.2	3.3	3.4	3.5	3.6	3.7	3.8	3.9	4.0
0	.0450	.0408	.0369	.0334	.0302	.0273	.0247	.0224	.0202	.0183
1	.1397	.1304	.1217	.1135	.1057	.0984	.0915	.0850	.0789	.0733
2	.2165	.2087	.2008	.1929	.1850	.1771	.1692	.1615	.1539	.1465
3	.2237	.2226	.2209	.2186	.2158	.2125	.2087	.2046	.2001	.1954
4	.1734	.1781	.1823	.1858	.1888	.1912	.1931	.1944	.1951	.1954
5	.1075	.1140	.1203	.1264	.1322	.1377	.1429	.1477	.1522	.1563
6	.0555	.0608	.0662	.0716	.0771	.0826	.0881	.0936	.0889	.1042
7	.0246	.0278	.0312	.0348	.0385	.0425	.0466	.0508	.0551	.0595
8	.0095	.0111	.0129	.0148	.0169	.0191	.0215	.0241	.0269	.0298
9	.0033	.0040	.0047	.0056	.0066	.0076	.0089	.0102	.0116	.0132

TABLE B-2 (Cont.)

x	λ 3.1	3.2	3.3	3.4	3.5	3.6	3.7	3.8	3.9	4.0
10	.0010	.0013	.0016	.0019	.0023	.0028	.0033	.0039	.0045	.0053
11	.0003	.0004	.0005	.0006	.0007	.0009	.0011	.0013	.0016	.0019
12	.0001	.0001	.0001	.0002	.0002	.0003	.0003	.0004	.0005	.0006
13	.0000	.0000	.0000	.0000	.0001	.0001	.0001	.0001	.0002	.0002
14	.0000	.0000	.0000	.0000	.0000	.0000	.0000	.0000	.0000	.0001

x	λ 4.1	4.2	4.3	4.4	4.5	4.6	4.7	4.8	4.9	5.0
0	.0166	.0150	.0136	.0123	.0111	.0101	.0091	.0082	.0074	.0067
1	.0679	.0630	.0583	.0540	.0500	.0462	.0427	.0395	.0365	.0337
2	.1393	.1323	.1254	.1188	.1125	.1063	.1005	.0948	.0894	.0842
3	.1904	.1852	.1798	.1743	.1687	.1631	.1574	.1517	.1460	.1404
4	.1951	.1944	.1933	.1917	.1898	.1875	.1849	.1820	.1789	.1755
5	.1600	.1633	.1662	.1687	.1708	.1725	.1738	.1747	.1753	.1755
6	.1093	.1143	.1191	.1237	.1281	.1323	.1362	.1398	.1432	.1462
7	.0640	.0686	.0732	.0778	.0824	.0869	.0914	.0959	.1002	.1044
8	.0328	.0360	.0393	.0428	.0463	.0500	.0537	.0575	.0614	.0653
9	.0150	.0168	.0188	.0209	.0232	.0255	.0280	.0307	.0334	.0363
10	.0061	.0071	.0081	.0092	.0104	.0118	.0132	.0147	.0164	.0181
11	.0023	.0027	.0032	.0037	.0043	.0049	.0056	.0064	.0073	.0082
12	.0008	.0009	.0011	.0014	.0016	.0019	.0022	.0026	.0030	.0034
13	.0002	.0003	.0004	.0005	.0006	.0007	.0008	.0009	.0011	.0013
14	.0001	.0001	.0001	.0001	.0002	.0002	.0003	.0003	.0004	.0005
15	.0000	.0000	.0000	.0000	.0001	.0001	.0001	.0001	.0001	.0002

x	λ 5.1	5.2	5.3	5.4	5.5	5.6	5.7	5.8	5.9	6.0
0	.0061	.0055	.0050	.0045	.0041	.0037	.0033	.0030	.0027	.0025
1	.0311	.0287	.0265	.0244	.0225	.0207	.0191	.0176	.0162	.0149
2	.0793	.0746	.0701	.0659	.0618	.0580	.0544	.0509	.0477	.0446
3	.1348	.1293	.1239	.1185	.1133	.1082	.1033	.0985	.0938	.0892
4	.1719	.1681	.1641	.1600	.1558	.1515	.1472	.1428	.1383	.1339
5	.1753	.1748	.1740	.1728	.1714	.1697	.1678	.1656	.1632	.1606
6	.1490	.1515	.1537	.1555	.1571	.1584	.1594	.1601	.1605	.1606

TABLE B-2 (Cont.)

					λ					
x	5.1	5.2	5.3	5.4	5.5	5.6	5.7	5.8	5.9	6.0
7	.1086	.1125	.1163	.1200	.1234	.1267	.1298	.1326	.1353	.1377
8	.0692	.0731	.0771	.0810	.0849	.0887	.0925	.0962	.0998	.1033
9	.0392	.0423	.0454	.0486	.0519	.0552	.0586	.0620	.0654	.0688
10	.0200	.0220	.0241	.0262	.0285	.0309	.0334	.0359	.0386	.0413
11	.0093	.0104	.0116	.0129	.0143	.0157	.0173	.0190	.0207	.0225
12	.0039	.0045	.0051	.0058	.0065	.0073	.0082	.0092	.0102	.0113
13	.0015	.0018	.0021	.0024	.0028	.0032	.0036	.0041	.0046	.0052
14	.0006	.0007	.0008	.0009	.0011	.0013	.0015	.0017	.0019	.0022
15	.0002	.0002	.0003	.0003	.0004	.0005	.0006	.0007	.0008	.0009
16	.0001	.0001	.0001	.0001	.0001	.0002	.0002	.0002	.0003	.0003
17	.0000	.0000	.0000	.0000	.0000	.0000	.0001	.0001	.0001	.0001

					λ					
x	6.1	6.2	6.3	6.4	6.5	6.6	6.7	6.8	6.9	7.0
0	.0022	.0020	.0018	.0017	.0015	.0014	.0012	.0011	.0010	.0009
1	.0137	.0126	.0116	.0106	.0098	.0090	.0082	.0076	.0070	.0064
2	.0417	.0390	.0364	.0340	.0318	.0296	.0276	.0258	.0240	.0223
3	.0848	.0806	.0765	.0726	.0688	.0652	.0617	.0584	.0552	.0521
4	.1294	.1249	.1205	.1162	.1118	.1076	.1034	.0992	.0952	.0912
5	.1579	.1549	.1519	.1487	.1454	.1420	.1385	.1349	.1314	.1277
6	.1605	.1601	.1595	.1586	.1575	.1562	.1546	.1529	.1511	.1490
7	.1399	.1418	.1435	.1450	.1462	.1472	.1480	.1486	.1489	.1490
8	.1066	.1099	.1130	.1160	.1188	.1215	.1240	.1263	.1284	.1304
9	.0723	.0757	.0791	.0825	.0858	.0891	.0923	.0954	.0985	.1014
10	.0441	.0469	.0498	.0528	.0558	.0588	.0618	.0649	.0679	.0710
11	.0245	.0265	.0285	.0307	.0330	.0353	.0377	.0401	.0426	.0452
12	.0124	.0137	.0150	.0164	.0179	.0194	.0210	.0227	.0245	.0264
13	.0058	.0065	.0073	.0081	.0089	.0098	.0108	.0119	.0130	.0142
14	.0025	.0029	.0033	.0037	.0041	.0046	.0052	.0058	.0064	.0071
15	.0010	.0012	.0014	.0016	.0018	.0020	.0023	.0026	.0029	.0033
16	.0004	.0005	.0005	.0006	.0007	.0008	.0010	.0011	.0013	.0014
17	.0001	.0002	.0002	.0002	.0003	.0003	.0004	.0004	.0005	.0006
18	.0000	.0001	.0001	.0001	.0001	.0001	.0001	.0002	.0002	.0002
19	.0000	.0000	.0000	.0000	.0000	.0000	.0000	.0001	.0001	.0001

TABLE B-2 (Cont.)

	λ									
x	7.1	7.2	7.3	7.4	7.5	7.6	7.7	7.8	7.9	8.0
0	.0008	.0007	.0007	.0006	.0006	.0005	.0005	.0004	.0004	.0003
1	.0059	.0054	.0049	.0045	.0041	.0038	.0035	.0032	.0029	.0027
2	.0208	.0194	.0180	.0167	.0156	.0145	.0134	.0125	.0116	.0107
3	.0492	.0464	.0438	.0413	.0389	.0366	.0345	.0324	.0305	.0286
4	.0874	.0836	.0799	.0764	.0729	.0696	.0663	.0632	.0602	.0573
5	.1241	.1204	.1167	.1130	.1094	.1057	.1021	.0986	.0951	.0916
6	.1468	.1445	.1420	.1394	.1367	.1339	.1311	.1282	.1252	.1221
7	.1489	.1486	.1481	.1474	.1465	.1454	.1442	.1428	.1413	.1396
8	.1321	.1337	.1351	.1363	.1373	.1382	.1388	.1392	.1395	.1396
9	.1042	.1070	.1096	.1121	.1144	.1167	.1187	.1207	.1224	.1241
10	.0740	.0770	.0800	.0829	.0858	.0887	.0914	.0941	.0967	.0993
11	.0478	.0504	.0531	.0558	.0585	.0613	.0640	.0667	.0695	.0722
12	.0283	.0303	.0323	.0344	.0366	.0388	.0411	.0434	.0457	.0481
13	.0154	.0168	.0181	.0196	.0211	.0227	.0243	.0260	.0278	.0296
14	.0078	.0086	.0095	.0104	.0113	.0123	.0134	.0145	.0157	.0169
15	.0037	.0041	.0046	.0051	.0057	.0062	.0069	.0075	.0083	.0090
16	.0016	.0019	.0021	.0024	.0026	.0030	.0033	.0037	.0041	.0045
17	.0007	.0008	.0009	.0010	.0012	.0113	.0015	.0017	.0019	.0021
18	.0003	.0003	.0004	.0004	.0005	.0006	.0006	.0007	.0008	.0009
19	.0001	.0001	.0001	.0002	.0002	.0002	.0003	.0003	.0003	.0004
20	.0000	.0000	.0001	.0001	.0001	.0001	.0001	.0001	.0001	.0002
21	.0000	.0000	.0000	.0000	.0000	.0000	.0000	.0000	.0001	.0001

	λ									
x	8.1	8.2	8.3	8.4	8.5	8.6	8.7	8.8	8.9	9.0
0	.0003	.0003	.0002	.0002	.0002	.0002	.0002	.0002	.0001	.0001
1	.0025	.0023	.0021	.0019	.0017	.0016	.0014	.0013	.0012	.0011
2	.0100	.0092	.0086	.0079	.0074	.0068	.0063	.0058	.0054	.0050
3	.0269	.0252	.0237	.0222	.0208	.0195	.0183	.1071	.0160	.0150
4	.0544	.0517	.0491	.0466	.0443	.0420	.0398	.0377	.0357	.0337
5	.0882	.0849	.0816	.0784	.0752	.0722	.0692	.0663	.0635	.0607
6	.1191	.1160	.1128	.1097	.1066	.1034	.1003	.0972	.0941	.0911
7	.1378	.1358	.1338	.1317	.1294	.1271	.1247	.1222	.1197	.1171
8	.1395	.1392	.1388	.1382	.1375	.1366	.1356	.1344	.1332	.1318
9	.1256	.1269	.1280	.1290	.1299	.1306	.1311	.1315	.1317	.1318

TABLE B-2 (Cont.)

	λ									
x	8.1	8.2	8.3	8.4	8.5	8.6	8.7	8.8	8.9	9.0
10	.1017	.1040	.1063	.1084	.1104	.1123	.1140	.1157	.1172	.1861
11	.0749	.0776	.0802	.0828	.0853	.0878	.0902	.0925	.0948	.0970
12	.0505	.0530	.0555	.0579	.0604	.0629	.0654	.0679	.0703	.0728
13	.0315	.0334	.0354	.0374	.0395	.0416	.0438	.0459	.0481	.0504
14	.0182	.0196	.0210	.0225	.0240	.0256	.0272	.0289	.0306	.0324
15	.0098	.0107	.0116	.0126	.0136	.0147	.0158	.0169	.0182	.0194
16	.0050	.0055	.0060	.0066	.0072	.0079	.0086	.0093	.0101	.0109
17	.0024	.0026	.0029	.0033	.0036	.0040	.0044	.0048	.0053	.0058
18	.0011	.0012	.0014	.0015	.0017	.0019	.0021	.0024	.0026	.0029
19	.0005	.0005	.0006	.0007	.0008	.0009	.0010	.0011	.0012	.0014
20	.0002	.0002	.0002	.0003	.0003	.0004	.0004	.0005	.0005	.0006
21	.0001	.0001	.0001	.0001	.0001	.0002	.0002	.0002	.0002	.0003
22	.0000	.0000	.0000	.0000	.0001	.0001	.0001	.0001	.0001	.0001

	λ									
x	9.1	9.2	9.3	9.4	9.5	9.6	9.7	9.8	9.9	10
0	.0001	.0001	.0001	.0001	.0001	.0001	.0001	.0001	.0001	.0000
1	.0010	.0009	.0009	.0008	.0007	.0007	.0006	.0005	.0005	.0005
2	.0046	.0043	.0040	.0037	.0034	.0031	.0029	.0027	.0025	.0023
3	.0140	.0131	.0123	.0115	.0107	.0100	.0093	.0087	.0081	.0076
4	.0319	.0302	.0285	.0269	.0254	.0240	.0226	.0213	.0201	.0189
5	.0581	.0555	.0530	.0506	.0483	.0460	.0439	.0418	.0398	.0378
6	.0881	.0851	.0822	.0793	.0764	.0736	.0709	.0682	.0656	.0631
7	.1145	.1118	.1091	.1064	.1037	.1010	.0982	.0955	.0928	.0901
8	.1302	.1286	.1269	.1251	.1232	.1212	.1191	.1170	.1148	.1126
9	.1317	.1315	.1311	.1306	.1300	.1293	.1284	.1274	.1263	.1251
10	.1198	.1210	.1219	.1228	.1235	.1241	.1245	.1249	.1250	.1251
11	.0991	.1012	.1031	.1049	.1067	.1083	.1098	.1112	.1125	.1137
12	.0752	.0776	.0799	.0822	.0844	.0866	.0888	.0908	.0928	.0948
13	.0526	.0549	.0572	.0594	.0617	.0640	.0662	.0685	.0707	.0729
14	.0342	.0361	.0380	.0399	.0419	.0439	.0459	.0479	.0500	.0521
15	.0208	.0221	.0235	.0250	.0265	.0281	.0297	.0313	.0330	.0347
16	.0118	.0127	.0137	.0147	.0157	.0168	.0180	.0192	.0204	.0217
17	.0063	.0069	.0075	.0081	.0088	.0095	.0103	.0111	.0119	.0128
18	.0032	.0035	.0039	.0042	.0046	.0051	.0055	.0060	.0065	.0071
19	.0015	.0017	.0019	.0021	.0023	.0026	.0028	.0031	.0034	.0037

TABLE B-2 (Cont.)

	λ									
x	9.1	9.2	9.3	9.4	9.5	9.6	9.7	9.8	9.9	10
20	.0007	.0008	.0009	.0010	.0011	.0012	.0014	.0015	.0017	.0019
21	.0003	.0003	.0004	.0004	.0005	.0006	.0006	.0007	.0008	.0009
22	.0001	.0001	.0002	.0002	.0002	.0002	.0003	.0003	.0004	.0004
23	.0000	.0001	.0001	.0001	.0001	.0001	.0001	.0001	.0002	.0002
24	.0000	.0000	.0000	.0000	.0000	.0000	.0000	.0001	.0001	.0001

TABLE B-3

The Standard Normal Distribution Function

$$\phi(x) = \int_{-\infty}^{x} \frac{1}{\sqrt{(2\pi)}} e^{-t^2/2} dt$$

x	0	1	2	3	4	5	6	7	8	9
.0	.5000	.5040	.5080	.5120	.5160	.5199	.5239	.5279	.5319	.5359
.1	.5398	.5438	.5478	.5517	.5557	.5596	.5636	.5675	.5714	.5753
.2	.5793	.5832	.5871	.5910	.5948	.5987	.6026	.6064	.6103	.6141
.3	.6179	.6217	.6255	.6293	.6331	.6368	.6406	.6443	.6480	.6517
.4	.6554	.6591	.6628	.6664	.6700	.6736	.6772	.6808	.6844	.6879
.5	.6915	.6950	.6985	.7019	.7054	.7088	.7123	.7157	.7190	.7224
.6	.7257	.7291	.7324	.7357	.7389	.7422	.7454	.7486	.7517	.7549
.7	.7580	.7611	.7642	.7673	.7704	.7734	.7764	.7794	.7823	.7852
.8	.7881	.7910	.7939	.7967	.7995	.8023	.8051	.8078	.8106	.8133
.9	.8159	.8186	.8212	.8238	.8264	.8289	.8315	.8340	.8365	8389
1.0	.8413	.8438	.8461	.8485	.8508	.8531	.8554	.8577	.8599	.8621
1.1	.8643	.8665	.8686	.8708	.8729	.8749	.8770	.8790	.8810	.8830
1.2	.8849	.8869	.8888	.8907	.8925	.8944	.8962	.8980	.8997	.9015
1.3	.9032	.9049	.9066	.9082	.9099	.9115	.9131	.9147	.9162	.9177
1.4	.9192	.9207	.9222	.9236	.9251	.9365	.9279	.9292	.9306	.9319
1.5	.9332	.9345	.9357	.9370	.9382	.9394	.9406	.9418	.9429	.9441
1.6	.9452	.9463	.9474	.9484	.9495	.9505	.9515	.9525	.9535	.9545
1.7	.9554	.9564	.9573	.9582	.9591	.9599	.9608	.9616	.9625	.9633
1.8	.9641	.9649	.9656	.9664	.9671	.9678	.9686	.9693	.9699	.9706
1.9	.9713	.9719	.9726	.9732	.9738	.9744	.9750	.9756	.9761	.9767
2.0	.9772	.9778	.9783	.9788	.9793	.9798	.9803	.9808	.9812	.9817
2.1	.9821	.9826	.9830	.9834	.9838	.9842	.9846	.9850	.9854	.9857
2.2	.9861	.9864	.9868	.9871	.9875	.9878	.9881	.9884	.9887	.9890

TABLE B-3 (Cont.)

x	0	1	2	3	4	5	6	7	8	9
2.3	.9893	.9896	.9898	.9901	.9904	.9906	.9909	.9911	.9913	.9916
2.4	.9918	.9920	.9922	.9925	.9927	.9929	.9931	.9932	.9934	.9936
2.5	.9938	.9940	.9941	.9943	.9945	.9946	.9948	.9949	.9951	.9952
2.6	.9953	.9955	.9956	.9957	.9959	.9960	.9961	.9962	.9963	.9964
2.7	.9965	.9966	.9967	.9968	.9969	.9970	.9971	.9972	.9973	.9974
2.8	.9974	.9975	.9976	.9977	.9977	.9978	.9979	.9979	.9980	.9981
2 9	.9981	.9982	.9982	.9983	.9984	.9984	.9985	.9985	.9986	.9986
3.0	.9987	.9987	.9987	.9988	.9988	.9989	.9989	.9989	.9990	.9990
3.1	.9990	.9991	.9991	.9991	.9992	.9992	.9992	.9992	.9993	.9993
3.2	.9993	.9993	.9994	.9994	.9994	.9994	.9994	.9995	.9995	.9995
3.3	.9995	.9995	.9995	.9996	.9996	.9996	.9996	.9996	.9996	.9997
3.4	.9997	.9997	.9997	.9997	.9997	.9997	.9997	.9997	.9997	.9998

TABLE B-4
The Gamma Function

$\Gamma(\alpha)$

α	0	1	2	3	4	5	6	7	8	9
1.0	1.0000	.9943	.9888	.9835	.9784	.9735	.9687	.9642	.9597	.9555
1.1	.9514	.9474	.9436	.9399	.9364	.9330	.9298	.9267	.9237	.9209
1.2	.9182	.9156	.9131	.9108	.9085	.9064	.9044	.9025	.9007	.8990
1.3	.8975	.8960	.8946	.8934	.8922	.8912	.8902	.8893	.8885	.8876
1.4	.8873	.8868	.8864	.8860	.8858	.8857	.8856	.8856	.8857	.8859
1.5	.8862	.8866	.8870	.8876	.8882	.8859	.8896	.8905	.8914	.8924
1.6	.8935	.8947	.8959	.8972	.8986	.9001	.9017	.9033	.9050	.9068
1.7	.9086	.9106	.9126	.9147	.9168	.9191	.9214	.9238	.9262	.9288
1.8	.9314	.9341	.9368	.9397	.9426	.9456	.9487	.9518	.9551	.9584
1.9	.9618	.9652	.9688	.9724	.9761	.9799	.9837	.9877	.9917	.9958
2.0	1.0000	1.0043	1.0086	1.0131	1.0176	1.0222	1.0269	1.0316	1.0365	1.0415
2.1	1.0465	1.0546	1.0568	1.0621	1.0675	1.0730	1.0786	1.0842	1.0900	1.0959
2.2	1.1018	1.1078	1.1140	1.1202	1.1266	1.1330	1.1395	1.1462	1.1529	1.1598
2.3	1.1667	1.1738	1.1809	1.1882	1.1956	1.2031	1.2107	1.2184	1.2262	1.2341
2.4	1.2422	1.2503	1.2586	1.2670	1.2756	1.2842	1.2930	1.3019	1.3109	1.3201

TABLE B-4 (Cont.)

α	0	1	2	3	4	5	6	7	8	9
2.5	1.3293	1.3388	1.3483	1.3580	1.3678	1.3777	1.3878	1.3981	1.4084	1.4190
2.6	1.4296	1.4404	1.4514	1.4625	1.4738	1.4852	1.4968	1.5085	1.5204	1.5325
2.7	1.5447	1.5571	1.5696	1.5824	1.5953	1.6084	1.6216	1.6351	1.6487	1.6625
2.8	1.6765	1.6907	1.7051	1.7196	1.7344	1.7494	1.7646	1.7799	1.7955	1.8113
2.9	1.8274	1.8436	1.8600	1.8767	1.8936	1.9108	1.9281	1.9457	1.9636	1.9816
3.0	2.000	2.019	2.037	2.057	2.076	2.095	2.115	2.136	2.156	2.177
3.1	2.198	2.219	2.240	2.262	2.284	2.307	2.330	2.353	2.376	2.400
3.2	2.424	2.448	2.473	2.498	2.524	2.549	2.575	2.602	2.629	2.056
3.3	2.683	2.711	2.740	2.768	2.798	2.827	2.857	2.888	2.918	2.950
3.4	2.981	3.013	3.046	3.079	3.112	3.146	3.181	3.216	3.251	3.287
3.5	3.323	3.360	3.398	3.436	3.474	3.513	3.553	3.593	3.634	3.675
3.6	3.717	3.760	3.803	3.846	3.891	3.936	3.981	4.028	4.075	4.122
3.7	4.171	4.220	4.269	4.320	4.371	4.423	4.476	4.529	4.583	4.638
3.8	4.694	4.751	4.808	4.867	4.926	4.986	5.047	5.108	5.171	5.235
3.9	5.299	5.365	5.431	5.499	5.567	5.637	5.707	5.779	5.851	5.925
4.0	6.000	6.076	6.153	6.231	6.311	6.391	6.473	6.556	6.640	6.726
4.1	6.183	6.901	6.990	7.081	7.173	7.267	7.362	7.458	7.556	7.656
4.2	7.757	7.850	7.963	8.069	8.176	8.285	8.396	8.508	8.622	8.738
4.3	8.855	8.975	9.096	9.219	9.344	9.471	9.600	0.731	0.864	0.999
4.4	10.136	10.275	10.417	10.561	10.707	10.855	11.005	11.158	11.314	11.471
4.5	11.632	11.795	11.960	12.128	12.299	12.472	12.648	12.827	13.009	13.194
4.6	13.381	13.572	13.766	13.962	14.162	14.366	14.572	14.782	14.995	15.211
4.7	15.431	15.655	15.882	16.113	16.348	16.586	16.829	17.075	17.325	17.579
4.8	17.84	18.10	18.37	18.64	18.91	19.20	19.48	19.77	20.06	20.36
4.9	20.67	20.98	21.20	21.61	21.94	22.27	22.60	22.04	23.20	23.64

TABLE B-5

Values of e^{-x}

x	e^{-x}	x	e^{-x}	x	e^{-x}	x	e^{-x}
0.00	1.00000	0.50	.60653	1.00	.36788	1.50	.22313
0.01	0.99005	0.51	.60050	1.01	.36422	1.51	.22091
0.02	.98020	0.52	.59452	1.02	.30659	1.52	.21871

TABLE B-5 (Cont.)

x	e^{-x}	x	e^{-x}	x	e^{-x}	x	e^{-x}
0.03	.97045	0.53	.58860	1.03	.35701	1.53	.21654
0.04	.96079	0.54	.58275	1.04	.35345	1.54	.21438
0.05	.95123	0.55	.57695	1.05	.34994	1.55	.21225
0.06	.94176	0.56	.57121	1.05	.34646	1.56	.21014
0.07	.93239	0.57	.56553	1.07	.34301	1.57	.20805
0.08	.92312	0.58	.55990	1.08	.33960	1.58	.20508
0.09	.91393	0.59	.55433	1.09	.33622	1.59	.20393
0.10	.90484	0.60	.54881	1.10	.33287	1.60	.20190
0.11	.89583	0.61	.54335	1.11	.32956	1.61	.19989
0.12	.88692	0.62	.53794	1.12	.32628	1.62	.19790
0.13	.87810	0.63	.53259	1.13	.32303	1.63	.19593
0.14	.86936	0.64	.52729	1.14	.31982	1.64	.19398
0,15	.86071	0.65	.52205	1.15	.31664	1.65	.19205
0.16	.85211	0.66	.51685	1.16	.31349	1.66	.19014
0.17	.84366	0.67	.51171	1.17	.31037	1.67	.18825
0.18	.83527	0.68	.50662	1.18	.30728	1.68	.18637
0.19	.82696	0.69	.50158	1.19	.30422	1.69	.18452
0.20	.81873	0.70	.49659	1.20	.30119	1.70	.18268
0.21	.81058	0.71	.49164	1.21	.29820	1.71	.18087
0.22	.80252	0.72	.48675	1.22	.29523	1.72	.17907
0.23	.79453	0.73	.48191	1.23	.29229	1.73	.17728
0.24	.78663	0.74	.47711	1.24	.28938	1.74	.17552
0.25	.77880	0.75	.47237	1.25	.28650	1.75	.17377
0.26	.77105	0.76	.46767	1.26	.28365	1.76	.17204
0.27	.76338	0.77	.46301	1.27	.28083	1.77	.17033
0.28	.75578	0.78	.45841	1.28	.27804	1.78	.16864
0.29	.74826	0.79	.45384	1.29	.27527	1.79	.16696
0.30	.74082	0.80	.44933	1.30	.27253	1.80	.16530
0.31	.73345	0.81	.44486	1.31	.26982	1.81	.16365
0.32	.72615	0.82	.44043	1.32	.26714	1.82	.16203
0.33	.71892	0.83	.43605	1.33	.26448	1.83	.16041
0.34	.71177	0.84	.41371	1.34	.26185	1.84	.15882
0.35	.70469	0.85	.42741	1.35	.25924	1.85	.15724
0.36	.69768	0.86	.42316	1.36	.25666	1.86	.15567
0.37	.69073	0.87	.41895	1.37	.25411	1.87	.15412
0.38	.68386	0.88	.41478	1.38	.25158	1.88	.15259
0.39	.67706	0.89	.41066	1.39	.24908	1.89	.15107

TABLE B-5 (Cont.)

x	e^{-x}	x	e^{-x}	x	e^{-x}	x	e^{-x}
0.40	.67032	0.90	.40657	1.40	.24660	1.90	.14957
0.41	.66365	0.91	.40252	1.41	.24414	1.91	.14808
0.42	.65705	0.92	.39852	1.42	.24171	1.92	.14661
0.43	.65051	0.93	.39455	1.43	.23931	1.93	.14515
0.44	.64404	0.94	.39063	1.44	.23693	1.94	.14370
0.45	.63763	0.95	.38674	1.45	.23457	1.95	.14227
0.46	.63128	0.96	.38289	1.46	.23224	1.96	.14086
0.47	.62500	0.97	.37908	1.47	.22993	1.97	.13916
0.48	.61878	0.98	.37531	1.48	.22764	1.98	.13807
0.49	.61263	0.99	.37158	1.49	.22537	1.99	.13670
2.00	.13534	2.40	.09072	2.80	.06081	4.00	.01832
2.01	.13399	2.41	.08982	2.81	.06020	4.10	.01657
2.02	.13266	2.42	.08892	2.82	.05961	4.20	.01500
2.03	.13134	2.43	.08804	2.83	.05001	4.30	.01357
2.04	.13003	2.44	.08716	2.84	.05843	4.40	.01228
2.05	.12873	2.45	.08629	2.85	.05784	4.50	.01111
2.06	.12745	2.46	.08544	2.86	.05727	4.60	.01005
2.07	.12619	2.47	.08458	2.87	.05670	4.70	.00010
2.08	.12493	2.48	.08374	2.88	.05613	4.80	.00823
2.09	.12369	2.49	.08291	2.89	.05558	4.90	.00745
2.10	.12246	2.50	.08208	2.90	.05502	4.00	.00674
2.11	.12124	2.51	.08127	2.91	.05448	5.10	.00610
2.12	.12003	2.52	.08046	2.92	.05393	5.20	.00552
2.13	.11884	2.53	.07966	2.93	.05340	5.30	.00499
2.14	.11765	2.54	.07887	2.94	.05287	5.40	.00452
2.15	.11648	2.55	.07808	2.95	.05234	5.50	.00400
2.16	.11533	2.56	.07730	2.96	.05182	5.60	.00370
2.17	.11418	2.57	.07654	2.97	.05130	5.70	.00335
2.18	.11301	2.58	.07577	2.98	.05079	5.80	.00303
2.19	.11192	2.59	.07502	2.99	.05029	5.90	.00274
2.20	.11080	2.60	.07427	3.00	.04979	6.00	.00248
2.21	.10970	2.61	.07353	3.05	.04736	6.25	.00193
2.22	.10861	2.62	.07280	3.10	.04505	6.50	.00150
2.23	.10753	2.63	.07208	3.15	.04285	6.75	.00117
2.24	.10646	2.64	.07136	3.20	.04076	7.00	.00001
2.25	.10540	2.65	.07065	3.25	.03877	7.50	.00055
2.26	.10435	2.66	.06995	3.30	.03688	8.00	.00034

TABLE B-5 (Cont.)

x	e^{-x}	x	e^{-x}	x	e^{-x}	x	e^{-x}
2.27	.10331	2.67	.06925	3.35	.03508	8.50	.00020
2.28	.10228	2.68	.06856	3.40	.03337	9.00	.00012
2.29	.10127	2.69	.06788	3.45	.03175	9.50	.00007
2.30	.10026	2.70	.06721	3.50	.03020	10.00	.00005
2.31	.09926	2.71	.06654	3.55	.02872		
2.32	.09827	2.72	.06587	3.60	.02732		
2.33	.09730	2.73	.06522	3.65	.02599		
2.34	.09633	2.74	.06457	3.70	.02472		
2.35	.09537	2.75	.06393	3.75	.02352		
2.36	.09442	2.76	.06329	3.80	.02237		
2.37	.09348	2.77	.06266	3.85	.02128		
2.38	.09255	2.78	.06204	3.90	.02024		
2.30	.09163	2.79	.06142	3.95	.01925		

Answers to Some Problems

Chapter 1

2. $\{(1, 2, 3), (1, 3, 2), (2, 1, 3), (2, 3, 1), (3, 1, 2), (3, 2, 1)\}$ 6, 2

4. a. 1
   b. .2

6. 10^5

8. $10 \cdot 9 \cdot 8 \cdot 7 \cdot 6 = 30{,}240$

11. 30, 1/7

13. a. 183/1174
    b. 467/1174
    c. 183/467
    d. Yes

17. 1/3

19. 3/5

Chapter 2

2. a. .1
   b. .39
   c. .74

4. b. .18
   c. .56

6. a. .044
   b. .005

9. $\mu_3' - 3\mu\mu_2' + 2\mu^3$

12. $2np^3 - 3np^2 + np$

15. λ

17. $\dfrac{n^3}{4} + \dfrac{n^2}{2} + \dfrac{n}{4}$

22. $\dfrac{\beta^4 - \alpha^4}{4(\beta - \alpha)}$

28. $3\mu\sigma^2 + \mu^3$

30. c. .36

31. c. .01

Chapter 3

1. 39.9, 2.7

4. 9, 108, 1620, 27

6. 6, 54, 648, 18

9. $\lambda = \dfrac{\ln 2}{T}$

13. $\sqrt{(\pi/2)}$, 2, $3\sqrt{(\pi/2)}$, $2-\pi/2$

16. 3/5, 2/5, 2/7, 1/25

19. 5/4, 25/6, 125/4, 125/48

Chapter 4

1. a. 30
   b. 21
   c. 50
   d. 25
   e. -15

3. $\binom{10}{x}\binom{10-x}{y}(.3)^x(.2)^y(.5)^{10-x-y}$

5. a. $n(1-p-q)$
   b. $n(1-p-q)(p+q)$
   c. $-n(1-p-q)p$
   d. $-n(1-p-q)q$

8. a. n
   b. 0

10. a. 0
    b. $1/4\pi^2$
    c. 0
    d. 1/2
    e. $1/4\pi$

12. a. $\frac{3}{5}(100-y)$
    b. $\frac{6}{52}(100-y)$
    c. $\frac{5}{7}(100-x)$
    d. $\frac{10}{49}(100-x)$
    e. 9
    f. 12

12. g. $\dfrac{25\cdot 91}{49}$

    h. $\dfrac{10\cdot 70}{49}$

16. a. $\dfrac{(n-z)p}{p+q}$

b. $\dfrac{(n-x)(1-p-q)}{1-p}$

c. $\dfrac{(n-z)pq}{(p+q)^2}$

d. $\dfrac{(n-x)(1-p-q)q}{(1-p)^3}$

e. $\dfrac{n(1-p-q)p^2}{(p+q)}$

f. $\dfrac{n(1-p-q)^2p}{1-p}$

g. $\dfrac{npq}{p+q}$

h. $\dfrac{n(1-p-q)q}{1-p}$

Chapter 5

2. $\lambda^3+3\lambda^2+\lambda$

3. $$\frac{r(r+1)(r+2)(1-p)^3}{p^3}+\frac{3r(r+1)(1-p)^2}{p^2}+\frac{r(1-p)}{p}$$

5. $3\sigma^2\mu+\mu^3$

7. $(1-t/\lambda)^{-1}$

10. 2, 6

11. $[(1-p-q)e^t+(p+q)]^n$

12. $2\rho\sigma_x\sigma_y\mu_x+\sigma_x{}^2\mu_y+\mu_x{}^2\mu_y$

Chapter 6

1. $P(y) = \left(\dfrac{10}{\sqrt{y}}\right)(.5)^{10} \qquad y = 0, 1, 4, 9, \ldots$

5. $P(y) = \displaystyle\int_y^{y+1} \frac{1}{\sqrt{(2\pi)}}\, e^{-x^2/2}dx \qquad y = \ldots, -2, -1, 0, 1, 2, \ldots$

7. $f(y) = e^y \qquad 0 < y < \ln 2$
   $= 0 \qquad$ otherwise

11. $f(y) = \dfrac{1}{2\Gamma(\alpha)\beta^\alpha}\, y^{(\alpha/2)-1}e^{-\sqrt{(y)}/\beta} \qquad 0 < y < \infty$
    $0 \qquad$ otherwise

13. $f(y) = \dfrac{2}{\sqrt{(2\pi)}\sigma}\, e^{-y/2\sigma^2} \qquad 0 < y < \infty$
    $= 0 \qquad$ otherwise

14. $f(y) = \dfrac{2\Gamma\left(\dfrac{n+1}{2}\right)}{\Gamma\left(\dfrac{n}{2}\right)\sqrt{(n\pi)}}\left(1+\dfrac{y^2}{n}\right)^{-[(n+1)/2]}$ $\quad 0 < y < \infty$

$= 0$ $\quad$ otherwise

18. $f(y) = \dfrac{1}{9!}\,y^9 e^{-y}$ $\quad 0 < y < \infty$

$= 0$ $\quad$ otherwise

19. $f(u, v) = \dfrac{1}{4\pi}\,e^{(u^2+v^2)/4}$ $\quad -\infty < u, v < \infty$

20. $f(z) = \frac{1}{2}\,e^{-z/2}$ $\quad 0 < z < \infty$

$= 0$ $\quad$ otherwise

Chapter 7

1. $\lambda, \frac{1}{2}(\lambda^2+\lambda)$
3. $p\mu+(1-p)\lambda,\ p(\sigma^2+\mu^2)+(1-p)(\lambda^2+\lambda)-(p\mu+(1-p)\lambda)^2$
5. $\frac{1}{2}(1-\lambda t)^{-1}+\frac{1}{2}(e^{-\lambda}e^{\lambda e^t})$
9. $r\mu_x+(1-r)np,\ r(\sigma_x{}^2+\mu^2)+(1-r)(npq+n^2p^2)-[r\mu_x+(1-r)np)]^2$
11. $E(X \mid y) = r\left[\mu_x+\dfrac{\rho\sigma_x}{\sigma_y}(y-\mu_y)\right]+(1-r)(n-y)\dfrac{p}{1-q}$

$$\text{Var}\,(X \mid y) = r\left\{\sigma_x{}^2(1-\rho)+\left[\mu_x+\frac{\rho\sigma_x}{\sigma_y}(y-\mu_y)\right]^2\right\}$$

$$+(1-r)\left[(n-y)\frac{p(1-p-q)}{(1-q)^2}+(n-y)^2\frac{p^2}{(1-q)^2}\right]-[E(X \mid y)]^2$$

13. $r(\rho\sigma_x\sigma_y+\mu_x\mu_y)-(1-r)(npq+n^2pq)-(r\mu_x+(1-r)np)(r\mu_y+(1-r)nq)$
17. 1/9, .002

Chapter 8

1. a. .45
   b. .77
   c. .77
   d. 1
   e. 1
   f. 1
3. 1.4
5. a. Yes
   b. Yes
   c. No
   d. No

7. a. Yes
   b. Yes
   c. No
   d. No

9. a. Yes
   b. Yes
   c. Yes
   d. Yes
   e. Yes

Chapter 9

1. $\frac{1}{n}\sum_{i=1}^{n}\alpha_i\beta_i$

4. $\frac{1}{n}\sum_{i=1}^{n}m_ip_i, \frac{1}{n^2}\sum_{i=1}^{n}m_ip_i(1-p_i)$

7. $\frac{1}{n}\sum_{i=1}^{n}(\alpha_i\beta_i^2+\alpha_i^2\beta_i^2)-\left(\frac{1}{n}\sum_{i=1}^{n}\alpha_i\beta_i\right)^2-\frac{1}{n^2}\sum_{i=1}^{n}\alpha_i\beta_i^2$

9. $\frac{n-1}{n}\alpha\beta^2, \alpha\beta^2$

11. $\frac{(n-1)^2}{n^3}\beta^4(\alpha+3)(\alpha+2)(\alpha+1)\alpha-\frac{(n-1)(n-3)}{n^3}\alpha^2\beta^4$

    $\frac{1}{n}\beta^4(\alpha+3)(\alpha+2)(\alpha+1)\alpha-\frac{n-3}{n(n-1)}\alpha^2\beta^4$

13. $1/\lambda, 1/n\lambda^2$

15. $\frac{1}{n}\sigma^2\sum_{i=1}^{n}m_i, \frac{2\sigma^4}{n^2}\sum_{i=1}^{n}m_i$

17. $\frac{n+1}{2}, \frac{n+1}{2n}$

19. $f(\bar{x}) = \frac{n^{3n}}{\Gamma(3n)2^{3n}}\bar{x}^{3n-1}e^{-n\bar{x}/2} \qquad 0 < \bar{x} < \infty$

    $= 0 \qquad$ otherwise

20. $P(\bar{x} = 3) = \frac{e^{-3n}(3n)^{3n}}{(3n)!}$

25. $f(t) = \frac{1}{2\sqrt{2}}\left(1+\frac{t^2}{2}\right)^{-3/2} \qquad -\infty < t < \infty$

28. $f(y) = \frac{6y}{(1+y)^4} \qquad 0 < y < \infty$

    $= 0 \qquad$ otherwise

Chapter 10

1. $\hat{\theta}_a, \hat{\theta}_m, \hat{\theta}_0, \hat{\theta}_p$

3. $\hat{\theta}_p$

5. $\hat{\theta}_a, \hat{\theta}_f$

7. $\hat{\theta}_c, \hat{\theta}_d, \hat{\theta}_e, \hat{\theta}_h, \hat{\theta}_i$

9. $\sigma^4/n^2, \dfrac{2(n-1)}{n^2}\sigma^4, \dfrac{2n-1}{n^2}\sigma^4$

11. $\bar{X}_n$

13. $\bar{X}_1, \bar{X}_2, \bar{X}_3, \ldots$

15. X

18. $\bar{X}_n$

19. $\dfrac{1}{n}\sum_{i=1}^{n}(X_i-\mu)^2$

Chapter 11

1. $P_\theta(\phi) = \binom{nm}{\phi} p^\phi(1-p)^{nm-\phi} \qquad \phi = 0, 1, 2, \ldots, nm$

3. $$f_\theta(\phi) = \frac{1}{\Gamma(n\alpha)\theta^{n\alpha}}\phi^{n\alpha-1}e^{-\phi/\theta} \qquad 0 < \phi < \infty$$
$$= 0 \qquad \text{otherwise}$$

13. $\bar{X}_n$

16. mpq/n

17. θ^2/n

20. $\theta^2/2$

Chapter 12

2. .516

4. .025

6. *accept* if $-2.96 < x < .96$
reject otherwise

8. *reject* if $\sum_{i=1}^{10} x_i > 13$
accept if $\sum_{i=1}^{10} x_i < 13$
reject with probability .54 if $\sum_{i=1}^{10} x_i = 13$

10. .17

14. *reject* if $\sum_{i=1}^{n} x_i < c$

accept otherwise

where c is given by

$$\frac{\theta_0^n}{F(n)} \int_0^c x^{n-1} e^{-\theta_0 x} dx = \alpha$$

17. *reject* if $x > -\ln .05$

accept otherwise

18. *reject* if $x < -\dfrac{\ln \cdot 95}{2}$

accept otherwise

Chapter 13

1. $\bar{x} - 1.04 < \theta < \bar{x} + 1.04$

4. $\theta \leqq x / -\ln .95$

10. Same as in Problem 1

12. $$f(\theta \mid x) = \frac{\Gamma(n+\alpha+\beta)}{\Gamma(x+\alpha)\Gamma(n-x+\beta)} \theta^{x+\alpha-1}(1-\theta)^{n-x+\beta-1} \qquad 0 < \theta < 1$$
$$= 0 \qquad \text{otherwise}$$

16. Same as in Problem 1

20. Same as in Problem 1.

Index